REPRODUCTIVE ECOLOGY OF TROPICAL FOREST PLANTS

MAN AND THE BIOSPHERE SERIES

Series Editor: J.N.R. Jeffers

VOLUME 7

REPRODUCTIVE ECOLOGY OF TROPICAL FOREST PLANTS

Edited by
K.S. Bawa and M. Hadley

PUBLISHED BY

PARIS

AND

The Parthenon Publishing Group
International Publishers in Science, Technology & Education

Published in 1990 by the United Nations Educational Scientific and Cultural Organization,
7 place de Fontenoy, 75700 Paris, France — Unesco ISBN 92-3-10266-9

and

The Parthenon Publishing Group Limited
Casterton Hall, Carnforth,
Lancs LA6 2LA, UK — ISBN 1-85070-268-3

and

The Parthenon Publishing Group Inc.
120 Mill Road, Park Ridge
New Jersey 07656, USA — ISBN 0-929858-22-0

Composed by Ryburn Typesetting Ltd, Halifax, England
Printed in Great Britain by Antony Rowe Ltd, Chippenham, Wiltshire

British Library Cataloguing in Publication Data

Reproductive ecology of tropical forest plants.
 1. Tropical plants. Ecology
 I. Bawa, K. II. Hadley, M. (Malcolm) III. UNESCO IV. Series
 581.52623

 ISBN 1-85070-268-3

Library of Congress Cataloging-in-Publication Data

Reproductive ecology of tropical forest plants / edited by K. Bawa and M. Hadley
      p.   cm. — (Man and the biosphere series : v. 7)
   Includes bibliographical references and index.
   ISBN 0-929858-22-0 : $65.00
   1. Rain forest plants — Physiological ecology. 2. Rain forest plants —
 Reproduction. 3. Forest flora—Tropics—Physiological ecology. 4. Forest
 flora—Tropics—Reproduction. 5. Rain forest ecology. 6. Forests and
 forestry—Tropics. 7. Pollination—Tropics. I. Bawa, K. (Kamal) II. Hadley, M.
 (Malcolm) III. Series: MAB (Series) : 7.
 QK938.R34R47   1990
 582'.016'09152—dc20
                                                              90-44130
                                                                 CIP

Unesco's Man and the Biosphere Programme

Improving scientific understanding of natural and social processes relating to man's interactions with his environment, providing information useful to decision-making on resource use, promoting the conservation of genetic diversity as an integral part of land management, enjoining the efforts of scientists, policymakers and local people in problem-solving ventures, mobilizing resources for field activities, strengthening of regional co-operative frameworks. These are some of the generic characteristics of Unesco's Man and Biosphere Programme.

Unesco has a long history of concern with environmental matters, dating back to the fledgling days of the organization. Its first Director General was biologist Julian Huxley, and among the earliest accomplishments was a collaborative venture with the French Government which led to the creation in 1948 of the International Union for the Conservation of Nature and Natural Resources. About the same time, the Arid Zone Research Programme was launched, and throughout the 1950s and 1960s this programme promoted an integrated approach to natural resources management in the arid and semi-arid regions of the world. There followed a number of other environmental science programmes in such fields as hydrology, marine sciences, earth sciences and the natural heritage, and these continue to provide a solid focus for Unesco's concern with the human environment and its natural resources.

The Man and Biosphere (MAB) Programme was launched by Unesco in the early 1970s. It is a nationally based, international programme of research, training, demonstration and information diffusion. The overall aim is to contribute to efforts for providing the scientific basis and trained personnel needed to deal with problems of rational utilization and conservation of resources and resource systems, and problems of human settlements. MAB emphasizes research for solving problems: it thus involves research by interdisciplinary teams on the interactions between ecological and social systems; field training; and applying a systems approach to understanding the relationships between the natural and human components of development and environmental management.

MAB is a decentralized programme with field projects and training activities in all regions of the world. These are carried out by scientists and technicians from universities, academies of sciences, national research laboratories and other research and development institutions, under the auspices of more than a hundred MAB National Committees. Activities are undertaken in co-operation with a range of international governmental and non-governmental organizations.

Further information on the MAB Programme is contained in *A Practical Guide to MAB, Man Belongs to the Earth*, a biennial report, a twice-yearly newsletter *InfoMAB*, MAB technical notes, and various other publications. All are available from the MAB Secretariat in Paris.

Man and the Biosphere Book Series

The Man and the Biosphere Series has been launched with the aim of communicating some of the results generated by the MAB Programme to a wider audience than the existing Unesco series of technical notes and state-of-knowledge reports. The series is aimed primarily at upper level university students, scientists and resource managers, who are not necessarily specialists in ecology. The books will not normally be suitable for undergraduate text books but rather will provide additional resource material in the form of case studies based on primary data collection and written by the researchers involved; global and regional syntheses of comparative research conducted in several sites or countries; and state-of-the-art assessments of knowledge or methodological approaches based on scientific meetings, commissioned reports or panels of experts.

The series spans a range of environmental and natural resource issues. Currently available are reviews on such topics as control of eutrophication in lakes and reservoirs, structure and function of a nutrient-stressed Amazonian ecosystem, ecological and social effects of large-scale logging of tropical forest in the Gogol Valley (Papua New Guinea), sustainable development and environmental management in small islands, rain forest regeneration and management, the role of land/inland water ecotones in landscape management and restoration. In press or in preparation are volumes on ecosystem redevelopment, assessment and control of non-point source pollution, research for improved land-use in arid northern Kenya, changing land-use patterns in the European Alps.

The Editor-in-Chief of the series is John Jeffers, until recently the Director of the Institute of Terrestrial Ecology, in the United Kingdom, who has been associated with MAB since its inception. He is supported by an Editorial Advisory Board of internationally-renowned scientists from different regions of the world and from different disciplinary backgrounds: E.G. Bonkoungou (Burkina Faso), Gonzalo Halffter (Mexico), Otto Lange (Federal Republic of Germany), Li Wenhua (China), Gilbert Long (France), Ian Noble (Australia), P.S. Ramakrishnan (India), Vladimir Sokolov (USSR) and Anne Whyte (Canada). Bernd von Droste and Malcolm Hadley of Unesco's Division of Ecological Sciences are *ex officio* members of the Board.

A publishing rhythm of three or four books per year is envisaged. Books in the series will be published initially in English, but special arrangements will be sought with different publishers for other language versions on a case-by-case basis.

Reproductive ecology of tropical forest plants

The last couple of decades has seen a fair number of innovative and meticulous studies on the reproductive ecology of tropical forest plants. These studies have highlighted some of the special features of these plants. Pollination is almost exclusively made by animals. Seed dispersal in a vast majority of species involves a wide variety of birds and mammals as vectors. Seeds, before and after dispersal, suffer heavy mortality from animals and pathogens. Studies on the nature of the interactions between plants and their pollinators, seed dispersers and seed predators have shed light on the structure and dynamics of tropical forest ecosystems, and have contributed to the clarification of issues related to the regeneration, management and conservation of tropical forest resources. At the same time, important differences have emerged in the seasonality of reproduction, as well as the nature of plant–animal interactions, in different parts of the world.

It was with a view to presenting research from different regions and providing a pan-tropical review of recent developments in the field, that an international workshop on the reproductive ecology of tropical forest plants was held at the Universiti Kebangsaan Malaysia in Bangi, Malaysia, from 8–12 June 1987. The workshop brought together some 150 scientists from over a score of countries, from as far afield as Australia, Brazil, China, Costa Rica, Paraguay, Uganda and Venezuela. There were over a dozen participants from each of France, India and the USA, with some 40 specialists from Malaysia, reflecting the particular interest of the scientific communities of these countries in the topic at hand. The workshop was organized as a joint venture of the MAB Programme of Unesco and the Decade of the Tropics of the International Union of Biological Sciences (IUBS), in co-operation with the Malaysian MAB National Committee and the Universiti Kebangsaan Malaysia. It was based on 20 invited papers and some 50 offered contributions, in the form of both oral and poster presentations.

The present volume represents the principal substantive output of the Bangi workshop. It presents a review of recent research in plant reproductive ecology, defined to include all stages of reproduction from the initiation of flowering to seedling establishment. The focus is on lowland tropical rain forests of Asia, Africa, Australia and the Americas. The book explores the implications of recent findings to improved understanding of forest structure and functioning, and examines how insights gained from reproductive ecology can be helpful in the management and conservation of tropical forest resources. As such, the book complements another title in the Man and the Biosphere Series, that on *Rain forest regeneration and management.*

The eight sections in the book comprise invited and a few, selected, contributed papers. The first section consists of the introductory chapter, which emphasizes the practical application of research in plant reproductive ecology. The next six sections on Phenology, Plant-pollinator interactions, sexual systems and gene flow, Seed and fruit dispersal, Seed physiology, seed germination and seedling ecology, Regeneration, and Reproductive biology in relation to tree improvement programmes, are each preceded by a commentary which seeks to provide a general perspective for the papers and highlight their main points. The final section consists of a concluding statement about the workshop and the general state of tropical forestry.

Warm thanks are due to all those who contributed to the organization of the Bangi workshop, and to the preparation and review of the present volume. Noraini Tamin was chairperson of the local organizing committee, with Kamaljit Bawa as head of the scientific programme committee. Otto Solbrig, Peter Ashton and Henry Howe assisted in the planning for the workshop. Fakhri Bazazz, Lucinda McDade, Jim Hamrick, Richard Primack and Nathaniel Wheelwright helped in reviewing manuscripts. Carol Goodwillie provided editorial assistance. Financial support for the workshop was provided by Unesco, IUBS and the US National Science Foundation, among others.

The views expressed in this volume are those of the editors and individual contributors, and are not necessarily shared by Unesco and their host institutions. Moreover, the designations employed in the book concerning the legal or constitutional status of any country, territory, city or area of its authorities, or concerning the delimitation of its frontiers or boundaries do not imply the expression of any opinion whatsoever on the part of Unesco.

CONTENTS

Contents

LIST OF CONTRIBUTORS

A.W.W.L. Abeygunasekera
Department of Botany
University of Peradeniya
Peradeniya
Sri Lanka

S. Appanah
Forest Research Institute Malaysia
Kepong
Selangor
52109 Kuala Lumpur
Malaysia

Joseph E. Armstrong
Department of Biological Sciences
Illinois State University
Normal
Illinois 61761
USA

Peter S. Ashton
Department of Organismic and
   Evolutionary Biology
Harvard University
22 Divinity Avenue
Cambridge
Massachusetts 02138
USA

D.N.C. Attygalla
Department of Botany
University of Peradeniya
Peradeniya
Sri Lanka

Carol K. Augspurger
Department of Biology
University of Illinois
505 S. Goodwin
Urbana
Illinois 61801
USA

John F. Barthell
Department of Entomological
   Sciences
University of California
Berkeley
California 94720
USA

Kamaljit S. Bawa
Department of Biology
University of Massachusetts-Boston
Harbor Campus
Boston
Massachusetts 02125
USA

Arno Brune
Mision Forestal Alemana
Casilla de Correos 471
Asuncion
Paraguay

H. T. Chan
Forest Research Institute Malaysia
Kepong
Selangor
42109 Kuala Lumpur
Malaysia

David B. Clark
Estación Biológica La Selva
Organization for Tropical Studies
Apartado 676
2050 San Pedro
Costa Rica

Fernando H. Cornejo
Smithsonian Tropical Research
   Institute
Apartado 2072
Balboa
Panama

S. Dayanandan
Department of Botany
University of Peradeniya
Peradeniya
Sri Lanka

Robin B. Foster
Smithsonian Tropical Research
   Institute
Apartado 2072
Balboa
Panama
and
Botany Department
Field Museum of Natural History
Chicago
Illinois 60605
USA

Gordon W. Frankie
Department of Entomological
   Sciences
University of California
Berkeley
California 94720
USA

Jutta K. Frankie
Friends of Lomas Barbudal, Inc.,
691 Colusa Avenue
Berkeley
California 94707
USA

K.N. Ganeshaiah
Department of Agricultural Botany
University of Agricultural Sciences
Bangalore 560065
India

Annie Gautier-Hion
Station Biologique de Paimpont
35380 Plénan le Grand
France

A. Rod Griffin
CSIRO Division of Forestry and
   Forest Products
P.O. Box 4008
Canberra ACT 2600
Australia

C.V.S. Gunatilleke
Department of Botany
University of Peradeniya
Peradeniya
Sri Lanka

I.A.U.N. Gunatilleke
Department of Botany
University of Peradeniya
Peradeniya
Sri Lanka

William A. Haber
Monteverde Conservation League
Apartado 10165
San José
Costa Rica

Malcolm Hadley
Division of Ecological Sciences
Unesco
7, Place de Fontenoy
75700 Paris
France

Annette Hladik
Laboratoire d'Ecologie Générale
Muséum National d'Histoire
   Naturelle
4, Avenue du Petit Chateau
91800 Brunoy
France

Henry F. Howe
Department of Biological Sciences
University of Illinois at Chicago
P.O. Box 4348
Chicago
Illinois 60680
USA

Stephen P. Hubbell
Princeton University
Department of Biology
Princeton
New Jersey 08544
USA

Anthony K. Irvine
CSIRO Tropical Forest Research
   Centre
P.O. Box 780
Atherton
Queensland 4883
Australia

Paulo Yoshio Kageyama
Forest Science Department
ESALQ University of São Paulo
CP 9, Piracicaba
SP 13400
Brazil

T.N. Koshoo
Tata Energy Research Institute
7, Jor Bagh
New Delhi 110003
India

R.R.B. Leakey
Institute of Terrestrial Ecology
Bush Estate
Penicuik EH26 0QB
United Kingdom

K.A. Longman
Institute of Terrestrial Ecology
Bush Estate
Penicuik EII26 0QB
United Kingdom

R.M. Manurung
Institute of Terrestrial Ecology
Bush Estate
Penicuik EH26 0QB
United Kingdom
Present address:
Agricultural Research Centre
P.O. Box 977
Kuching
Sarawak
Malaysia

Sophie Miquel
Laboratoire d'Ecologie Générale
Muséum National d'Histoire
    Naturelle
4, Avenue du Petit Chateau
91800 Brunoy
France

Linda E. Newstrom
Department of Entomological
    Sciences
University of California
Berkeley
California 76720
USA

F.S.P. Ng
Forest Research Institute Malaysia
Kepong
Selangor
52109 Kuala Lumpur
Malaysia

A. Orozco-Segovia
Centro de Ecologia
UNAM
Apartado 70–275
Deleg. Coyoacan
04510 Mexico D.F.
Mexico

Richard B. Primack
Biology Department
Boston University
2 Cummington Street
Boston
Massachusetts 02215
USA

Salleh Mohd. Nor
Forest Research Institute Malaysia
Kepong
Selangor
52109 Kuala Lumpur
Malaysia

George E. Schatz
Missouri Botanical Garden
P.O. Box 299
St. Louis
Missouri 63166
USA

R. Uma Shaanker
Department of Crop Physiology
University of Agricultural Sciences
Bangalore 560065
India

John Terborgh
Duke University
Center for Tropical Conservation
Wheeler Building
3705 Erwin Road
Durham
North Carolina 27705
USA

Carlos Vázquez-Yanes
Centro de Ecologia
UNAM
Apartado 70–275
Deleg. Coyoacan
04510 Mexico D.F.
Mexico

S. B. Vinson
Department of Entomology
Texas A & M University
College Station
Texas 77843
USA

S. Joseph Wright
Smithsonian Tropical Research
    Institute
Apartado 2072
Balboa
Panama

S. K. Yap
Forest Research Institute Malaysia
Kepong
Selangor
52109 Kuala Lumpur
Malaysia

Helen J. Young
Department of Botany
Duke University
Durham
North Carolina 27706
USA

Section 1

Introduction

CHAPTER 1

REPRODUCTIVE ECOLOGY OF TROPICAL FOREST PLANTS: MANAGEMENT ISSUES

Kamaljit S. Bawa, Peter S. Ashton and Salleh Mohd. Nor

INTRODUCTION

The unabated transformation and devastation of tropical wildlands has become one of the most pressing issues of our times. Not only are the rates of deforestation very high, but also approximately 40% of the existing forest areas have been degraded in recent times. It is estimated that tropical rain forests will largely be logged or converted in about 30 years time, except for those that might be conserved as nature reserves. Obviously there is a need for greater investment in tropical rain forest management world-wide.

Indeed, over the past few years the continuing loss of tropical rain forests has become an area of international concern. A number of programmes and action plans have reflected an increase in activities towards conservation and management of tropical forest resources. Among these are the IUCN-UNEP-WWF sponsored World Conservation Strategy, the WRI sponsored Tropical Forests: A Call for Action, and the FAO sponsored Tropical Forestry Action Plan, as well as work within the Decade of the Tropics of the International Union of Biological Sciences (IUBS) and the Man and the Biosphere (MAB) Programme of Unesco.

It was within such a context that the international workshop which gave rise to the present volume was organized in Bangi (Malaysia). The general consensus of participants was that the workshop was timely and addressed an important subject. Although the primary purpose of the workshop was to explore recent scientific advances in the reproductive ecology of tropical rain forest plants, there was at the same time a need to examine the practical applications of these findings. Indeed, many contributions in this volume review the application of current knowledge to practical problems in management. Our purpose here is to briefly

3

consider some areas of scientific research critical to the management of forest resources on a sustained basis.

There are three crucial interrelated issues that the manager of indigenous forests must address, degradation and depletion of forest resources, regeneration and restoration of forest ecosystems, and conservation of genetic resources. The principal cause of depletion of forest resources is changes in land-use patterns. In addition the conversion of forested land to agriculture and excessive logging as well as grazing have led to the reduction in forest stock within the forested areas. The challenges generated by the reduction and degradation of forest cover can be adequately met only if serious attempts are made to restore forest ecosystems. Restoration inevitably must involve improved reforestation of degraded lands through plantation of native species, and the extension of forest boundaries by artificial and natural regeneration. Finally, coupled with reforestation, conservation of existing genetic resources is of high priority. The resources to be conserved and the manner in which they ought to be conserved are serious issues requiring strong scientific input.

Most research on the reproductive ecology of tropical forest plants, from flowering to regeneration, has had strong theoretical underpinnings. The test of predictions emerging from hypotheses relating to coevolution and the structure, organization and dynamics of communities has been a major impetus for much of the work. Nevertheless, many types of basic research in reproductive ecology have strong practical applications.

REPRODUCTIVE COST IN RELATION TO STAND STRUCTURE AND PLANTATION DESIGN

In Asia, the great majority of fruit trees are components of the mature phase of the forest in the main canopy (e.g. mangoes, rambutans) and the understorey (e.g. mangosteens, also the neotropical *Annona* fruits). The principal timber trees are emergents, both forest gap and building phase species producing light industrial hardwood often lacking heartwood (e.g. *Albizia, Dyera, Alstonia*; also *Hevea* and *Ceiba*), and quality timber species of the mature phase (*Shorea*, and the principal leguminous, meliaceous, and lauraceous timbers). Most of these timber species have dry fruits and seeds, often wind dispersed or gyrating. Dioecy (separate sexes) in tropical trees is associated with fleshy fruits (Bawa, 1980; Givnish, 1980). It is interesting that Ashton (1969) observed an increase in the representation of dioecious individuals from less than 5% in the emergent stratum of Far Eastern mixed dipterocarp forest to more than 30% in the understorey, the large representation of emergent juveniles in the latter not withstanding. Forest fruit and timber trees therefore substantially avoid competition for space.

4

This fact provides opportunities, long known to subsistence farmers in the tropics but only recently entering into commercial plantation practice, of increasing profitability by more efficient use of space through multiple species, multiple product, plantation. A notable advantage of this approach is that a much earlier return can be made on investment in quality hardwood timber plantation, by interplanting with rattan and fruit trees which can be culled from 6–10 years of age onwards. Others are that such plantations are well suited to small-holders and increased labour intensity. They are therefore socio-politically more acceptable than pure timber plantations, and the timber trees included in them are hence more secure.

The genus *Parkia* is unusual as it includes relatively fast-growing trees of the building phase which not only provide light shade favourable to quality hardwood regeneration, but also highly nutritious fruit. Likewise, the durians (*Durio* section *Durio*) are mature phase emergents yielding both fruit and quality timber. There are some 20 species of durian, and up to six species are cultivated in some ancient centres of settled agriculture such as Brunei Darrussalm. Different species have different soil preferences, several occurring in nature on infertile podsolized soils, thus providing improvement opportunities for agricultural diversification through breeding and their use for rootstock and for grafting.

In general, though, genetic improvement must be directed to increasing the yield of a single commodity. Plants survive by performing at their maximum potentiality for their site and genotype. Increase in yield of fruit by one species can therefore only be achieved at the cost of reduced wood production, and vice versa. Thus, Primack (in prep.) has found evidence that wood increment declines drastically in the occasional mast fruiting years during which the merantis and kapurs (*Shorea, Dryobalanops,* Dipterocarpaceae), prime timber trees, reproduce in western Malesia. This may be because these trees produce inflorescences instead of a seasonal leaf flush, thus reducing their leaf area by perhaps as much as half. Interestingly, Dayanandan *et al.* (this volume) have found that the exceptionally fast growing tiniya dun (*Shorea trapezifolia*) of Sri Lanka not only flowers annually, but produces inflorescences and a new leaf flush simultaneously. These properties identify tiniya dun, with its readily available seed (albeit lacking dormancy), and its rapid growth, as a plantation species of unusual promise. The possibility of transferring the gene responsible for its simultaneous reproductive and vegetative growth to other *Shorea* also arises.

The mangosteen (*Garcinia mangostana*), well known for its slow growth rate, belongs to a genus in which flowers and fruit are presented in the shade of the forest understorey. Jamuluddin (1978) and Ashton and Hall (in prep.) have evidence that members of the understorey guild, which often start flowering early in life, can manifest exceptionally low maximum girth

5

growth rates. These small trees may include some of the oldest individuals in the forest. Here, it seems, natural selection may have already favoured fruit over wood production. This needs to be taken into account in selecting new species for introduction, and in breeding programmes.

The mangosteen is dioecious, but the male tree is unknown in cultivation and the tree reproduces apomictically. Bawa (1980) and Givnish (1980) hypothesized that dioecy may be causally associated with seed dispersal by vertebrates, that is with large seeds and fleshy fruits. In this case, knowledge of the breeding system is essential to enable increases in fruit production, because the number, if any, of male trees required to maximize fruit trees in a stand has to be balanced against the space for fruit production which must instead be allocated to males.

There is growing evidence of site-related differences in fecundity among tropical trees. There is evidence of reduction in average fruit size and nutritional value in mixed-species stands with decline in soil fertility (Ashton, unpublished data). Wood (1956) implied that dipterocarps in peat swamps may flower less frequently than in more fertile dry land sites, and this has been confirmed in an unpublished phenological report by the silvicultural staff of the Sarawak Forest Department. Burgess (1972) found that *Shorea leprosula*, a fast growing species of mesic sites, flowers more frequently than others in its section in Peninsular Malaysia. C.V.S. and I.A.U.N. Gunatilleke and their students (in prep.) have observed that *S. trapezifolia*, *S. disticha* and *S. worthingtonii*, which respectively occupy the mesic intermediate and xeric parts of the catena in Sinharaja forest in the wet lowland of south-western Sri Lanka, flower in declining frequency and intensity. These observations imply that poor sites can be expected to yield less timber and also less fruit than favourable sites.

PHENOLOGY

A vast body of knowledge about leafing, flowering and fruiting periodicity of tropical forest plants, both at the level of communities and of individual species, has been developed during the last 2 decades. This information has revealed considerable spatial and temporal variation in phenological patterns. Species differ with respect to timing, duration and frequency of flowering and fruiting. Moreover, communities differ in terms of overall phenological patterns. For example, the type of mass flowering that has been observed in the South-east Asian rain forests (Yap and Chan, this volume) has not been noted in the neotropics. Our understanding of the factors that regulate initiation, periodicity and frequency of flowering mostly remains obscure (though see Ashton *et al.*, 1988; Ashton 1989). It is thus not surprising that despite 2 decades of research in tropical plant phenology, for the most part we are far from developing predictive models of phenological events.

The effect of logging on phenological patterns is not known, but is an area that should be of primary concern to the forest manager. Logging may change the environmental regime and the spacing patterns of the conspecific trees. Both changes may influence the amount of flowering and fruit and seed set. Altered spacing and phenological patterns may also change the mating relationships with unknown genetic consequences.

Characterization of phenological patterns at the level of species-populations is of utmost importance to the tree breeder. In several species, individuals within a population mature seeds asynchronously. Seeds collected at only one point in time in such populations may not adequately represent the genetic diversity of the population.

POLLINATION AND SEXUAL SYSTEMS

Attempts to gather information about pollination of large canopy trees in tropical rain forests are still in a very preliminary stage. For many commercially important species, we have virtually no knowledge about the mode of pollination or the extent to which there is a species–specific relationship between the pollen vector and the plant species. Our knowledge about the dynamics of pollinator populations in tropical forests is also poorly developed.

At the community level, pollination mechanisms of tropical rain forest trees involve a wide variety of vertebrates and invertebrates as pollen vectors. Species specificity in pollination mechanisms is rare and each species of pollen vector may service many species of plants either at the same or at different times. Thus the maintenance of a particular plant species within an ecosystem may be contingent upon the presence of other plant species which serve as a continuing resource for its pollinators. However, little is known about the extent to which the perturbation of species diversity in an ecosystem might influence specific plant–pollinator interactions.

Tropical forest trees, because of their diverse floral morphologies and pollination systems, also present problems for controlled pollinations. The flowers of canopy species are often technically difficult to manipulate. Flowering is generally highly synchronized within populations and frequently of short duration. Chan and Appanah (1980) found individual flowers of *Shorea* section *Mutica*, which are small, though produced abundantly, to undergo anthesis during a single night, and to be extremely easily damaged. Pollen yield in flowers of these trees is small. Stigmas are generally small, too. These features, common to many mature phase trees of

7

the emergent and main canopy strata, demand great deftness to achieve successful crosses.

A further problem for breeders, in many and perhaps the majority of above-canopy species, is the high level of post-anthesis mortality (Bawa and Webb, 1984). Reasons remain unclear but may include inability of the tree to sustain development of all flowers into ripe fruit, post-anthesis abortion of unpollinated flowers, or selective mortality of selfed over outcrossed flowers. This problem is particularly acute in those species, such as mangoes, in which fruit is large. Above-canopy species generally produce large numbers of flowers on each inflorescence. Successful pollination experiments may require time-consuming removal of all unmanipulated flowers.

A notable exception to the above generalizations among Old World tropical evergreen forest trees is the group of species pollinated by vertebrates. Bird and bat pollinated flowers are universally robust and larger, with large frequently numerous stamens producing abundant pollen (Faegri and van der Pijl, 1971; Start and Marshall, 1976). Interestingly the sunbirds, flower peckers and megachiropteran bats that are pollinators of Asian tropical trees are not attracted to understorey flowers, though they may visit or even confine themselves to forest gaps. They can however detect ramiflorous flowers, such as those of durians, in the branches of emergent crowns, and can be lured from there down the trunk in species which are also cauliforous. This poses a major difference with the New World, where some species of both hummingbird and microchiropteran bat pollinators are understorey specialists. It is possible that the understorey flowering trees which they visit would not set fruit if grown in plantation in the Old World.

Yap (1982) found that Malaysian understorey species can broadly be divided into two classes. One, which is epitomized by *Xerospermum noronhianum* (Sapindaceae), a relative of the rambutan studied by Yap (1980) and by Appanah (1982), flowers annually but with low to moderate intensity. Flowers are borne in small numbers over long periods, with relatively low synchrony within populations. Male flowers in this androdioecious species and also in other, truly dioecious, species may be short-lived. Post-anthesis mortality among successfully pollinated female flowers is relatively low. Such species appear to attract low numbers of generalist pollinators. Their long period of flowering and low post-anthesis mortality make these trees relatively amenable to experimentation.

Other understorey tree species flower synchronously and at intervals of varying duration. This flowering syndrome is associated with the production of powerful olfactory lures, which is the only means by which pollinators can be attracted in the leafy understorey from long distances. Olfactory lures are prevalent in the archaic families Annonaceae and Myristicaceae.

Fascinatingly, at least some of these trees, in spite of the clumsiness of their primitive beetle and fly pollinators, are highly oligolectic (S. Rogstad, in prep). Here too, introduction of such species and their exotic relatives, may pose problems for successful pollination.

The flowering characteristics of pioneer species appear more obviously related to their fruit and seed characteristics. The trees are well known to start reproduction early in life, to fruit abundantly and frequently, and to have small seeds, sometimes with dormancy, which can be effectively dispersed by wind or small animals. Some flower and fruit continuously (for example, see Croat, 1978), others in synchronized flushes. Flowers are generally small and numerous, but apparently vary with respect to length of anthesis and pollinator specificity.

Self-incompatibility and self-sterility are other topics of considerable practical importance, but subjects about which not much is known. Nevertheless, enough is known concerning breeding systems of tropical rain forest trees in Asia to make some broad generalizations, albeit tentative. Besides Bawa and his associates in Central America (Bawa, 1979; Bawa *et al.*, 1985), Chan (1981) and Yap (1980) in the Far East, the Gunatillekes and their co-workers (in prep.) in South Asia and others have demonstrated, contrary to all expectation, that rain forest trees are overwhelmingly outbreeders. This seems to be so whether they are emergents or in the understorey, climax species or pioneers. The likely widespread occurrence of apomixis through adventive embryony, that is asexual embryogenesis (Kaur, 1977; Kaur *et al.*, 1978, 1986), which is confirmed in a few taxa and inferred in a wide range of genera and families including important timber genera such as *Shorea*, and fruit trees which include *Citrus*, mangoes (*Mangifera*) and *Garcinia*, has been a surprise, and contrary to theoretical expectations. So far, apomixis has only been found in trees of the forests' mature phase, but pioneer and gap phase species have been little investigated. The percentage of apomixis is variable in some species but may be uniformly high in others (Kaur, 1977). In the absence of confirmatory evidence for adventive embryony, which is time-consuming to obtain and therefore impractical on a community basis, it is impossible to distinguish between self-compatibility and psuedogamy (that is, adventive embryony in which pollination is a necessary stimulant to embryogenesis). Estimates for self-compatibility among tropical trees, already low, may therefore still be exaggerated.

We need to measure the degree of selfing in individual trees, the extent of variation in selfing rates and the degree of inbreeding within a population. Fortunately, electrophoretic procedures utilizing genetic markers are now available, and to some extent have been utilized to estimate quantitatively the degree of outcrossing in tropical trees (Bawa and O'Malley, 1987; O'Malley *et al.*, 1988).

GENE FLOW, EFFECTIVE POPULATION SIZE AND GENETIC VARIATION

Information about gene flow and effective population size is of critical importance in designing suitable breeding strategies for tree improvement programmes. Also, appropriate measures for conservation of forest genetic resources are contingent upon the knowledge of the breeding structure of populations. Despite much work on pollination and seed dispersal on one hand, and the density and dispersion patterns of the trees on the other, we have no knowledge about the effective population size for tropical trees.

Central to any breeding programme is knowledge of the pattern and degree of genetic variability within and between species populations. Data on population genetic structure also provide the basis for adequate sampling for *ex situ* conservation and for suitable design of reserves for *in situ* conservation of forest genetic resources. High levels of outbreeding imply high genetic variability within populations. Here patterns of genetic variation in populations are largely determined by pollen and seed dispersal patterns. In Asian rain forest trees, pollen is animal dispersed except among some gregarious river bank and ridge top species, as is seed except among many emergents. High levels of apomixis imply low genetic variability within populations, though without the increase in gene fixation which accompanies self-compatibility. Gene fixation is also increased where breeding populations are small however, and this is substantially determined by pollen and seed dispersal distances.

Patterns of genetic variation should be apparent in patterns of phenotypic variation. Ashton (1969, 1984) has indicated that this might be so among rain forest trees species in Asia. Striking is the general tendency for taxa to manifest extraordinary morphological uniformity throughout their geographical range, which can often be large. Sympatric, closely related species, differ morphologically in small ways which are nevertheless constant throughout their ranges. In Dipterocarpaceae, taxa in which geographical subspecies are recognized, and which have been examined, have been found to be facultatively apomictic, suggesting that facultative apomixis may serve to fix favourable genotypes and thus increase the rate of allopatric differentiation in outbreeders. In the Far Eastern sapindaceous monoecious genera *Pometia*, *Allophyllus* and *Nephelium*, which are known to be highly self-compatible, a complex reticulate pattern of local and regional morphological variation is manifested, often accomplished by ecotypic specialization, which defies narrow species definition (Leenhouts, 1968, 1986).

In summary, tropical tree populations are expected to show a wide variety of population genetic structures (Kageyama, this volume). We are now just beginning to describe these structures and explore their implications for conservation and management of forest resources.

SEED BIOLOGY

Surprisingly, there is an absence of very basic information about seed biology. For many species of commercial importance, seed maturation times and seed vectors are not known. In the absence of knowledge about seed dispersal agents, one cannot assess their role in forest dynamics via their interactions with other plant species. Seed dispersal studies in South American forests indicate that at times of food scarcity, fruits and seeds of certain species may provide resources critical to the survival of many fruit-eating animals (Terborgh, 1986). Identification of such species and how their elimination due to logging may influence vertebrates should be accorded high priority in research. At an even more fundamental level, little is known about factors regulating seed viability, seed germination and seed dormancy for many of the commercially important species. Techniques for storage and handling of seeds and seedlings are other very basic activities that need attention. For many species whose seeds yield secondary forest products, large amounts are collected by humans, but the impact of such collection on regeneration of the constituent species remains unknown.

The role of seed predators in influencing the distribution of host species is well known (Janzen, 1970). In many South-east Asian forests, fruiting is supra-annual. Identification of hosts of seed predators during non-fruiting years may make it possible to manipulate the frequency of such hosts in order to control the population of seed predators.

REPRODUCTION AND POPULATION GENETIC STRUCTURE

Much of the research in plant reproduction, regeneration and population genetics has been done in isolation from each other. Genetic consequences of various phenological patterns, breeding systems and pollination and seed dispersal mechanisms are virtually unknown in tropical plants. Spatial and temporal variation in reproductive output has generally not been correlated with seedling establishment or the variation in genetic diversity of the resulting seedlings. For example, if recruitment in tropical forest trees is episodic, regeneration may mostly occur after periods of heavy fruit production concomitant with the release of large amounts of genetic variability. Obviously there is a need to explore correlations among the quantity of seed produced, the amount of recruitment and the level of genetic variation among cohorts produced from different reproductive episodes. Thus not only ought there to be greater integration of various types of ecological research such as phenology and regeneration, but ecological studies must also be meshed with genetic studies to understand the dynamics of regeneration in tropical rain forests.

CONCLUSIONS

Clearly many areas of plant reproductive ecology, defined in a broad sense, offer opportunities for the application of basic research to practical problems in forestry management. This viewpoint is echoed in many contributions in this volume. However, by no means do we imply that our ability to restore and rehabilitate, and to conserve tropical ecosystems must await biological information, because we already know enough to proceed. Further studies on phenology, pollination and breeding systems, seed biology, regeneration and population genetics, can certainly refine our ability to manage effectively tropical ecosystems for conservation and sustained yield of forest products. This is particularly true for species of known commercial value, for many of which we lack information about basic biology. At a time when many of these species or the habitats they occupy are being threatened, it is opportune to ask what we can learn about them and how we can use the information so gained to propagate, utilize and conserve such species.

REFERENCES

Appanah, S. (1982). Pollination of androdioecious *Xerospermum intermedium* Radlk. (Sapindaceae) in a rain forest. *Botanical Journal of the Linnean Society*, **18**, 11–14

Ashton, P. S. (1969). Speciation among tropical forest trees: Some deductions in the light of recent evidence. In Lowe-McConnell, R.H. (ed.) *Speciation in Tropical Environments*, pp. 155–96. (London: Academic Press)

Ashton, P. S. (1984). Biosystematics of tropical forest plants: A problem of rare species. In Grant, W.F. (ed.) *Plant Biosystematics*, pp. 497–518. (New York: Academic Press, for International Organization of Plant Biosystematics)

Ashton, P.S. (1989). Dipterocarp reproductive biology. In Lieth, H. and Werger, M.J.A. (eds) *Tropical Rain Forest Ecosystems. Ecosystems of the World*, Volume 14B, pp. 219–40 (Amsterdam: Elsevier)

Ashton, P.S., Givnish, T. and Appanah, S. (1988). Staggered flowering in the Dipterocarpaceae: New insights into floral induction and the evolution of mast flowering in the aseasonal tropics. *American Naturalist*, **132**, 44–66

Bawa, K.S. (1979). Breeding systems of trees in a wet tropical forest. *New Zealand Journal of Botany*, **17**, 521–4

Bawa, K.S. (1980). Evolution of dioecy in flowering plants. *Annual Review of Ecology and Systematics*, **11**, 15–39

Bawa, K.S. and O'Malley, D.M. (1987). Estudios geneticos y de sistemas de cruzamiento en algunase especies arboreas de bosques tropicales. *Review of Tropical Biology*, **35** (Supplement 1), 177–88

Bawa, K.S. and Webb, C.J. (1984). Flower, fruit and seed abortion in tropical trees: Implications for the evolution of paternal and maternal reproduction patterns. *American Journal of Botany*, **71**, 736–51

Bawa, K.S., Perry, D. and Beach, J.H. (1985). Reproductive biology of tropical lowland rain forest trees. I. Sexual systems and incompatability mechanisms. *American Journal of Botany*, **27**, 331–45

Burgess, P.F. (1972). Studies on the regeneration of the hill forests of Malay Peninsula: The phenology of dipterocarp. *Malayan Forester*, **35**, 103–23

Chan, H.T. (1981). Reproductive biology of some Malaysian dipterocarps. III. Breeding systems. *Malaysian Forester*, **44**, 28–36

Chan, H.T. and Appanah, S. (1980). Reproductive biology of some Malaysian dipterocarps. I. Flowering biology. *Malaysian Forester*, **43**, 132–43

Croat, T.B. (1978). *Flora of Barro Colorado Island*. (Stanford: Stanford University Press)

Faegri, K. and van der Pijl, L. (1971). *The Principles of Pollination Ecology*. Second edition. (Oxford and New York: Pergamon)

Gan, Y.Y. (1976). *Population and phylogenetic studies on species of Malaysian rain forest trees*. Ph.D. Thesis. University of Aberdeen, Aberdeen.

Givnish, T. (1980). Ecological constraints on the evolution of breeding systems in seed plants: Dioecy and dispersal in gymnosperms. *Evolution*, **34**, 959–72

Jamaluddin bin Basharuddin. (1978). *Ecological studies of variation in dynamics and species distribution in relation to habitat in some mixed dipterocarp forests of Sarawak, East Malaysia*. Masters Thesis. University of Aberdeen, Aberdeen.

Janzen, D.H. (1970). Herbivores and the number of tree species in tropical forests. *American Naturalist*, **104**, 501–28

Kaur, A. 1977. Embryological and cytological studies of some member of the Dipterocarpaceae. Ph.D. Thesis. University of Aberdeen, Aberdeen.

Kaur, A., Ha, C.O., Jong, K., Sands, V.E., Chan, H.T., Soepadmo, E. and Ashton, P.S. (1978). Apomixis may be widespread among trees of the climax rain forest. *Nature*, **271**, 440–41

Kaur, A., Jong, K., Sands, V.E. and Soepadmo, E. (1986). Cytoembryology of some Malyasian dipterocarps, with some evidence of dapomixis. *Botanical Journal of the Linnean Society*, **92**, 75–88

Leenhouts, P.F. (1968). A conspectus of the genus *Allophylus* (Sapindaceae). A problem of the complex species. *Blumea*, **15**, 301–58

Leenhouts, P.F. (1986). A taxomic revision of *Nephalium* (Sapindaceae). *Blumea*, **31**, 373–436

O'Malley, D.M., Buckley, D.P., Prance, G.T. and Bawa, K.S. (1988). Genetics of Brazil nut (*Bertholletia excelsa* Humb. Bonpl.: Lecythidaceae). 2. Mating system. *Theoretical and Applied Genetics*, **76**, 929–32

Start, A.N. and Marshall, A.G. (1976). Nectarivorous bats as pollinators of trees in West Malaysia. In Burley, J. and Styles, P.T. (eds.) *Tropical Trees: Variation, Breeding and Conservation*, pp. 141–50. (London: Academic Press)

Terborgh, J. (1986). Keystone plant resources in the tropical environment. In Soulé, M.E. (ed.) *Conservation Biology: The Science of Scarcity and Diversity*, pp. 330–44. (Sunderland, Massachusetts: Sinauer Associates)

Wood, G.H.S. (1956). The dipterocarp flowering season in North Borneo, 1955. *Malayan Forester*, **19**, 193–201

Yap, S.K. (1980). Phenological behaviour of some fruit tree species in a lowland dipterocarp forest of West Malaysia. In Furtado, J.I. (ed.) *Tropical Ecology and Development. Proceedings of V International Symposium on Tropical Ecology*, pp. 161–67. (Kuala Lumpur: International Society of Tropical Ecology)

Yap, S.K. (1982). The phenology of some forest species in a lowland dipterocarp forest. *Malaysian Forester*, **45**, 21–35

Section 2

Phenology

CHAPTER 2

PHENOLOGY – COMMENTARY

Kamaljit S. Bawa and F.S.P. Ng

Phenology of tropical rain forest plants raises a number of interesting questions. In a seemingly aseasonal climate, what cues do plants use for the initiation of vegetative and reproductive growth? Given the lack of variation in climate, why do different species initiate vegetative growth and reproduction at different times? What accounts for tremendous variation in patterns of leaf flushing and flowering among species? Why do some species flower more than once a year, others once a year and still others every 2 or more years? How is the phenology of plants correlated with the phenology of pollinators and herbivores? How does selection from such diverse forces as herbivores, pollinators, seed dispersal agents and seed predators influence patterns of leafing, flowering and fruiting? Answers to such questions require characterization of phenological phases with respect to timing, duration and frequency at the level of species. In recent years, a number of phenological patterns have been described in tropical forest plants but the possible factors underlying these patterns largely remain obscure. Two out of the three papers in this section, one from Malaysia and the other from Costa Rica, summarize data on the phenology of trees, and the third describes the results of an empirical study undertaken in Panama aimed at elucidating factors responsible for the initiation of flowering.

General mass flowering at irregular intervals is a notable feature of many aseasonal forests in South-east Asia. This flowering pattern is characterized by supra-annual flowering and may involve one species, a group of related species or a majority of species in the community. Yap and Chan describe community-wide general flowering in dipterocarp forests. They observed 310 trees belonging to 16 species of *Shorea* over an 11-year period at four sites. Mass flowering occurred in the years 1976, 1981 and 1983. The proportion of species and individuals that participated in mass flowering varied from one episode to another. Moreover, Yap and Chan show considerable site specific variation in phenological responses of species. Not only was the intensity of flowering different at the four sites, but also some species flower at one site but not at the other(s).

Yap and Chan's study also shows that mass flowering can occur at different times of the year in different episodes. For example, mass flowering in Malaysian forests has been generally recorded to occur in April–May period (Burgess, 1972; Ng, 1977). However, in 1981, the mass flowering occurred in September–October. Ng (1981) has shown two leaf-flushing peaks in April and October in dipterocarps of Peninsular Malaysia. Generally, the flowering of dipterocarps is associated with leaf-flushing in April, but in 1981, it apparently was associated with leaf-flushing in October. Dayanandan *et al.* (this volume) also note two periods of flowering for dipterocarps in Sri Lanka, in April–May and November–December.

There is no documentation of the response of pollinator populations to mass flowering. Appanah (this volume) remarks that there is general abundance of insect pollinators during periods of mass flowering. In 1976, Ng (unpublished observations) noted a marked increase in the number of pollen-collecting bees. One might assume that population densities of pollinators decline during off years. Yap and Chan have observed that flowering in off years generally does not result in fruiting. Lack of fruit-set could be due to insufficient pollinators or resource depletion from the previous mast fruiting episode.

Janzen (1974) has attributed the evolution of mass fruiting to the pressure from seed predators. According to Janzen, production of large quantities of seeds after intervals of more than 1 year results in the satiation of seed predators. Satiation allows the escape of many more seeds from the predators than would be the case if trees were to flower every year and produce smaller quantities of fruits. Ashton *et al.* (1988) have suggested that the cue for floral induction in mass fruiting species is a drop of approximately 2°C or more in minimum night-time temperature for 3 or more nights.

Frankie *et al.* summarize the results of their comprehensive studies of phenology and plant pollinator interactions. In contrast to the Malaysian dipterocarp forests, most tree species in neotropical lowland rain forests in Central America flower annually, though some species do flower in alternate years (Frankie *et al.*, 1974). In the neotropics, phenology of various species at the population level has also been examined (Bawa, 1983 and references therein). Studies at the population level show considerable year to year quantitative variation in flowering and fruiting.

In order to understand the co-evolution between flowers and their pollinators, studies of flowering phenology ought to be coupled with studies of the phenology of the associated pollinators. Frankie *et al.* also briefly describe their comprehensive investigations of the biology of bees, including their nesting behaviour, feeding and mating ecology and population dynamics. It is apparent that our knowledge of the behavioural ecology and population biology of tropical pollinators is rather limited, yet crucial for the conservation and management of forest resources.

It has often been suggested that water availability is a critical factor in the initiation of flowering in many tropical trees. The suggestion is based on the observation that many species in neotropical forests initiate flowering in the dry season. Wright and Cornejo describe the results of an unusual experiment conducted to determine if moisture stress is indeed responsible for the timing of flowering. They continuously irrigated forested areas in Panama during the dry season to maintain water level at a certain threshold. They found that irrigation had no effect on the flowering periodicity. Wright and Cornejo conclude that water availability is not the proximal cue for flowering for many species, but emphasize that long-term observations are required for a firm conclusion.

Clearly phenological patterns of tropical forest trees are diverse and complex. Equally complex are the factors that regulate these patterns. It is thus not surprising that despite considerable research on phenology in recent years, we are still far from developing any predictive models of flowering or fruiting. Because the patterns of leaf-flushing, flowering and fruiting influence populations of herbivores, pollinators and seed dispersal agents respectively, an understanding of phenology – patterns as well as the underlying factors – is basic to the understanding of a wide variety of species interactions in tropical forests.

The year to year variation in seed and fruit set is also likely to influence population recruitment. Moreover, if the number of mating individuals varies greatly from one flowering episode to another, different cohorts may also differ in the amount of genetic diversity contained within the cohorts. As mentioned earlier, the consequences of temporal variation in seed output on population recruitment and the generation of genetic diversity have not been examined.

A detailed knowledge of flowering and fruiting patterns is also critical for the successful management of forest genetic resources. Information about seed and fruit set and seedling establishment schedules is required for *in situ* management of forest stands for conservation as well as production. Adequate sampling for *ex situ* collections requires collection of seeds in the years when the maximum number of individuals participate in the reproductive episode.

REFERENCES

Ashton, P.S, Givnish, T.J. and Appanah, S. (1988). Staggered flowering in the Dipterocarpaceae: New insights into floral induction and the evolution of mast fruiting in the aseasonal tropics. *American Naturalist*, **132**, 44–60

Bawa, K.S. (1983). Patterns of flowering in tropical plants. In Jones, C.E. and Little, R.J. (Eds) *Handbook of Experimental Pollination Biology*, pp. 394–410. (New York: Van Nostrand Reinhold)

Burgess, P.F. (1972). Studies on the regeneration of the hill forests of the Malay Peninsula: The phenology of dipterocarps. *Malayan Forester*, **35**, 103–23

Frankie, G.W., Baker, H.G. and Opler, P.A. (1974). Comparative phenological studies of trees in tropical wet and dry forests in the lowlands of Costa Rica. *Journal of Ecology*, **62**, 881–919

Janzen, D.H. (1974). Tropical blackwater rivers, animals and mast fruiting by the Dipterocarpaceae. *Biotropica*, **4**, 69–103

Ng, F.S.P. (1977). Gregarious flowering of dipterocarps in Kepong. *Malaysian Forester*, **40**, 126–37

Ng, F.S.P. (1981). Vegetative and reproductive phenology of dipterocarps. *Malaysian Forester*, **44**, 197–221

CHAPTER 3

PHENOLOGICAL BEHAVIOUR OF SOME SHOREA SPECIES IN PENINSULAR MALAYSIA

S.K. Yap and H.T. Chan

ABSTRACT

Flowering and fruiting behaviour of 310 mature trees belonging to 16 species of Shorea *was studied in natural forests as well as in a plantation at Kepong. Results show that flowering of* Shorea *is more frequent than previously reported. In addition to the gregarious flowering occurring once in a number of years and the sporadic, irregular, flowering activity, there is also an intermediate intensity of flowering which appears more regularly. Besides variation in the flowering patterns in different sites, interspecific and intraspecific variation has also been observed. Members of the Red Meranti Group are able to flower more regularly than the other groups. A high proportion of trees produce fruits during good flowering years, but most trees do not produce fruits during sporadic flowering.*

INTRODUCTION

Many dipterocarp tree species as well as species of many other families flower at intervals of several years. Such a flowering pattern has been termed as "general or gregarious flowering" (Ashton, 1969; Burgess, 1972; Medway, 1972; Ng, 1977; Wood, 1956). Both Ridley (1901) and Foxworthy (1932) had observed exceptionally heavy flowering years and almost annual flowering of some species of the Dipterocarpaceae. Wood (1956) reported that approximately two-thirds of the then-known 200 dipterocarp species in Sabah flowered during 1955. Medway (1972) observed 63% and 78% of dipterocarp species flowering during 1963 and 1968, respectively, in Selangor. Between these general episodes of flowering, intermittent, sporadic or isolated flowering involving several individuals of a species or sometimes several species, has also been reported (Ashton, 1969; Whitmore, 1976). A general flowering is normally followed by a general or mast fruiting (Ashton, 1969; Janzen, 1974; Ng, 1988; Wood, 1956). Fruit set following sporadic flowering is often poor (Burgess, 1972; McClure, 1966).

Most of the prior observations of the phenological behaviour of dipterocarps have been based on trees in the arboreta or small populations within a forest site. In the present study, many trees of each species of *Shorea* were individually selected and kept under regular surveillance. Furthermore, the phenological behaviour of these species was followed at four sites over a period of 11 years (1973 to 1983).

Members of the Dipterocarpaceae are the main timber producing species in Malaysia. In the early management system of the Department of Forestry, logging activities were planned such that subsequent harvests could depend on natural regeneration (Wyatt-Smith, 1963). With the depletion of the lowland forests, logging activities are now concentrated on the hills. Because of the steep terrain in these forests, natural regeneration can be poor. In many areas, the stock of seedlings is too low for producing the next crop. Thus, enrichment planting with selected species has to be done. To produce the required planting stock, large quantities of seeds are required. In view of the irregularity of the dipterocarp flowering, an organized monitoring system is essential.

STUDY SITES AND METHODS

A total of 300 mature trees belonging to 16 species of *Shorea* at the Ampang and Gombak Forest Reserves in Selangor, the Pasoh Forest Reserve in Negri Sembilan, and those planted at the Forest Research Institute Malaysia in Kepong, were selected (Figure 3.1). They were individually tagged and their phenological behaviour was recorded at monthly intervals. Some of these trees were lost through natural death and wind throw during the period of observation.

A detailed study of the flowering and fruiting behaviour of *Shorea* populations (section *Mutica*) was carried out during the 1976 general flowering at Pasoh. For this study, a total of 100 trees of *Shorea leprosula, S. macroptera, S. lepidota, S. acuminata, S. dasyphylla* and *S. parvifolia* within a 15 ha transect were selected. Observations were undertaken twice weekly throughout the flowering and fruiting period.

FLOWERING BEHAVIOUR OF *SHOREA* SPECIES

Flowering was observed every year from 1973 to 1983 in some of the 16 selected *Shorea* species at the study sites (Figure 3.2). Even during lean

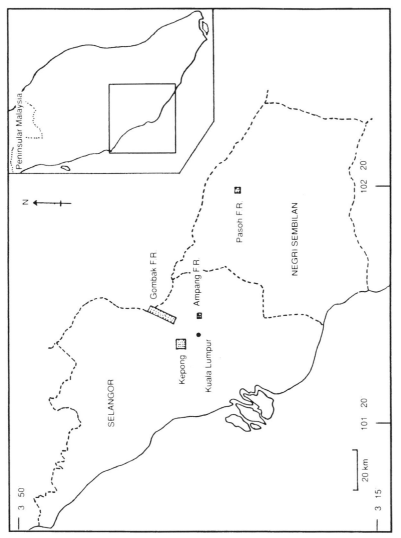

Figure 3.1 Location of the four study sites

23

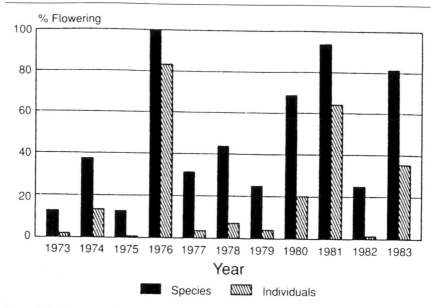

Figure 3.2 The proportion of species and individuals flowering collectively at the four study sites (1973–1983)

years, such as 1973 and 1975, 12.5% of the species, involving less than 2% of the trees, flowered. General flowering was observed during 1976, 1981 and 1983 at all the four sites. These, flowering episodes respectively involved 100%, 93.8% and 81.3% of all the species and 83.2%, 64.2% and 35.5% of all the individuals. For intermediate flowering years, such as 1974 and 1980, flowering of intermediate intensity with 37.4% and 68.8% of the species, and 13.2% and 20.1% of individuals respectively was observed.

FLOWERING AT DIFFERENT SITES

Although most species flower during general flowering years, trees at different sites behave differently. In the general flowering of 1976, trees of *S. maxwelliana* and *S. multiflora* flowered in Ampang, Gombak and Pasoh but not in Kepong. In 1981, trees of *S. curtisii* in Gombak did not bear any flowers, although trees in Kepong and Ampang flowered heavily. In the same year, trees of *S. macroptera* were sterile in Gombak and Pasoh but flowered in Kepong and Ampang.

Variation in terms of the proportion of trees that flowered for each of the four sites is illustrated in Figure 3.3. Flowering was detected in all the years

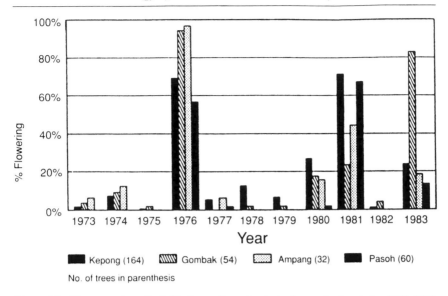

Figure 3.3   The proportion of individual trees flowering in the four study sites during 1973–1983

for trees in Kepong, while those in Pasoh had 6 years (1973, 1974, 1975, 1978, 1979 and 1982) of non-flowering. No flowering was observed in Ampang in 1975, 1978, 1979 and 1982; while, only in 1977, trees in Gombak did not flower. During the general flowering of 1976, the proportion of flowering individuals in Kepong and Pasoh was conspicuously lower than that of other sites. In contrast, the proportion of individuals flowering in Kepong and Pasoh was high in 1981; while that in Gombak far exceeded the other areas in 1983.

Species tended to flower somewhat synchronously among the study sites, particularly during general flowering years (Figure 3.4). In 1976, most of the species flowered in March and April at all the four sites. However, in 1983, peak flowering was observed in April and May for trees in Kepong and Gombak, but in May and June at Ampang and Pasoh. In 1981, five species in Gombak and Ampang began flowering around June, two of them (*S. leprosula* and *S. acuminata*) flowered again during the September to October general flowering. Only one tree of *S. acuminata* was involved in both flowerings. For *S. sumatrana*, flowering peaks at Kepong were observed in May and September in 1981 and different trees were involved in both the flowerings. Flowering in September–October was also observed in 1978, when some trees of *S. curtisii*, *S. platyclados* and *S. parvifolia* flowered in Kepong.

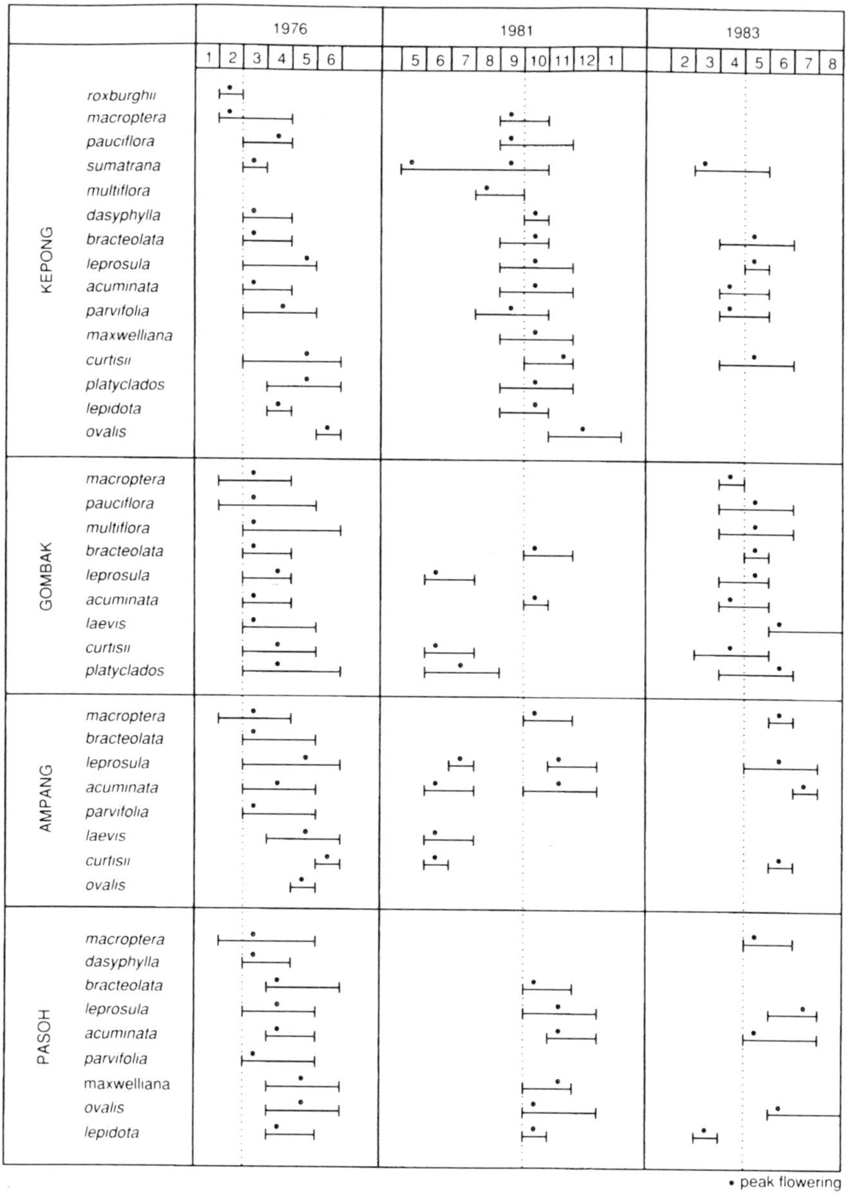

Figure 3.4 Duration of flowering of *Shorea* species in the study sites during the general flowering of 1976, 1981 and 1983

Table 3.1   Flowering frequency and periodicity of the *Shorea* species studied

Species	No. of individuals observed	No. of times flowering observed	No. of trees that flowered during consecutive years
Red Meranti Group			
Section *Brachypterae*			
S. platyclados	28	8	5
S. pauciflora	6	4	1
Section *Ovalis*			
S. ovalis	10	5	1
Section *Mutica*			
S. acuminata	19	7	6
S. curtisii	22	6	5
S. dasyphylla	4	3	1
S. lepidota	20	3	0
S. leprosula	49	8	10
S. macroptera	26	7	3
S. parvifolia	33	7	8
Damar Hitam Group			
Section *Richetioides*			
S. multiflora	6	5	1
Balau Group			
Section *Shorea*			
S. laevis	10	3	0
S. maxwelliana	28	5	1
S. sumatrana	19	7	12
White Meranti Group			
Section *Anthoshorea*			
S. bracteolata	16	5	1
S. roxburghii	4	5	1

VARIATION IN FLOWERING FOR DIFFERENT SPECIES

The frequency of flowering varies with different species, as shown in Table 3.1. With the exception of *S. sumatrana*, all the seven species that flowered six times or more during the observation period were members of the sections *Mutica* and *Brachypterae* of the Red Meranti Group of *Shorea*.

During the period of observation, some of the individuals from 14 out of the 16 selected species flowered for 2 consecutive years (Table 3.1). The proportion of individuals capable of this flowering pattern varies from 10% to 63% depending on species. Some trees of *S. acuminata* (31.6%), *S.bracteolata* (6.3%), *S. curtisii* (22.7%), *S. leprosula* (20.4%), *S. macroptera* (11.5%), *S. parvifolia* (24.2%), *S. platyclodos* (17.9%) and *S. sumatrana* (63.2%) flowered for 3 consecutive years.

FLOWERING WITHIN A SITE

All the 100 adult individuals of *Shorea* (section *Mutica*) flowered during the 1976 general flowering at Pasoh. *Shorea macroptera* was the first to come into flower, followed by *S. dasyphylla*, *S. lepidota*, *S. parvifolia*, *S. acuminata* and *S. leprosula* respectively (Figure 3.5). The blooming times of these closely related species were staggered, with some overlapping between the terminal stages of an earlier species and the initial phases of a later species. The duration of flowering for the whole section was about 13 weeks.

The duration of anthesis of each species was remarkably short, ranging from 2 to 3.5 weeks, with an apparent prolongation among later flowering species. The period, from first recognition of inflorescence initiation to the onset of anthesis, seemed also to be longer for the later flowering species, ranging from 2.5 weeks in *S. macroptera* to 6.5 weeks in *S. leprosula*. The disparity in the timing of bud initiation and development resulted in the individual species being at the height of flowering at different times.

FRUITING BEHAVIOUR OF *SHOREA* SPECIES

Flowering is usually followed by fruiting, but not all trees which flower bear fruits. The proportion of trees which failed to produce fruits varies between years and among individuals of a species. In the general f lowering years of 1976, 1981 and 1983, a high proportion of trees that flowered produced fruits (87%, 79% and 87%, respectively). Also, 10 out of 16 species set fruit on all the flowering trees. This was not true for the sporadic flowering years. In 1973, only one species fruited (out of two species that flowered), while in 1977, three out of the six flowering species did not produce fruits. The difference was more striking in 1982, when none of the four species that flowered bore fruits (Table 3.2). However, species such as *S. laevis*, *S. multiflora*, *S. ovalis* and *S. sumatrana* were the exceptions in that all the flowering trees set fruits.

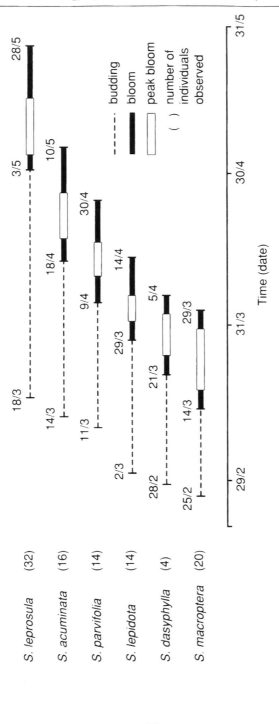

Figure 3.5 Relative flowering times of six *Shorea* species (section *Mutica*) (after Chan and Appanah, 1980)

Table 3.2 Number of *Shorea* trees flowering and fruiting in the study sites during the period 1973–1983

Species		No. of trees observed*	73	74	75	76	77	78	79	80	81	82	83
S. acuminata:	flowering	19	0	0	0	19	2	1	1	8	15	0	4
	fruiting		0	0	0	19	1	0	1	8	15	0	4
S. bracteolata:	flowering	16	0	1	0	14	0	0	0	1	14	0	3
	fruiting		0	1	0	14	0	0	0	1	14	0	2
S. curtisii:	flowering	22	0	0	0	18	0	4	0	5	12	0	16
	fruiting		0	0	0	16	0	4	0	4	4	0	16
S. dasyphylla:	flowering	4	0	0	0	3	0	0	0	0	2	1	0
	fruiting		0	0	0	3	0	0	0	0	2	0	0
S. laevis:	flowering	10	0	0	0	8	0	0	0	0	1	0	1
	fruiting		0	0	0	8	0	0	0	0	1	0	1
S. lepidota:	flowering	20	0	0	0	9	0	0	0	0	9	0	1
	fruiting		0	0	0	9	0	0	0	0	8	0	0
S. leprosula:	flowering	49	2	5	0	47	1	0	1	16	28	0	16
	fruiting		0	2	0	41	0	0	1	11	21	0	15
S. macroptera:	flowering	26	0	2	1	26	0	0	0	2	10	1	5
	fruiting		0	1	1	26	0	0	0	2	10	0	4
S. maxwelliana:	flowering	28	0	0	0	17	1	0	0	1	24	1	0
	fruiting		0	0	0	16	0	0	0	1	23	0	0
S. multiflora:	flowering	6	0	0	0	5	1	1	0	0	1	0	5
	fruiting		0	0	0	5	0	1	0	0	1	0	5
S. ovalis:	flowering	10	0	2	0	6	0	0	0	1	5	0	1
	fruiting		0	2	0	6	0	0	0	1	5	0	1
S. parvifolia:	flowering	33	0	8	0	28	0	4	1	8	29	0	21
	fruiting		0	6	0	25	0	4	0	8	29	0	19
S. pauciflora:	flowering	6	0	0	0	6	0	0	0	1	1	0	3
	fruiting		0	0	0	4	0	0	0	0	1	0	3
S. platyclados:	flowering	28	3	0	0	26	4	8	0	1	13	1	7
	fruiting		3	0	0	26	3	7	0	1	12	0	7
S. roxburghii:	flowering	4	0	0	1	1	1	1	0	0	0	0	0
	fruiting		0	0	1	0	1	1	0	0	0	0	0
S. sumatrana:	flowering	19	0	1	0	1	0	3	9	15	11	0	5
	fruiting		0	1	0	1	0	3	9	15	11	0	5

* Some trees were lost through natural death and wind throw during the period of observation

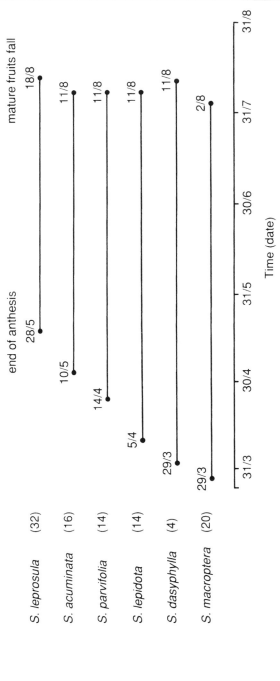

Figure 3.6   Relative rates of fruit development of six *Shorea* species (after Chan, 1981)

31

During the 1976 general flowering in Pasoh, heavy shedding of abortive immature fruits occurred during the first 2 weeks following anthesis. Out of the 100 trees observed to have started fruiting, 18 failed to produce mature fruits. Despite the disparity in flowering time among the six closely related species mentioned above, ripe fruits of these species were observed at around the same time during the first 3 weeks of August (Figure 3.6). The period of flowering to fruit-ripening of the earlier flowering species (*S. macroptera*) was longer than that of the later flowering species (*S. leprosula*).

DISCUSSION

Flowering of dipterocarps has been described as general (intense and gregarious flowering which is synchronized at irregular intervals of several years) or isolated (sporadic flowering which involves several individuals of a species or several species and occurs during intermittent years). From this study, it is evident that flowering of *Shorea* species occurs every year, with years of discernible increased flowering activity. There is, however, no marked distinction between general and isolated flowering in the number of species and individuals. A gradation of flowering intensities can be observed. General flowering intensities can range from 100% to 81.3% of the species and 83.2% to 35.5% of individuals flowering around the same period, while isolated flowering intensities can range from 68.8% to 12.5% of the species and from 20.1% to 0.7% of individuals. It appears, therefore, that there are years that should be rightly termed "intermediate flowering years" in addition to the acclaimed general and sporadic flowering years.

Burgess (1972) noted that often less than 50% of mature individual dipterocarp trees flower even during good flowering years. This was also supported by Cockburn (1975). Appanah (1985) observed 63% of dipterocarp trees flowering during the 1976 general flowering at Pasoh. Our observations of 83.2%, 64.25% and 35.5% individuals flowering at Ampang, Gombak, Kepong and Pasoh during the 1976, 1981 and 1983 general flowering, therefore, reaffirm that general flowering is widespread and can vary in intensity. A similar trend has also been observed in other families at Pasoh (Yap, 1982).

Trees at different sites (all within a 100 km radius) behave differently. Even in the good flowering year of 1976, more trees in Gombak and Ampang flowered in comparison to the other sites. In 1981, more intense flowering was observed at Kepong and Pasoh, while, in 1983, only trees in Gombak flowered intensely. Within a species, trees may flower at one site but not at another. It is therefore important that several populations be monitored before drawing general conclusions about the flowering characteristics of a species.

Flowering of dipterocarps is more regular in the seasonal evergreen forest. Species found in the northern belt of Peninsular Malaysia (Zone 1 in Burgess, 1972), which is at the very edge of the seasonal forest, behave differently from those south of it. Annual flowering had been recorded by Symington (1943) in *Hopea ferrea* and *Shorea roxburghii* (*S. talura*) found in this region.

Besides variation in flowering patterns in different climatic zones, Wood (1956) observed that nearly all the 20 species which did not flower during the gregarious year of 1955 occurred in freshwater swamps or on mountains above 1000 m elevation. This out-of-phase flowering of swamp species has also been documented by Wyatt-Smith (1963) in *S. rugosa*.

In terms of flowering frequency, members of the section *Mutica* of the Red Meranti Group of *Shorea* flower more regularly and have the ability to flower during consecutive years. The flowering behaviour as displayed by these trees appears to contradict the postulation that there is a need for a gradual internal accumulation of assimilates for the initiation of the flowering process among dipterocarps (Ashton, 1982; Burgess, 1972; Palmer, 1979; Wycherley, 1973).

Although flowering appears to be somewhat synchronous among species within and between sites, as exemplified during the general flowering years, detailed observations on closely related species showed that flowering peaks are actually temporally isolated, although intraspecific flowering is synchronous. This staggering in flowering times among closely related species may be a useful mechanism for reducing competition for similar pollinators and may explain the paucity of natural hybrids among the genus *Shorea* (see also Ashton *et al.*, 1988).

Wood (1956) recorded a delay in flowering of species in the west coast of Sabah when compared to that in the east coast. A distinct delay was also detected in the flowering of the same species at higher altitudes. This disparity of flowering time has also been recorded by Sasaki *et al.* (1979) in the west and east of the Main Range in Peninsular Malaysia.

Both Burgess (1972) and Ng (1977) reported that gregarious flowering of dipterocarps tends to occur during the second quarter of the year, at around April to May, although there are some trees flowering every month of the year. Through analysis of 952 herbarium collections in Sabah, Cockburn (1975) found the flowering peak in May. In contrast, flowering during the last quarter of the year was reported by Yap (1987) in 4 out of the 12 years of observation. This second peak corresponds with that of some understorey fruit tree species such as *Xerospermum noronhianum* (*X. intermedium*), as observed by Yap (1982). This suggests that gregarious flowering can occur at any month of the year. In the 1981 gregarious flowering, an earlier flowering of five species was observed in June at Gombak and Ampang while most of the peak flowering was during September. Interestingly, *S. acuminata* and *S. leprosula* participated in both flowering episodes. This out-of-phase flowering prior to the September general flowering suggests that the stimulus in the earlier part of the year was not strong enough to trigger off a general flowering during the second quarter of the year until the later stage. The intensity of flowering every year may be controlled by the strength of the stimulus having different influences on different species.

Though general flowering of dipterocarps is normally followed by mast fruiting, not all trees set fruits. Our observations on successful fruit-set ranged from 79% to 87% following the 1976, 1981 and 1983 general flowering. The percentage of trees which set fruit was drastically reduced following isolated flowering, particularly those of species belonging to the section *Mutica* of the Red Meranti Group. This may be the result of low pollination success, because many species of this section have been found to be highly self-incompatible (Chan, 1981) and out-crossing among isolated trees during an out-of-phase flowering would be most unlikely. On the other hand, consistently successful fruit set in some species (*S. ovalis*, *S. laevis*, *S. multiflora* and *S. sumatrana*) may be due to their apomictic mode of reproduction. This observation has been confirmed by Kaur (1977) for *S. ovalis*.

In summary, this study has shown that flowering of *Shorea* species is more frequent than previously reported, and there are some that flower almost annually. In addition to the well-known general flowering that occurs once in many years and the isolated flowering, there appears to be also an intermediate intensity of flowering activity which appears frequently. Besides variation in the flowering patterns in different sites, interspecific and intraspecific variation has been observed in the 16 species studied. Members of Red Meranti Group are able to bear flowers more regularly compared with the other groups. This result emphasizes the importance of having a number of trees of different populations monitored before a description of the flowering pattern of a species can be made. Although there is a staggering in the flowering times of different species in an area, ripe fruits are found at around the same period. A high proportion of flowering trees bear fruits during the good flowering years, while most trees do not produce fruits during sporadic flowering years. All these phenological characteristics are of great importance in rain forest management and silvicultural prescription, as *Shorea* trees are the most important commercial timber species in Malaysia.

REFERENCES

Appanah, S. (1985). General flowering in the climax rain forests of South-east Asia. *Journal of Tropical Ecology*, 1, 225 40

Ashton, P.S. (1969). Speciation among tropical forest trees: some deductions in the light of recent evidence. *Biological Journal of the Linnean Society*, 1, 155–96

Ashton, P.S. (1982). Dipterocarpaceae. *Flora Malesiana* Series I, **Vol.9, Part 2**, 237–600

Ashton, P.S., Givnish, T.J. and Appanah, S. (1988). Staggered flowering in the Dipterocarpaceae: new insights into floral induction and the evolution of mast fruiting in the aseasonal tropics. *American Naturalist*, **132**, 44–66

Burgess, P.F. (1972). Studies on the regeneration of the hill forests of the Malay Peninsula: The phenology of dipterocarps. *Malayan Forester*, **35**, 103–23

Chan, H.T. (1981). Reproductive biology of some Malaysian dipterocarps: III. Breeding systems. *Malaysian Forester*, **44**, 28–36

Cockburn, P.F. (1975). Phenology of dipterocarps in Sabah. *Malaysian Forester*, **38**, 160–70

Foxworthy, F.W. (1932). Dipterocarpaceae of the Malay Peninsula. *Malayan Forest Records*, **10**.

Janzen, D.H. (1974). Tropical blackwater rivers, animals and mast fruiting by the Dipterocarpaceae. *Biotropica*, **6**, 69–103

Kaur, A. (1977). Embryological and cytological studies on some members of the Dipterocarpaceae. Ph.D. Thesis. University of Aberdeen, Aberdeen.

McClure, H.E. (1966). Flowering, fruiting, and animals in the canopy of a tropical rain forest. *Malayan Forester*, **29**, 182–203

Medway, Lord (1972). Phenology of a tropical rain forest in Malaya. *Biological Journal of the Linnean Society*, **4**, 117–46

Ng, F.S.P. (1977). Gregarious flowering of dipterocarps in Kepong, 1976. *Malaysian Forester*, **40**, 126–37

Ng, F.S.P. (1988). Forest tree biology. In Earl of Cranbrook (ed.) *Key Environments: Malaysia.*, pp. 102–25. (Oxford: Permagon Press)

Palmer, J.R. (1979). Gregarious flowering of dipterocarps in Kepong, 1976. Letter to Editor. *Malaysian Forester*, **42**, 74–5

Ridley, H.N. (1901). The timbers of the Malay Peninsula. *Agricultural Bulletin, Straits Settlement and Federated Malay States*, **1**, 53

Sasaki, S., Tan, C.H. and Zulfatah, A.R. (1979). Some observations on unusual flowering and fruiting of dipterocarps. *Malaysian Forester*, **42**, 38–45

Symington, C.F. (1943). *Foresters' Manual of Dipterocarps*. Malayan Forest Records **16**.

Whitmore, T.C. (1976). *Tropical Rain Forests of the Far East*. (Oxford: Clarendon Press)

Wood, G.H.S. (1956). Dipterocarp flowering season in North Borneo, 1955. *Malayan Forester*, **19**, 195–201

Wyatt-Smith, J. (1963). *Manual of Silviculture for Inland Forests*. **Vol. 1**. Malayan Forest Records **23**.

Wycherley, P.R. (1973). Phenology of plants in the humid tropics. *Micronesica*, **9**, 21–35

Yap, S.K. (1982). Phenology of some fruit tree species in a lowland dipterocarp forest. *Malaysian Forester*, **45**, 21–35

Yap, S.K. (1987). Gregarious flowering of dipterocarps: observations based on fixed tree populations in Selangor and Negri Sembilan, Malay Peninsula. In Kostermans, A.J.G.K. (ed.) *Proceedings of the Third Round Table Conference on Dipterocarps*. Samarinda, Indonesia. 16–20 April 1985. pp. 305–317. (Jakarta: Unesco)

CHAPTER 4

PLANT PHENOLOGY, POLLINATION ECOLOGY, POLLINATOR BEHAVIOUR AND CONSERVATION OF POLLINATORS IN NEOTROPICAL DRY FOREST

Gordon W. Frankie, S.B. Vinson, Linda E. Newstrom,
John F. Barthell, William A. Haber and Jutta K. Frankie

ABSTRACT

A review of 20 years of research in Costa Rica from 1968 to the present, on phenology and plant–pollinator interactions is presented within the context of four principal time phases. Most of the early research was based on plant phenological studies that established the flower and fruit resource base in lowland dry and wet forests. After 1974, research was conducted almost entirely in the dry forest on plant–pollinator interactions. Once it was determined that bees and moths were the most important pollinators, attention shifted to the biology and ecology of these two insect groups. Most of the recent research in the dry forest has centered on the anthophorid bees in the genus Centris, *with an emphasis on their nesting biology, chemical ecology, habitat preferences and population ecology.* Centris *bees were singled out because of their numerical and functional importance as pollinators of a high proportion of the tree and climber species. An examination of factors limiting their populations offers explanation as to why* Centris *(and other) bees have greatly declined in numbers during recent years. Such an examination also suggests possible courses of action that biologists may be obliged to pursue to conserve pollinator faunas.*

INTRODUCTION

During the past 20 years, we have conducted a long series of related studies on the reproductive ecology of plants and their insect pollinators in the lowland wet and dry forests of Costa Rica. Early work focused on surveys of general community-wide patterns of plant phenology and plant–pollinator interactions. As the investigations developed, experimental studies were added, which increasingly focused on floral behaviour, bee pollinator behaviour, nesting biology and chemical ecology. Ongoing work now consists of a mixture of experimental, population and conservation biology studies. Throughout the entire 20-year period, research directions were dictated by the organisms themselves and what they revealed to the research group.

In this paper, we briefly review our past work and the most significant findings. We also review relevant findings of other colleagues who have worked on related problems in plant reproductive biology at the same site. Past work is described during four major periods: 1968–74, 1975–80, 1981–85 and 1986–present. In the last period, we describe ongoing work that is mostly concerned with behaviour, population ecology and chemical ecological studies on native solitary bees. Finally, we outline recommendations for the future conservation and management of bees and other pollinators.

STUDY SITES

Studies were conducted in lowland wet and dry forest sites in Costa Rica. The wet forest, La Selva Biological Station (Heredia Province), is the internationally known teaching and research site of the Organization for Tropical Studies. The La Selva forest has been described in Frankie *et al.* (1974*a*) and Janzen (1983).

The dry forest study area consisted of several sites within a 20 000 ha region of southern Guanacaste Province (see descriptions and maps in Frankie *et al.*, 1974*a*, 1974*b*, 1983). The site has since been divided into a complex with different owners and administrations, consisting of the Lomas Barbudal Biological Reserve, Palo Verde Wildlife Refuge, government agricultural land, and private land. Despite the land divisions and varying management policies, many of the dry forest sites are still being studied by biologists.

PHASE I RESEARCH: WET AND DRY FORESTS, 1968–1974

The original and general goal of this phase, which began in 1968, was to learn as much as possible about the reproductive ecology of lowland wet and dry forest plants. Intensive surveys were designed to gather this information. Trees were the early focus because of ongoing co-operative studies with other research groups in the same forests.

The research started with systematic phenological studies of trees, treelets and shrubs. The work was conducted in both forests until 1974, after which it was continued only in the wet forest, until 1981. The first papers on these studies (Frankie *et al.*, 1974*a*, *b*; Opler *et al.*, 1976, 1980) provided broad overviews on leaf, flower and fruit phenological patterns of the above life forms. For example, with regard to leaf patterns, most trees in both forests lost leaves in the dry season; leaf-flushing periodicity depended on individual tree species and on seasonal rainfall. Flowering of trees in both forests was concentrated in the dry season; however, substantial flowering was also recorded in both forests during the wet season. Mature fruit production was mostly observed in the dry forest during the dry season; in the wet forest, it occurred primarily in the wet season. This work also revealed many ecological characteristics of the flower and fruit resources for studies that followed on plant reproductive ecology.

Wet forest observations were continued uninterruptedly from 1968–1981. The resulting 12.5-year data set, which focused on trees, is currently being analysed for long-term individual tree patterns and general community patterns. Early analyses have revealed several distinctive long-term flowering patterns of individual trees: continuous, seasonal, episodic and supra-annual (Newstrom *et al.*, 1990). During this first phase, general observations were also gathered on floral behaviour and pollinators in both forests (fruit and seed studies were de-emphasized after 1970). An analysis of the observations indicated that bees and moths accounted for the vast majority of pollinator types in each forest (Frankie, 1975).

Bawa's classic work (1974) on controlled pollinations, which demonstrated that most dry forest trees were obligate outcrossers, placed great importance on pollinator groups suspected of moving between trees. Experimental studies in 1972 on bee foraging behaviour in the dry forest (Frankie *et al.*, 1976) identified several bee taxa that made intertree movements. By the end of 1974, it was clear that far more work could be accomplished in dry than wet forest because of the ease of working with shorter trees. Further, the lower tree diversity (by three times) in the dry forest meant that forest-wide patterns could be unravelled sooner. Thus, most research after 1974 shifted to the dry forest.

PHASE II RESEARCH: DRY FOREST, 1975–1980

A large portion of the work during the second phase focused on developing detailed information on floral biology and pollinator behaviour of most tree species and, to a lesser extent, other life forms in the dry forest. The first paper to emerge from this work (Frankie *et al.*, 1983) emphasized characteristics of the large bee pollination system, which is found most commonly among trees and climbers. The work also emphasized the distinction between large-bee and small-bee flowers. Studies on the hawkmoth (Haber, 1983, 1984; Haber and Frankie, 1982, 1989) and small-moth pollination systems (in prep.), also common in the dry forest, provided a wealth of information on nocturnal pollination. Early work by Heithaus *et al.* (1975) demonstrated the importance of bats and bat flowers in the dry forest; however, this pollination system is encountered far less frequently than those involving bees and moths. Finally, a study on beetle, butterfly and hummingbird pollination indicated that these systems were relatively rare among dry forest tree species (Haber and Frankie, in prep.).

Because bees were determined to be numerically and functionally the most important pollinator group in the dry forest, research focused on this insect group. It is estimated that 250 + species of bees occur throughout the dry forest; the vast majority are solitary in habit.

Intensive work on bee foraging in 1972, using capture-mark-recapture methods, demonstrated that several *Centris* species, *Gaesischia exul* (both Anthophoridae) and a megachilid bee moved rapidly and over long distances among flowering individuals of *Andira inermis*, a common tree in the Fabaceae (Frankie *et al.*, 1976). Pilot work, which preceded this study, indicated that *Centris* species, *G. exul* and *Trigona* species also forage widely among trees of *Byrsonima crassifolia*, *Pterocarpus rohrii* and vines of *Securidaca sylvestris* (G. Frankie, P. Opler and K. Bawa, unpub.). Another study demonstrated that large bees preferred to forage higher in the crowns and canopies of trees than small bees (Frankie, 1975; Frankie and Coville, 1979), which corresponded with the observations that large-bee flowers are mostly in the canopy of dry forests. Observations on a similar stratification of bee flowers have been made by Perry (1984) and Bawa *et al.* (1985) in the La Selva wet forest on the Atlantic lowland side of the country. However, one group of large bees, the Euglossini (Apidae), are also known to be specialists on understorey flowers in both forests.

The foraging behaviour of bees was found to be influenced by chemical, spatial and temporal variations in floral resources among individual plants of a given species (Frankie *et al.*, 1982; Frankie and Haber, 1983). Studies of nectar chemistry (especially sugars and amino acids) of many dry forest species by Baker and Baker (1982, 1983) are directly related to this research. These studies document in detail the chemical diversity of floral nectars.

During this research phase, studies were also initiated on nesting biology, male territoriality, mating behaviour and chemical ecology of selected *Centris* bee species. These bees were selected because of their importance as pollinators and their high numerical abundance in many intact dry forest locations (Frankie and Baker, 1974; Frankie *et al.*, 1976; Heithaus, 1979*a*, 1979*b*, 1979*c*). The nesting biology of *C. aethyctera*, a common ground nesting species was described (Vinson and Frankie, 1977). Male territorial and mating behaviour of *C. adani* (=*C. adanae*), also a common ground nester (Frankie *et al.*, 1980), and the chemicals associated with scent-marking in *C. adanae* (Vinson *et al.*, 1982) were also described.

PHASE III RESEARCH: DRY FOREST, 1981–1985

An intensive effort was made to study territorial behaviour in most of the 16 or so *Centris* species that occur in the dry forest. Information was gathered on 13 of the species, and much of it is reported in Coville *et al.* (1986) and Frankie *et al.* (1989). Associated chemical ecology of *Centris* revealed two types of scent-marking glands, mandibular and leg glands, and their respective chemical compositions (Vinson *et al.*, 1982, 1984; Williams *et al.*, 1984; Coville *et al.*, 1986). Chemicals from both types of glands are used to scent mark territories, which in turn appeared to attract females. This period of research was assisted by a new taxonomic treatment of *Centris* and the closely related *Epicharis* by Snelling (1984).

Casual observations from 1978–1983 on carpenter bees (*Xylocopa*, a related genus in the Anthophoridae) revealed that these large bees had much in common with *Centris* from the standpoint of male territoriality and mating behaviour. They are also important pollinators despite their low densities as compared to *Centris*. Because these reproductive traits were easier to study in *Xylocopa* than *Centris*, studies on the two largest species, *X. fimbriata* and *X. qualanensis*, were initiated. A heretofore unknown scent-marking gland in the males of both species, which they apparently use for attracting females for mating, was described and the scents characterized (Vinson *et al.*, 1986; Williams *et al.*, 1987; see also Andersen *et al.*, 1988). Details of mating have been worked out for these two *Xylocopa* species (Vinson and Frankie, 1990); however, only a partial mating story has been developed to date for *Centris*.

Several studies on the nesting biologies of most dry forest *Centris* species (16) were conducted during Phase III. Two major groups, ground nesters with eight species and tree hole nesters with seven to eight species, were recognized. Emphasis was placed on ground nesters first because of the ease of finding and observing these *Centris* at ground level. Information on nest architecture, preferred soil types and parasites of ground nesters is found in Vinson and Frankie (1977), Coville *et al.* (1983, 1986) and Vinson *et al.* (1987). The tree hole species were more difficult to study until an experimental method was developed to lure them to artificial traps for careful study (see next section).

PHASE IV RESEARCH: 1986–PRESENT

During the past 4 years, our efforts have focused on nesting biology, habitat preferences and population ecology of several *Centris* species, with an emphasis on understanding factors limiting populations.

Nesting biology

Nesting biologies of most ground – and tree – hole nesting *Centris* are now known. Early work on *C. aethyctera* (Vinson and Frankie, 1977) provided guidance for more detailed studies on other ground nesters. In the case of *C. segregata* (Coville *et al.*, 1983), routine nest excavations and rearings led to two important findings. First, two distinct types of males were discovered in this species ("beta" and "regular" males in the sense of Alcock *et al.*, 1976, 1977). Second, evidence was gathered to suggest that *C. segregata* is actually a dimorphic form of *C. inermis* (see discussion in Snelling, 1984). Work on *C. heithausi* and *C. flavofasciata* nesting provided information on mortality due to parasites and possible alternate-year emergences of adults in the case of *C. flavofasciata* (Coville *et al.*, 1986; Vinson *et al.*, 1987). The *C. flavofasciata* study also revealed beta and regular males (GWF and SBV, unpub.). Nesting biologies of tree hole nesting *Centris* have been worked out for most dry forest *Centris*. A series of nest construction characteristics readily separates each species (Frankie *et al.*, 1989).

Habitat preferences

Survey and experimental studies have indicated habitat preferences for both ground and tree hole nesters. A variety of soil and habitat preferences for ground nesters are indicated in the previously mentioned papers. With regard to tree hole nesters, studies have shown an overall preference by most species for shaded subsites in mesic and riparian forests. *Centris nitida*, however, displays a strong tendency to nest in oak forests (Frankie *et al.*, 1988). One extremely rare species, *C. lutea*, may prefer abandoned nests of carpenter bees in savanna vegetation (GWF and SBV, unpub.). Overall, the occurrence of up to 12 sympatric species (ground and tree hole nesters) in a given block of forest is related to the diverse nesting habitats found in this genus (Frankie *et al.*, 1989). (See also following paragraphs on limiting factors).

Population ecology

Work on nesting and habitat preferences has led logically to studies on the population ecology of *Centris* bees. These studies are concerned with long-term population monitoring and identification and quantification of mortality factors. An annual forest-wide monitoring programme of tree hole nesting *Centris* was begun in 1986. The monitoring was aimed at:

(1) Developing density information on *Centris* in representative habitats;

(2) Providing supplemental data on habitat preferences; and,

(3) Rearing natural enemies.

To date, characteristic densities and general habitat preferences for four of six tree hole *Centris* have been determined (Frankie *et al.*, 1988). Parasitic bees, flies and beetles have been reared, and most taxa have been identified to species.

Limiting factors

Important limiting factors have been determined for both ground inhabiting and tree hole *Centris*. With regard to ground nesters, surveys indicate that certain soil types are preferred for nesting by most species. For example, *C. flavifrons*, *C. fuscata* and *C. inermis* appear to prefer sandy loam soils (Vinson and Frankie, 1988). Nests constructed by the bees are shallow and thus easily disturbed by farmers, who also prefer this soil type for agricultural purposes. Extensive ploughing in the dry season over many years appears to have been responsible for much of the devastation of ground nesting bee populations (GWF and SBV, unpub.). Censusing since 1972 suggests an overall decrease of 80 + % in ground nesting *Centris*, as well as other anthophorids: *Gaesischia exul* and several *Epicharis* species. *Centris adanae* and *C. aethyctera* exist at slightly higher densities than other ground nesting *Centris*, and this may be related to the more general soil requirements of these species (SBV and GWF, unpub.). Finally, a marked preference for selected beach strand soils (almost pure sand) greatly limits the distribution of two species, *C. aethiocesta* and *C. flavofasciata* (Vinson *et al.*, 1987; Vinson and Frankie, 1988). Populations of these two species are threatened in areas where beach front development is inevitable.

Fire is another important limiting factor. In the case of ground nesting *Centris*, fire burns off vegetation that these species may use for orientation to nest sites and cover for protection against parasitization. Preliminary surveys indicate that parasites can more readily locate *Centris* nests in burned versus unburned sites. When parasite activity is high in an area, females tend to construct fewer cells per nest (GWF and SBV, unpub.).

Fire also limits the number of available nest sites to tree nesters by its destruction of dead branches on live trees. This dead wood contains cavities made by larvae of wood-boring beetles that several *Centris* species use for nesting. Fire also burns hot in some patches, thus opening up forests, which then become invaded by grass, especially the highly combustible jaragua grass from Africa. If fires continue to burn these patches year after year, large openings or savannas will eventually result. This kind of progressive degradation leads to hotter, drier forest environments that are unsuitable for tree hole nesting *Centris* bees (Frankie *et al.*, 1988). Overall, fire kills trees that provide important pollen and nectar resources for all bees. Some of the best food trees for bees are also the most susceptible species to fire damage and include three *Tabebuia* species (Bignoniaceae) and *Dalbergia retusa* (Fabaceae).

Riparian, mesic and well-developed oak forests are poorly represented in most conserved dry forests of Guanacaste. Because of their poor representation and the propensity of tree hole *Centris* to nest in these habitats, these habitats are considered limiting.

Finally, we have identified one limiting factor common to both ground nesters and tree hole nesters – the oil-producing flower of the Malpighiaceae. Flowers of *Stigmatophyllon* vines (Anderson, 1987) and especially trees of *Byrsonima crassifolia* provide essential oil for nest construction and/or brood provisioning for *Centris* and *Epicharis* species (Buchmann, 1987; Vinson *et al.*, 1991). The oil is considered a key resource of these bees. Areas where members of this family are rare, are noticeably lacking in these bee taxa.

CONSERVATION IMPLICATIONS AND CONSIDERATIONS

During the development of our research programme, it became apparent that one of our dry forest research sites, Lomas Barbudal, supported a high diversity of bees, especially in the genus *Centris*. At that time, it was also realized that without an appropriate management effort the site would be burned and probably converted to cattle pasture. Once the site (2300 ha) was established as an official reserve of the Costa Rican national park system in 1986, our research group became actively involved with its management. A large part of this involvement was spent aggressively pursuing projects designed to protect critical habitats for bees (and other pollinator groups). The following list of recommendations reflects the variety of activities that have proved useful for protecting bee habitats at the Lomas Barbudal reserve.

(1) Conduct biological assessments to characterize as much of the habitat diversity and selected bee species diversity as possible. Be prepared to assess other important organism groups as well;

(2) Conduct biological assessments of non-represented habitats that lie just outside or in the near vicinity of the conserved area, with the possibility that these non-conserved areas may function as important reservoirs of

target pollinator groups and that, one day, the opportunity may arise to conserve these areas with the proper biological justification;

(3)   Conduct research to demonstrate the importance or preference of bees for particular habitats. Publish results of scientific findings in journals that allow for summaries in the language of the host country where research was conducted (if other than English);

(4)   Present findings of research to knowledgeable biologists of the host country for their suggestions as to introducing the assessment information into the decision-making administrative structure of the local/regional/national conservation community; and,

(5)   Be prepared to participate in a variety of hands-on conservation projects to see research findings implemented in some effective manner. Be prepared to serve on technical committees in host countries where decision-making on conservation matters requires input from biologists.

The steps outlined above represent considerations and courses of action that we have found necessary to follow since 1979. The luxury of conducting research in a wide variety of field sites no longer exists as it did when we first began our work in 1968.

ACKNOWLEDGEMENTS

We thank the National Science Foundation and the National Geographic Society for major support. The University of California Research Expedition Programme, and the California/Texas Agricultural Experiment Stations provided support for selected studies. Several individuals offered technical and/or logistical assistance during all or most research phases; these are Paul Opler, Herbert and Irene Baker, Lili and Werner Hagnauer and the David Stewart family. Rita Alfaro, Mario Boza and Luis Gomez provided moral and administrative assistance on matters related to conservation biology.

REFERENCES

Alcock, J., Jones, C.E. and Buchmann, S.L. (1976). Location before emergence of the female bee, *Centris pallida*, by its male (Hymenoptera: Anthophoridae). *Journal of Zoology, London*, **179**, 189–99

Alcock, J., Jones, C.E. and Buchmann, S.L. (1977). Male mating strategies in the bee *Centris pallida* Fox (Anthophoridae: Hymenoptera). *American Naturalist*, **111**, 145–55

Andersen, J.F., Buchmann, S.L., Weisleder, D., Plattner, R.D. and Minckley, R.L. (1988). Identification of thoracic gland constituents from male *Xylocopa* spp. Latreille (Hymenoptera:

Anthophoridae) from Arizona. *Journal of Chemical Ecology*, **14**, 1153–62

Anderson, C. (1987). *Stigmaphyllon* (Malphighiceae) in Mexico, Central America, and the West Indies. *Contributions of University of Michigan Herbarium*, **16**, 1–48

Baker, H.G. and Baker, I. (1982). Chemical constituents of nectar in relation to pollination mechanisms and phylogeny. In Nitecki, M.H. (ed.). *Biochemical Aspects of Evolutionary Biology*, pp. 131–71. (Chicago: University of Chicago Press)

Baker, H.G. and Baker, I. (1983). Floral nectar sugar constituents in relation to pollinator type. In Jones, C.E. and Little, R.J. (eds). *Handbook of Experimental Pollination Biology*, pp. 117–41. (New York: Van Nostrand Reinhold))

Bawa, K.S. (1974). Breeding systems of tree species of a lowland tropical community. *Evolution*, **28**, 85–92

Bawa, K.S., Bullock, S.H., Perry, D.R., Coville, R.E. and Grayum, M.H. (1985). Reproductive biology of tropical lowland rain forest trees. II. Pollination systems. *American Journal of Botany*, **72**, 346–56

Buchmann, S.L. (1987). The ecology of oil flowers and their bees. *Annual Review of Ecology and Systematics*, **18**, 343–69

Coville, R.E., Frankie, G.W. and Vinson S.B. (1983). Nests of *Centris segregata* (Hymenoptera: Anthophoridae) with a review of the nesting habits of the genus. *Journal of Kansas Entomological Society*, **56**, 109–22

Coville, R.E., Frankie, G.W., Buchmann, S.L., Vinson, S.B. and Williams, H.J. (1986). Nesting and male behavior of *Centris heithausi* (Hymenoptera: Anthophoridae) in Costa Rica with chemical analysis of the hindleg glands of males. *Journal of Kansas Entomological Society*, **59**, 325–36

Frankie, G.W. (1975). Tropical forest phenology and pollinator plant coevolution. In Gilbert, L.E. and Raven, P.H. (eds). *Coevolution of Animals and Plants*, pp. 192–209. (Austin: University of Texas Press)

Frankie, G.W. (1976). Pollination of widely dispersed trees by animals in Central America, with an emphasis on bee pollination systems. In Burley, J. and Styles, B.T. (eds). *Variation, Breeding and Conservation of Tropical Forest Trees*, pp. 151–9. (London: Academic Press)

Frankie, G.W. and Baker, H.G. (1974). The importance of pollinator behavior in the reproductive biology of tropical trees. *Annals Institute Biologia Universidad, Mexico* 45, Serv. Bot., **(1)**, 1–10

Frankie, G.W. and Coville, R.E. (1979). An experimental study on the foraging behavior of selected solitary bee species in the Costa Rican dry forest. *Journal of Kansas Entomological Society*, **52**, 591–602

Frankie, G.W. and Haber, W.A. (1983). Why bees move among mass-flowering neotropical trees. In Jones, C.E. and Little, R.J. (eds). *Handbook of Experimental Pollination Biology*, pp. 361–72 (New York: Van Nostrand Reinhold)

Frankie, G.W., Baker, H.G. and Opler, P.A. (1974a). Comparative phenological studies of trees in tropical wet and dry forests in the lowlands of Costa Rica. *Journal of Ecology*, **62**, 881–919

Frankie, G.W., Baker, H.G. and Opler, P.A. (1974b). Tropical plant phenology: Applications for studies in community ecology. In Lieth, H. (ed.) *Phenology and Seasonality Modelling*, pp. 287–96. (New York: Springer-Verlag)

Frankie, G.W., Opler, P.A. and Bawa, K.S. (1976). Foraging behaviour of solitary bees: Implications for outcrossing of a neotropical forest tree species. *Journal of Ecology*, **64**, 1049–57

Frankie, G.W., Vinson, S.B. and Coville, R.E. (1980). Territorial behavior of *Centris adani* and its reproductive function in the Costa Rican dry forest (Hymenoptera: Anthophoridae). *Journal of Kansas Entomological Society*, **53**, 837–57

Frankie, G.W., Haber, W.A., Baker, H.G. and Baker, I. (1982). A possible chemical explanation for differential foraging by anthophorid bees on individuals of *Tabebuia rosea* in the Costa Rican dry forest. *Brenesia*, **19/20**, 397–405

Frankie, G.W., Haber, W.A., Opler, P.A. and Bawa, K.S. (1983). Characteristics and organization of the large bee pollination system in the Costa Rican dry forest. In Jones, E.C. and Little, R.J. (eds). *Handbook of Experimental Pollination Ecology*, pp. 411–47 (New York: Van Nostrand Reinhold)

Frankie, G.W., Vinson, S.B., Newstrom, L.E. and Barthell, J.F. (1988). Nest site and habitat preferences of *Centris* bees in the Costa Rican dry forest. *Biotropica*, **20**, 301–10

Frankie, G.W., Vinson, S.B. and Williams, H. (1989). Ecological and evolutionary sorting of 12 sympatric species of *Centris* bees in Costa Rican dry forest. In Bock, J.H. and Linhart, Y.B. (eds.) *The*

Evolutionary Ecology of Plants, pp. 535–49. (Boulder, San Francisco and London: Westview Press)

Haber, W.A. (1983). Insects: Checklist of Sphingidae. In Janzen, D.H. (ed.) *Costa Rican Natural History*, pp. 645–7. (Chicago: University of Chicago)

Haber, W.A. (1984). Pollination by deceit: *Plumeria rubra. Biotropica*, **16**, 269–75

Haber, W.A. and Frankie, G.W. (1982). Pollination ecology of *Luehea* (Tiliaceae) in Costa Rican deciduous forest. *Ecology*, **63**, 1740–50

Haber, W.A. and Frankie, G.W. (1989). A tropical hawkmoth community: Costa Rican dry forest Sphingidae. *Biotropica*, **21**, 155–72

Heithaus, E.R. (1979*a*). Community structure of neotropical flower visiting bees and wasps: diversity and phenology. *Ecology*, **60**, 190–202

Heithaus, E.R. (1979*b*). Flower-feeding specialization in wild bee and wasp communities in seasonal neotropical habitats. *Oecologia*, **42**, 179–94

Heithaus, E.R. (1979*c*). Flower visitation records and resource overlap of bees and wasps in northwest Costa Rica. *Brenesia*, **16**, 9–52

Heithaus, E.R., Fleming, T.H. and Opler, P.A. (1975). Foraging patterns and resource utilization in seven species of bats in a seasonal tropical forest. *Ecology*, **56**, 841–54

Janzen, D.H. (ed.) (1983). *Costa Rican Natural History*. (Chicago: University of Chicago Press)

Newstrom, L.E., Frankie, G.W., Baker, H.G. and Colwell, R.C. (1990). Phenology in the tropical rain forest at la Selva. In McDade, L.A., Bawa, K.S., Harshorn, G.S. and Hespenhelde, H.A. (eds.) *La Selva: Ecology and Natural History of a Neotropical Rain Forest.* (University of Chicago Press) (in review)

Opler, P.A., Frankie, G.W. and Baker, H.G. (1976). Rainfall as a factor in the release, timing and synchronization of anthesis by tropical trees and shrubs. *Journal of Biogeography*, **3**, 231–6

Opler, P.A., Frankie, G.W. and Baker, H.G. (1980). Comparative phenological studies of shrubs in tropical wet and dry forests in the lowlands of Costa Rica. *Journal of Ecology*, **68**, 167–88

Perry, D.R. (1984). The canopy of the tropical rain forest. *Scientific American*, **251**, 138–47

Snelling, P.R. (1984). Studies on the taxonomy and distribution of American Centridine bees (Hymenoptera: Anthophoridae). Contribution No. 347. Natural History Museum, Los Angeles.

Vinson, S.B. and Frankie, G.W. (1977). Nests of *Centris aethyctera* (Hymenoptera: Apoidea: Anthophoridae) in the dry forest of Costa Rica. *Journal of Kansas Entomological Society*, **50**, 301–11

Vinson, S.B. and Frankie, G.W. (1988). A comparative study of the ground nests of *Centris flavifrons* and *Centris aethiocesta* (Hymenoptera: Anthophoridae). *Entomological Experimental Application*, **49**, 181–7

Vinson, S.B. and Frankie, G.W. (1990). Territorial and mating behavior of *Xylocopa fimbriata F.* and *Xylocopa gualanensis* Cockerell from Costa Rica. *Journal of Insect Behavior*, **3**, 13–32

Vinson, S.B., Williams, H.J., Frankie, G.W., Wheeler, J.W., Blum, M.S. and Coville, R.E. (1982). Mandibular glands of male *Centris adani*, their function in scent marking and territorial behavior (Hymenoptera: Anthophoridae). *Journal of Chemical Ecology*, **8**, 319–27

Vinson, S.B., Williams, H. J., Frankie, G. W. and Coville, R.E. (1984). Comparative morphology and chemical contents of the mandibular glands of males of several *Centris* species (Hymenoptera: Anthophoridae) from Costa Rica. *Comparative Biochemistry and Physiology*, **77a**, 685–88

Vinson, S.B., Frankie, G.W. and Williams, H.J. (1986). Description of a new dorsal mesosomal gland in two *Xylocopa* species (Hymenoptera: Anthophoridae) from Costa Rica. *Journal of Kansas Entomological Society*, **59**, 185–9

Vinson, S.B., Frankie, G.W. and Coville, R.E. (1987). Nesting habits of *Centris flavofasciata* Friese (Hymenoptera: Apoidea: Anthophoridae) in Costa Rica. *Journal of Kansas Entomological Society*, **60**, 249–63

Vinson, S.B., Williams, H.J., Frankie, G.W. and Shrum, G. (1991). Floral lipid chemistry of floral lipids by *Centris* bees (Hymenoptera: Anthophoridae). *Biotropica*, (in review).

Williams, H.J., Vinson, S.B., Frankie, G.W., Coville, R.E. and Ivie, G.W. (1984). Description, chemical contents and function of tibial gland in *Centris nitida* and *Centris trigonoides subtarsata* (Hymenoptera: Anthophoridae) in the Costa Rican dry forest. *Journal of Kansas Entomological Society*, **57**, 50–4

Williams, H.J., Vinson, S.B. and Frankie, G.W. (1987). Chemical content of the dorsal mesosomal gland of two *Xylocopa* species (Hymenoptera: Anthophoridae) from Costa Rica. *Comparative Biochemistry and Physiology*, **86B**, 311–12

47

CHAPTER 5

SEASONAL DROUGHT AND THE TIMING OF FLOWERING AND LEAF FALL IN A NEOTROPICAL FOREST

S. Joseph Wright and Fernando H. Cornejo

ABSTRACT

Flowering and leaf fall are concentrated in the dry season in many neotropical forests. The hypothesis that moisture availability is the proximal cue was tested by augmenting water supplies on two 2.25 ha plots on Barro Colorado Island, Panama. Soil moisture was maintained at field saturation throughout the dry season. The timing of flowering (18 species) and leaf fall (43 species) was little affected. Moisture availability is not the proximal cue for flowering and leaf fall for most species in the tropical moist forests of Barro Colorado Island.

INTRODUCTION

Flowering and leaf fall are often concentrated in the dry season in Central American forests (Janzen, 1967; Daubenmire, 1972; Frankie *et al.*, 1974; Croat, 1978). Janzen (1967) suggested that reproduction in the dry season would facilitate pollination and seed dispersal and also free resources for vegetative growth in the wet season, thereby enhancing competitive ability. Borchert (1980) and Reich and Borchert (1982, 1984) eschew this adaptive scenario in favour of a purely physiological mechanism. They suggest that the timing of leaf fall and floral bud break are controlled by the water status of the plant.

Rates of leaf fall for tropical forest communities almost always peak during seasonal drought (see references compiled by Proctor, 1984). It is frequently inferred that avoidance of water stress is the ultimate cause of leaf abscission, and correlations are known to occur between leaf fall and the water status of individual plants (Daubenmire, 1972; Borchert, 1980;

49

Lieberman, 1982; Reich and Borchert, 1982, 1984). However, leaf abscission is a complex process. Many phenomena may affect the timing of abscission and water availability may not be a critical factor (Hopkins, 1966; Haines and Foster, 1977). For example, Bernhard-Reversat *et al.* (1972) found a strong correlation between incident radiation and leaf fall and no correlation with precipitation.

We report on the first year of an experiment designed to determine whether water availability is the proximal cue for leaf fall and flowering. Water availability was augmented by irrigation of two 2.25 ha plots of mature forest on Barro Colorado Island (BCI), Panama. The timing of leaf fall and flowering was determined from weekly collections from 60 litter traps. If water availability is the proximal cue, the timing of leaf fall and flowering should be altered by the manipulation.

STUDY SITE

Barro Colorado Island is the largest island (1500 ha) in Gatun Lake. Gatun Lake was flooded between 1911 and 1914 to complete the Panama Canal. Annual rainfall between 1977 and 1986 averaged 2540 mm. The dry season usually begins in December and ends in April. Median rainfall during the first 13 weeks of the calendar year is just 88 mm (Leigh and Wright, in press). Diurnal variation in average temperature (8 to 11°C) is four or five times larger than seasonal variation (2.2°C). The average annual temperature in the laboratory clearing is 27°C. Under the Holdridge Life-Zone System, the forest is tropical moist forest. Figure 5.1 presents weekly rainfall and incident radiation for the year of this study. Croat (1978) and Leigh *et al.* (1982) provide further descriptions of BCI.

Flowering and leaf fall are strongly seasonal on BCI. About 330 of the 652 native, woody species flower each month at the end of the dry season and the beginning of the wet season (March, April and May), while as few as 200 species flower each month in the late wet season (September, October and November; Croat, 1978). The rate of leaf fall for the community as a whole peaks early in the dry season (December and January) and is about twice as high during the remainder of the dry season as in the wet season (Figure 5.1, also see Haines and Foster, 1977; Leigh and Smythe, 1978). Most deciduous species lose their leaves in the dry season; however, 88% of the native tree species are evergreen (Croat, 1978), and the timing of leaf fall for individual evergreen species was not known prior to this study.

The study site is on the northern half of Poacher's Peninsula (9°08' N, 79°51' W). The forest is among the oldest on BCI (Robin Foster, pers. comm.) and has not been altered by man for more than 500 years (Piperno, in press). The terrain is flat or gently sloping and the canopy is 30 m tall. Common trees

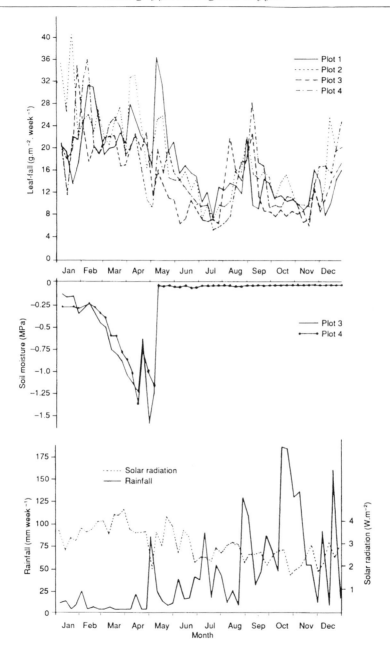

Figure 5.1 Weekly values of leaf fall for the four study plots (upper panel), of soil moisture for the two control plots (middle panel), and of rainfall and solar radiation (lower panel) in 1986. Plots 1 and 2 were irrigated. Rainfall was measured at the study site. Solar radiation is based on a 24-hour day and was measured 3 km from the study site at the laboratory clearing on BCI

include *Alseis blackiana, Apeiba membranacea, Beilschmiedia pendula, Dipteryx panamensis, Jacaranda copaia, Prioria copaifera, Protium panamense, Quararibea asterolepis, Tetragastris panamensis, Trichilia cipo,* and *Virola sebifera.* Nomenclature follows Croat (1978).

METHODS

Water availability was manipulated by irrigating two square plots of 2.25 ha each. Two similar plots serve as unirrigated controls. Water is delivered by PVC pipe to sprinklers mounted 1.8 m above the ground. Water is taken from Gatun Lake where nutrient concentrations are lower than in rain-water collected in the laboratory clearing (Gonzalez *et al.*, 1975, Robert Stallard pers. comm.). The sprinklers are hexagonally packed and 15.3 m apart. If there are no obstructions, water arcs 4 or 5 m into the air and lands at the bases of the six equidistant neighbouring sprinklers. In fact, trunks, branches, lianas and leaves almost invariably intercept and break up the stream of water before it reaches the ground.

The schedule for irrigation is determined by soil water potentials. The objective is to maintain soil water potentials at -0.03 MPa (-30 centibars) which is field saturation. Measurements were taken at eight randomly located stations in each plot. Tensiometers were placed 25 and 45 cm below the soil surface at each station in the irrigated plots. Dew-point psychrometers were placed at 25 cm at all stations, at 50 cm at four stations in each plot, and at 100 cm at two stations in each plot. Psychrometers were recorded weekly, and tensiometers were recorded three or more times each week during the dry season.

Irrigation will take place throughout the dry seasons of 1986, 1987 and 1988. This report covers 16 December 1985 through to 15 December 1986. Tensiometer measurements first fell below zero on 16 December 1985, signalling the beginning of the annual soil drying cycle. Irrigation began on 21 December 1985 and ended on 16 April 1986, when the wet season rains began. About 135 metric tons of water were delivered to each irrigated plot on 5 days each week until 3 March and on 6 days each week thereafter. Irrigation took place between 09.00 and 14.00 h and lasted between 1.5 and 2 h on each plot each day. Significant amounts of precipitation fell on 13 and 14 January (27 mm) and 31 March (18 mm) and wet the soil sufficiently to maintain soil water potentials without irrigation for the following 2 to 4 days.

Litter traps were located randomly but at least 15 m apart and at least 2 m from the palm *Oenocarpus panamanus.* The fronds of this palm redirect falling litter so that almost none falls beneath them. Ten traps were placed in each plot on 23 November 1985. This number was

52

increased to 15 per plot on 4 March 1986. The surface area of each trap is 0.25 m² and consists of plastic screening with 1.2 mm mesh supported 40 cm above the ground.

Trap contents were collected weekly and dried to a constant weight at 60°C. All leaves, flowers and fruit were identified to species and weighed. About 1% of the leaves (by weight) could not be identified to species. Fine litter and wood less than 2 cm in diameter were also dried and weighed. Only leaves and flowers are considered here.

Henceforth, records will refer to the presence of leaves or flowers in a trap. Records are independent events if the fall of each leaf or flower is assumed to be an independent event. Estabrook *et al.* (1982) advocate this assumption for analyses of the timing of phenological events. The analyses are restricted to the 43 species with 25 or more records for leaf fall in each treatment and the 18 species with five or more records for flowers in each treatment. Leaves and flowers are analysed separately for each species.

The null hypothesis is that the temporal distributions of records do not differ between treatments, or more exactly, that there is no treatment-census interval interaction. Contingency tables were constructed with main effects being treatment and census interval. Census intervals were obtained by grouping weeks so that each expected cell count of the contingency table exceeded five. Thus, the number of census intervals differed among species and was greater for species with more records. The analysis had two steps. First, plots were compared within treatments to test for between-plot, within-treatment effects. None were significant. Records were then summed over plots within treatments to test for treatment effects. Since numbers of records were relatively small for flowers, we also inspected the temporal distributions of flowering records directly.

RESULTS

Irrigation maintained soil water potentials at or just below field saturation throughout the dry season. The lowest daily average value recorded from tensiometers for either irrigated plot at either depth was –0.058 MPa. Some spatial heterogeneity existed in the irrigated plots. Consistently wet stations produced readings of –0.01 to –0.03 MPa throughout the dry season while dry stations fell to –0.06 to –0.09 MPa. This variability is inevitable and reflects local drainage patterns and irrigation shadows created by large trunks located close to sprinklers. In the control plots, average soil water potentials fell to –1.6 MPa (Figure 5.1)

The null hypothesis that the temporal distributions of records did not differ between treatments could rarely be rejected. For leaf fall, the null

Table 5.1 The temporal distribution of leaf fall for irrigated (IRR) and control (CTL) plots for two species of canopy trees

Astronium graveolens (Anacardiaceae)

IRR	0	1	1	1	2	0	0	1	1	4	1	0	0	0	0	0	0	1	1	0	1	1	2	1	6	3
CTL	0	0	0	0	0	0	0	1	1	1	2	4	1	3	2	3	3	3	1	4	4	4	4	4	2	2	2

Dipteryx panamensis (Leguminosae)

IRR	3	4	4	8	6	14	16	17	11	25	12	11	14	9	9	5	6	8	7	8	7	7	7	5	4	9
CTL	0	0	3	0	1	7	13	15	13	12	4	2	3	1	0	2	1	6	2	2	1	2	1	2	3	3

Entries are the number of traps in which leaves were recorded in each 2-week interval. The first and last colums represent the 2-week intervals beginning and ending on 16 December 1985 and 15 December 1986, respectively

hypothesis was rejected for two, of the 43 species analysed. Those species are *Dipteryx panamensis* ($p < 0.01$) and *Astronium graveolens* ($p < 0.001$). If the Bonferroni procedure is used, the appropriate alpha level for rejection is 0.05/43. Only *A. graveolens* qualifies.

Table 5.1 presents temporal distributions of records for leaf fall for *D. panamensis* and *A. graveolens*. Rows refer to treatments and columns to 2-week intervals. Entries are numbers of records summed over plots within treatments. Two-week intervals are used to facilitate presentation; the data were collected in 1-week intervals. Normally, individuals of *D. panamensis* are briefly deciduous toward the end of the dry season. This was the pattern in the control plots where leaf fall is concentrated in the late dry season; leaf fall was much less concentrated in the irrigated plots (Table 5.1)

Astronium graveolens is evergreen. Leaf fall for *A. graveolens* occurred throughout the wet season in the control plots, but was concentrated in the dry season and at the end of the following wet season in the irrigated plots (Table 5.1). This result is difficult to interpret as a response to the treatment (Table 5.1). Table 5.2 summarizes the analysis of leaf fall for all 43 species.

The null hypothesis that the temporal distribution of floral records does not differ between treatments could only be tested for *Anacardium excelsum* ($\chi^2 = 0.20$, $df = 3$, $p = 0.98$), *Apeiba membranacea* ($\chi^2 = 3.31$, $df = 4$, $p = 0.51$), *Guatteria dumetorum* ($\chi^2 = 1.81$, $df = 1$, $p = 0.17$), *Maripa panamensis* ($\chi^2 = 3.81$, $df = 1$, $p = 0.051$), *Phyrganocydia corymbosa* ($\chi^2 = 1.01$, $df = 1$, $p = 0.32$), *Prionostemma aspera* ($\chi^2 = 2.18$, $df = 1$, $p = 0.14$), and *Xylopia macrantha* ($\chi^2 = 16.1$, $df = 10$, $p = 0.10$). Table 5.3 presents flowering times for these species and for 11 additional species with five or more flowering records in each treatment. The format is the same as in Table 5.1. The boxes include all records for each species.

Table 5.2 Analyses of the timing of leaf fall

Species	Family	Life form	χ^2	df	p
Alseis blackiana	Rubiaceae	Tree	11.6	19	0.90
Anacardium excelsum	Anacardiaceae	Tree	3.4	16	1.00
Apeiba mambranacea	Tiliaceae	Tree	10.9	20	0.95
Astronium graveolens	Anacardiaceae	Tree	22.8	4	< 0.001
Banisteriopsis cornifolia	Malpighiaceae	Liana	17.9	11	0.09
Beilschmiedia pendula	Lauraceae	Tree	13.1	18	0.79
Colubrina glandulosa	Rhamnaceae	Liana	13.7	12	0.32
Croton bilbergianus	Euphorbiaceae	Treelet	10.8	5	0.06
Davilla nitida	Dilleniaceae	Liana	5.9	14	0.97
Dipteryx panamensis	Leguminosae	Tree	29.9	14	< 0.01
Doliocarpus olivaceus	Dillenaceae	Liana	17.1	13	0.19
Drypetes standleyi	Euphorbiaceae	Tree	5.5	12	0.94
Faramea occidentalis	Rubiaceae	Treelet	4.5	7	0.72
Guarea guidonia	Meliaceae	Treelet	6.6	15	0.97
Guatteria dumetorum	Annonaceae	Tree	18.1	19	0.52
Heisteria concinna	Olacaceae	Treelet	7.6	10	0.67
Hippocratea voluvilis	Hippocrateaceae	Liana	1.7	5	0.89
Hiraea sp.	Malpighiaceae	Liana	4.7	4	0.32
Hirtella triandra	Chrysobalanaceae	Treelet	5.7	4	0.22
Inga cocleensis	Leguminosae	Tree	2.6	7	0.92
Jacaranda copaia	Bignoniaceae	Tree	11.6	40	1.00
Licania platypus	Chrysobalanaceae	Tree	6.7	5	0.25
Luehea seemannii	Tiliaceae	Tree	7.5	6	0.28
Maquira costaricana	Moraceae	Tree	5.5	5	0.36
Maripa panamenses	Convolvulaceae	Liana	15.4	17	0.57
Mouriri myrtilloides	Melastomataceae	Shrub	3.7	5	0.60
Oenocarpus panamanus	Palmae	Palm	7.0	4	0.14
Phryganocydia corymbosa	Bignoniaceae	Liana	3.4	6	0.76
Pouteria unilocularis	Sapotaceae	Tree	8.8	12	0.72
Prionostemma aspera	Hippocrateaceae	Liana	12.7	18	0.81
Prioria copaifera	Leguminosae	Tree	5.3	12	0.95
Protium panamense	Burseraceae	Tree	9.6	9	0.38
Quararibea asterolepis	Bombacaceae	Tree	5.3	8	0.72
Sorocea affinis	Moraceae	Treelet	5.7	4	0.23
Stricnos dariense	Loganiaceae	Liana	13.5	14	0.49
Terminalia amazonica	Combretaceae	Tree	5.3	13	0.97
Tetragastris panamensis	Burseraceae	Tree	33.0	32	0.42
Trattinnickia aspera	Burseraceae	Tree	5.4	8	0.71
Trichilia cipo	Meliaceae	Tree	11.6	19	0.90
Uncaria tomentosa	Rubiaceae	Liana	2.0	4	0.74
Virola sebifera	Myristicae	Tree	8.3	15	0.91
Xylopia macrantha	Annonaceae	Tree	12.5	22	0.95
Unknown sp.	Bignoniaceae	Liana	8.4	14	0.87

Table 5.3 The temporal distribution of flowering for irrigated (IRR) and control (CTL) plots

Anacardium excelsum (Anacardiaceae)

```
IRR  0  0  0 |3  4  6  10  9  8  5  5  1  2| 0  0  0  0  0  0  0  0  0  0  0  0  0  0
CTL  0  0  0 |2  2  4  5  4  4  3  2  1  1| 0  0  0  0  0  0  0  0  0  0  0  0  0  0
```

Apeiba membranacea (Tiliaceae)

```
IRR  0  0  0  0  0  0  0  0  0 |3  4  4  3  5  4  3  4  2  2  0  0  1  0  0  0  0|
CTL  0  0  0  0  0  0  0  0  0 |3  6  7  5  5  6  7  6  5  2  2  2  3  0  2  1  1|
```

Beilschmiedia pendula (Lauraceae)

```
IRR |1| 0  0  0  0  0  0  0  0  0  0  0  0  0  0  0  0  0  0  0  0  0  0  0 |1  4  9|
CTL |0| 0  0  0  0  0  0  0  0  0  0  0  0  0  0  0  0  0  0  0  0  0  0  0 |0  5  6|
```

Cydista aequinoctalis (Bignoniaceae)

```
IRR  0  0  0  0  0  0  0  0  0  0 |2  6  5  3  2  1  2  0  1  0  1| 0  0  0  0  0
CTL  0  0  0  0  0  0  0  0  0  0 |0  2  0  1  1  0  2  0  0  1  0| 0  0  0  0  0
```

Dipteryx panamensis (Leguminosae)

```
IRR  0  0  0  0  0  0  0  0  0  0  0  0 |1  0  0  2  0  2  0| 0  0  0  0  0  0
CTL  0  0  0  0  0  0  0  0  0  0  0  0 |2  3  3  7  2  4  2| 0  0  0  0  0  0
```

Guatteria dumetorum (Annonaceae)

```
IRR |3  4  2  2  4  1  2  2  2  1  2  2  0  0  0  0  1  1  3  2  3  2  3  4  4  6|
CTL |0  0  0  0  0  0  1  0  0  0  0  0  0  0  0  0  1  0  2  1  2  1  0  0  1  1|
```

Heisteria concinna (Olacaceae)

```
IRR  0  0  0  0  0  0  0  0  0  0  0  0  0  0  0  0  0  0  0 |0  0  0  0  0  2  3|
CTL  0  0  0  0  0  0  0  0  0  0  0  0  0  0  0  0  0  0  0 |2  2  1  0  1  5  3|
```

Jacaranda copaia (Bignoniaceae)

```
IRR  0  0  0  0  0 |0  4  3  0| 0  0  0  0  0  0  0  0  0  0  0  0  0  0  0  0  0
CTL  0  0  0  0  0 |1  11  11  3| 0  0  0  0  0  0  0  0  0  0  0  0  0  0  0  0  0
```

Licania platypus (Chrysobalanaceae)

```
IRR  0  0  0  0  0  0 |0  2  2  1| 0  0  0  0  0  0  0  0  0  0  0  0  0  0  0  0
CTL  0  0  0  0  0  0 |2  4  4  3| 0  0  0  0  0  0  0  0  0  0  0  0  0  0  0  0
```

Luehea seemannii (Tiliaceae)

```
IRR  0 |0  3  3  1| 0  0  0  0  0  0  0  0  0  0  0  0  0  0  0  0  0  0  0  0
CTL  0 |1  2  2  0| 0  0  0  0  0  0  0  0  0  0  0  0  0  0  0  0  0  0  0  0
```

Maripa panamensis (Convolvulaceae)

```
IRR  0  0  0  0   6  7  5  1  2  0  1   0  0  0  0  0  0  0  0  0  0  0  0  0  0  0
CTL  0  0  0  0   5 10 11  6  5  5  3   0  0  0  0  0  0  0  0  0  0  0  0  0  0  0
```

Oenocarpus panamanus (Palmae)

```
IRR   0  0  0  0  0  1   0  0  0  0  0  0  0  0  0  0  0  0  0  0   2  1  0  0  1  7
CTL   0  0  1  2  1  1   0  0  0  0  0  0  0  0  0  0  0  0  0  0   1  0  0  1  0  2
```

Phyrganocydia corymbosa (Bignoniaceae)

```
IRR  0  0  0  0  0  0  0  0  0  0  0   0  0  0  0  1  0  2  3  1  1  1  1   0  0  0
CTL  0  0  0  0  0  0  0  0  0  0  0   1  1  2  0  1  2  4  3  2  3  2  1   0  0  0
```

Prionostemma aspera (Hippocrateaceae)

```
IRR   4  2  1  1  6  5  0  0  0  1  2  2  2  3  2  1  0  2  5  6  3  5  0  0  0  1
CTL   0  0  1  3  4  8  1  0  0  0  0  0  0  0  0  1  0  0  0  0  0  0  0  0  0  0
```

Quararibea asterolepis (Bombacaceae)

```
IRR  0  0  0  0  0  0  0  0  0  0  0   3  2  2  2  1  0  0  0  1  0  0   0  0  0  0
CTL  0  0  0  0  0  0  0  0  0  0  0   0  3  2  2  2  2  2  1  0  0  1   0  0  0  0
```

Tabebuia rosea (Bignoniaceae)

```
IRR  0  0  0  0   0  1  6  1  0  0   0  0  0  0  0  0  0  0  0  0  0  0  0  0  0  0
CTL  0  0  0  0   2  2  3  1  0  1   0  0  0  0  0  0  0  0  0  0  0  0  0  0  0  0
```

Tetragastris panamensis (Burseraceae)

```
IRR  0  0  0  0  0  0  0  0  0  0  0  0  0  0  0  0  0   1  6  7  2  0  1  0  0  0  4
CTL  0  0  0  0  0  0  0  0  0  0  0  0  0  0  0  0  0   0  2  3  2  1  0  0  0  0  1
```

Xylopia macrantha (Annonaceae)

```
IRR   0  0  1  0  0  0  0  0  1  0  0  1  0  2  4  7  7  8  8  8  7  6  6  3  3  5
CTL   2  2  2  1  1  1  2  0  0  2  0  1  3  3  4  5  4  4  4  4  4  4  4  4  4  3
```

Boxes include all records for each species. See caption to Table 5.1 for further explication

DISCUSSION

Maintaining soil water potentials at or very near to field saturation throughout the dry season had little effect on the timing of flowering and leaf fall. Flowers of most species were recorded for the same 8 to 24 weeks regardless of treatment (Table 5.3). The null hypothesis that the temporal distribution of leaf fall differs between control and irrigated plots was tested for 43 species. The distribution of probability values associated with the test statistic is essentially random; 81% of the values are greater than 0.25 and 4.7% are less than 0.05 (Table 5.2).

Four explanations will be considered for the general lack of response to the radical manipulation of soil moisture. First, it is possible that irrigation did not relieve water stress. Perhaps the reduced relative humidity and increased incident radiation characteristic of the dry season (Figure 5.1) induces water stress even when plants are rooted in irrigated, saturated soil.

This possibility can be discounted. Pre-dawn water potentials were essentially zero (−0.02 to −0.07 MPa) in the irrigated plots for five species of *Psychotria*, two species of *Piper*, and saplings of *Trichilia cipo* and *Prioria copaifera* in late March 1986. Values were significantly lower for all nine species in the control plots and were as low as −2.7 MPa (SJW, unpublished data). Midday water potentials were also recorded for 10 species of trees and treelets late in the 1987 dry season. The irrigation schedule of 1986 has been maintained throughout 1987. The midday water potentials were significantly higher in the irrigated plots for eight of the 10 species (Wright and Cornejo, 1990), and values in the irrigated plots were similar to wet season values for species that were also studied by Rundel and Becker (1987). We conclude that irrigation effectively removed dry season water stress.

A second possible explanation for the lack of response is that irrigation changed other factors whose effect compensated for relief from water stress. It seems improbable that nearly identical flowering times could be an outcome of hypothetical opposing factors. Still, irrigation might produce anaerobic conditions at the roots or change nutrient availability. Again, these possibilities can be discounted. Concentrations of O_2 and CO_2 in soil gases have been monitored weekly and monthly, respectively, since irrigation began. Values vary seasonally but did not differ significantly between treatments in 1986 (T. Kursar, unpublished data). Nutrient concentrations have also been monitored in irrigated and control plots since December 1985 and did not exhibit treatment effects in 1986 (J. Yavitt, K. Wieder, P. Vitousek and SJW, unpublished data).

A third possible explanation for the lack of response is that the analyses performed here lack power. This seems unlikely for two reasons. First, visual inspection suggests that the timing of flowering is remarkably similar for both treatments for most species (Table 5.3). No analysis will distinguish these distributions. Second, for the timing of leaf fall, Wright and Cornejo (1990) have performed analyses based on dry weights for the more common species. Their analyses confirm the generally negative results presented here. With few exceptions, the timing of leaf fall is unaffected by relief from the annual cycle of soil moisture availability.

The final possibility is that the timing of flowering and leaf fall are controlled by factors other than seasonal drought. Possible proximal cues for leaf fall and anthesis in tropical plants include day length (Njoku, 1958), abrupt drops in temperature (reviewed by Opler *et al.*, 1976), and changes in incident radiation (Bernhard-Reversat *et al.*, 1972). We cannot evaluate these possibilities, but we can evaluate the model of tropical plant phenology proposed by Borchert (1980) and Reich and Borchert (1982, 1984).

Reich and Borchert (1984) tentatively divided the observed phenological patterns of tropical trees into three main classes. Their class A includes species that abscise their leaves early in the dry season and flower and flush new leaves in response to rainfall. Their class B includes species that abscise their leaves during the middle or late dry season and experience bud break under continuously dry conditions and/or in response to rain. Their class C includes evergreen species. Reich and Borchert (1984) propose that water stress affects the timing of leaf fall for all three classes, although for evergreen species they state that "...the effects of water stress on the time of leaf fall and shoot emergence are moderate."

Twenty-three of the 43 species for which leaf fall was analysed are trees. Ten of these species are evergreen, and the remainder fit broadly into category B, although several abscise their leaves very early in the dry season. If water stress were the proximal cue for leaf fall as Reich and Borchert (1984) propose, leaf fall for the 13 species in class B should have been strongly affected by irrigation, and leaf fall for the 10 evergreen species may or may not have been affected. In fact, the timing of leaf fall was only affected for one evergreen species and one class B species, and, as discussed above, even these responses are of dubious statistical significance.

On the other hand, the response of the one class A species under study supports the predictions of Reich and Borchert (1984). *Tabebuia guayacan* normally becomes deciduous soon after the dry season begins and flowers after a dry season rain. This pattern has been drastically upset by irrigation (SJW, pers. obs.). The three individuals on the irrigated plots did not drop their leaves during the 1986 dry season. Instead, leaf drop occurred asynchronously among branches on the same individual in the first few months of the wet season. Flowering occurred branch by branch soon after leaf fall.

We conclude that the model of tropical tree phenology proposed by Reich and Borchert (1984) is applicable to a very limited number of species in tropical moist forest on BCI. For the vast majority of species, the proximal cues for leaf fall and flowering have not been identified. The results reported here indicate that water stress is rarely an important factor.

Two caveats remain. First, with two exceptions, the results reported here are for trees, treelets and lianas (Tables 5.2 and 5.3). Our conclusions should not be extrapolated to other life forms. In particular, preliminary analyses suggest that some understorey shrubs and herbs have responded to the treatment.

The second caveat concerns the complex life cycle of a flower or leaf. This includes bud initiation, bud break, senescence, and abscission. The seasonal rhythm of soil moisture availability may affect any of these stages. Our data bear most directly on the timing of senescence and abscission. In addition, since most flowers are short-lived, our results are probably also indicative of the timing of floral bud break. The timing of bud initiation and vegetative bud break are not addressed by our results. However, if the first year of irrigation affects these events, the timing of flowering and leaf fall may be changed in subsequent years. It is also possible that drought does synchronize phenological events and that several years without drought will be required to destabilize synchrony established in the last dry season before the manipulation. For these reasons, results from the second and third years of the experiment will be important.

ACKNOWLEDGEMENTS

Mitch Aide, Alan Smith and Neal Smith commented on the manuscript. Financial support came from the Environmental Sciences Program of the Smithsonian Institution and the Exxon Foundation (FHC).

REFERENCES

Bernhard-Reversat, F., Huttel, C. and Lemée, G. (1972). Some aspects of the seasonal ecological periodicity and plant activity in an evergreen rain forest of the Ivory Coast. In Golley, F.B. and Misra, R. (eds). *Papers from a Symposium on Tropical Ecology with an Emphasis on Organic Productivity*, pp. 217–34. International Society for Tropical Ecology, Athens.

Borchert, R. (1980). Phenology and ecophysiology of tropical trees: *Erythrina poeppigiana* O.F. Cook. *Ecology*, **61**, 1065–74

Croat, T.B. (1978). *Flora of Barro Colorado Island*. (Stanford: Stanford University Press)

Daubenmire, R. (1972). Phenology and other characteristics of tropical semi-deciduous forest in north-western Costa Rica. *Journal of Ecology*, **60**, 147–70

Estabrook, G.F., Winsor, J.A., Stephenson, A.G. and Howe, H.F. (1982). When are two phenological patterns different? *Botanical Gazette*, **143**, 374–8

Frankie, G.W., Baker, H.G. and Opler, P.A. (1974). Comparative phenological studies of trees in tropical wet and dry forests in the lowlands of Costa Rica. *Journal of Ecology*, **62**, 881–919

Gonzalez, A., Alvarado-Dufree, G. and Diaz, C.T. (1975). *Canal Zone Water Quality Study*. Final report. Water and Laboratories Branch, Panama Canal Company, Canal Zone.

Haines, B. and Foster, R.B. (1977). Energy flow through litter in a Panamanian forest. *Journal of Ecology*, **65**, 147–55

Hopkins, B. (1966). Vegetation of the Olokemeji Forest Reserve, Nigeria. IV. The litter and soil with special reference to their seasonal changes. *Journal of Ecology*, **54**, 687–703

Janzen, D.H. (1967). Synchronization of sexual reproductin of trees within the dry season in Central America. *Evolution*, **21**, 620–37

Leigh, E.G., Jr., and Smythe, N. (1978). Leaf production, leaf consumption and the regulation of folivory on Barro Colorado Island. In Montgomery, G.G. (ed.) *The Ecology of Arboreal Folivores*, pp. 33–50. (Washington, D.C.: Smithsonian Institution Press)

Leigh, E.G., Jr., Rand, A.S. and Windsor, D.M. (eds.) (1982). *The Ecology of a Tropical Forest. Seasonal Rhythms and Long-term Changes*. (Washington, D.C.: Smithsonian Institution Press)

Leigh, E.G., Jr. and Wright, S.J. The role of Barro Colorado Island in the understanding of tropical forest ecology. In Gentry, A.H. (ed.) *Four Neotropical Forests*. (New Haven, Connecticut: Yale

University Press) (in press).

Lieberman, D. (1982). Seasonality and phenology in a dry tropical forest in Ghana. *Journal of Ecology*, **70**, 791–806

Njoku, E. (1958). The photoperiodic reponse of some Nigerian plants. *Journal of the West African Science Association*, **4**, 99–111

Opler, P.A., Frankie, G.W. and Baker, H.G. (1976). Rainfall as a factor in the release, timing and synchronization of anthesis by tropical trees and shrubs. *Journal of Biogeography*, **3**, 231–6

Piperno, D.R. (1990). Fitolitos, arquelogia y cambios prehistoricos de la vegetacion en un lote de cincuenta hectareas de la isla de Barro Colorado. In Leigh, E.G. Jr., Rand, A.S. and Windsor, D.M. (eds.). *Ecologia de un bosque tropical*, pp. 153–6. (Washington, D.C.: Smithsonian Institution)

Proctor, J. (1984). Tropical forest litterfall. II: The data set. In Chadwick, A.C. and Sutton, S.L. (eds). *Tropical Rain Forest: The Leeds Symposium*. pp. 83–113. (Leeds: Leeds Philosophical and Literary Society)

Reich, P.B. and Borchert, R. (1982). Phenology and ecophysiology of the tropical tree, *Tabebuia neochrysantha* (Bignoniaceae). *Ecology*, **63**, 294–9

Reich, P.B. and Borchert, R. (1984). Water stress and tree phenology in a tropical dry forest in the lowlands of Costa Rica. *Journal of Ecology*, **72**, 61–74

Rundel, P.W. and Becker, P.F. (1987). Cambios estacionales en las relaciones hydricas y en la fenologia vegetativa de plantas del estrato bajo del bosque tropical de la Isla de Barro Colorado, Panama. *Revista de Biologia Tropical*, **35**, 71–84

Wright, S.J. and Cornejo, F.H. (1990). Seasonal drought and the timing of leaf fall in a tropical forest. *Ecology*, **71**, 1165–75

Section 3

Plant–pollinator interactions,
sexual systems and gene flow

CHAPTER 6

PLANT–POLLINATOR INTERACTIONS, SEXUAL SYSTEMS AND POLLEN FLOW – COMMENTARY

Kamaljit S. Bawa

Plant–pollinator interactions, sexual and breeding systems, and levels of gene flow in tropical forest trees are of interest for several reasons. Many species of trees in tropical rain forests have densities of one reproductively mature individual or less per hectare (Hubbell and Foster, 1983). Spatial isolation of conspecifics could result in limited pollen flow and inbreeding unless mating patterns are such that they allow considerable outcrossing. Thus for many years, the extent of inbreeding and outcrossing has been a central issue in the population biology of tropical forest trees (Ashton, 1969; Bawa, 1979). In recent years, a great diversity of pollination mechanisms and breeding systems have been documented in rain forest trees (Appanah, 1981; Bawa, Perry and Beach, 1985; Bawa et al., 1985; and references therein). The diversity of pollination mechanisms, coupled with taxonomic diversity in tropical rain forests, has also provided excellent opportunities to study the degree of coevolution between plants and their pollinators (Feinsinger, 1983). Finally, the study of the spatial and temporal distribution of various plant–pollinator interactions at the community level has provided insights into the role of such interactions in community structure (Stiles, 1975; Appanah, 1981; Bawa et al., 1985).

In the lowland wet tropics, pollination mechanisms and sexual and breeding systems have been studied most extensively at two sites: La Selva in Costa Rica and Pasoh in Malaysia. Schatz and Appanah respectively summarize the results from these two sites. Dayanandan et al. present results of comprehensive studies on the pollination ecology of the Dipterocarpaceae in Sinharaja, a premontane wet tropical forest in Sri Lanka. Irvine and Armstrong examine interactions between plants and beetle pollinators in an Australian rain forest. Young provides estimates of pollen flow in an aroid herb. Shaanker and Ganeshaiah review the relationship between patterns of pollen deposition and the number of seeds per fruit.

Schatz and Appanah note that, at both La Selva and Pasoh, tree species are largely outcrossed via self-incompatibility or by virtue of being dioecious. Apomixis has been reported for some species at Pasoh, but species at La Selva have not been examined from this point of view. Studies of herbaceous species at La Selva reveal a higher incidence of self-incompatibility than encountered in trees (Kress and Beach, 1989).

Pollination mechanisms at both sites are diverse. Bawa *et al.* (1985), reviewed in Schatz, have shown that the relative frequencies of various pollination systems are dissimilar in different strata of the forest. Appanah's qualitative observations in Malaysia confirm the quantitative trends noted for the Costa Rican rain forest. Studies at both sites show that the diversity of pollinators is greatest in the understorey. Schatz distinguishes between diurnal and nocturnal pollination systems. He points out that the former are driven by visual cues and the latter by odours. The diurnal pollination systems appear to be more common in the canopy, and the nocturnal in the understorey. Pollination "guilds" consisting of species sharing the same vectors have been studied extensively in the case of hummingbird pollinated plants (Stiles, 1975) and beetle pollinated plants (Schatz, Irvine and Armstrong, this volume). Irvine and Armstrong note that, in Australia, the nocturnal beetle pollination system is encountered in all life forms and at all levels of the forest. They also suggest that beetle pollination may be more common in Australian than in neotropical rain forests.

Dayanandan *et al.* report the results of their comprehensive studies of flowering, floral morphology, pollination mechanisms and breeding systems in the Dipterocarpaceae. They show that various species of *Shorea* differ in their flowering patterns. In *Vatteria copallifera*, flowering patterns vary among populations. Trees in open disturbed habitats flower more frequently than trees in closed, undisturbed forests. Dayanandan *et al.* show that species of *Shorea* and *Vatteria*, like other tropical rain forest trees, are mostly outcrossed. The principal pollen vectors are bees. Dipterocarpaceae are a dominant component of the canopy and many species are commercially exploited. The information on reproductive biology provided by Dayanandan *et al.* should be of considerable importance in the conservation and management of dipterocarps.

Young describes the reproductive biology, with particular reference to the movement of pollen flow, of an understorey aroid. Estimates of pollen flow and effective population size provide important insights into the dynamics of microevolutionary processes as well as conservation strategies. There is an urgent need to extend the type of study conducted by Young to other plants.

In many species of plants, all ovules do not mature into seeds. Many factors are involved in the abortion of ovules. Shaanker and Ganeshaiah examine the role of pollen deposition patterns in regulating the number of seeds. They note that, in many multi-ovulated species, a large fraction of ovules develop into seeds. Shaanker and Ganeshaiah show that the high level of seed set is due to the deposition of many grains on the stigma. Flowers receiving pollen grains fewer than the number of ovules are aborted. Shaanker and Ganeshaiah's research shows the existence of subtle pre-fertilization mechanisms employed by plants to regulate their reproductive output. Elucidation of such mechanisms helps us understand the evolution of plant reproductive strategies, as well as the factors limiting seed and fruit set.

The detailed investigations of specific pollination systems are just beginning, and much remains to be learned. As stressed by Schatz, Appanah, Irvine and Armstrong and others, comprehensive data on flowering patterns, floral rewards, and sexual systems are required to elucidate the structure and functioning of reproductive systems at the level of species, groups of related species, and communities.

Papers in this section reveal the diversity and complexity of reproductive systems of plants in tropical lowland rain forests. Community-wide studies indicate that the diversity of pollination mechanisms is greatest in the understorey, and that the maintenance of the understorey may be critical to the overall integrity of the interactions in the community. Within the community are the various "guilds". Some of these guilds, as for example the hummingbirds and their host plants, are well studied (Stiles, 1975); others such as the beetles are the targets of intensive studies, as pointed out by Schatz and by Irvine and Armstrong in their contributions. The number of pollinator as well as plant species involved in these guilds vary among the pollinator guilds as well as within geographical regions. Identification of the factors that limit the number of species of a guild is an important theoretical issue. The effect of removal of one or more species of plants on other plant species pollinated by the same group of vectors is a significant management issue. At this level, the specificity of plant–pollinator interactions is not well understood. Nor is the geographical variation in the interaction well documented. The extent of specificity, as well as geographical variation, have important theoretical and practical implications.

In terms of sexual and breeding systems, there is now overwhelming evidence that a majority of tree species in tropical rain forests are outcrossed. However, the presence of apomixis in several species indicates that uniparental reproduction occurs. The challenge is to estimate quantitatively the relative frequency of outcrossing, selfing, and apomixis among the progeny of the same individual or population. Genetic markers in the form of allozymes, recently utilized to estimate quantitatively the amount of outcrossing in tropical tree species, offer the potential to investigate the mating patterns and mating systems in more detail than hitherto possible (O'Malley and Bawa, 1987; O'Malley *et al.*, 1988).

Genetic markers are also expected to be used increasingly to estimate gene flow, effective size of populations, and the amount and patterns of genetic variation in populations (Bawa and Krugman, 1990). Despite rapid progress in understanding the reproductive modes of trees, information about their population genetic parameters remains very meagre. Yet such information is critical for designing effective strategies to maintain genetic diversity in nature reserves, and *ex situ* collections.

REFERENCES

Appanah, S. (1981). Pollination in Malaysian primary forests. *Malaysian Forester*, **44**, 37–42

Ashton, P.S. (1969). Speciation among tropical forest trees: some deductions in the light of recent evidence. *Biological Journal of the Linnean Society*, **1**, 55–96

Bawa, K.S. and Krugman, S. (1990). Reproductive biology and genetics of tropical trees in relation to conservation and management. In Gómez-Pompa, A., Whitmore, T.C. and Hadley, M. (eds). *Rain Forest Regeneration and Management*. Man and the Biosphere Series, vol. 6, pp. 119–136. (Carnforth: Parthenon Publishing and Paris: Unesco)

Bawa, K.S. (1979). Breeding systems of trees in a tropical wet forest. *New Zealand Journal of Botany*, **17**, 521–4

Bawa, K.S., Perry, D.R. and Beach, J.H. (1985). Reproductive biology of tropical lowland rain forest trees. I. Sexual systems and self-incompatibility mechanisms. *American Journal of Botany*, **72**, 331–45

Bawa, K.S., Perry, D.R., Bullock, S.H., Covile, R.E. and Grayum, M.H. (1985). Reproductive biology of tropical lowland rain forest trees. II. Pollination mechanisms. *American Journal of Botany*, **72**, 346–56

Feinsinger, P. (1983). Coevolution and pollination. In Futuyma, D.J. and Slatkin, M. (eds). *Coevolution*, pp. 282–310. (Sunderland, Massachusetts: Sinauer Associates)

Hubbell, S.P. and Foster, R.B. (1983). Diversity of canopy trees in a neotropical forest and implications for conservation. In Sutton, S.L., Whitmore, T.C. and Chadwick, S. (eds). *Tropical Rain Forest: Ecology and Management*, pp. 25–41. (Oxford: Blackwell Scientific Publications)

Kress, J. and Beach, J.H. (1989). Flowering Plant reproductive systems at La Selva Biological Stations. Manuscript.

O'Malley, D.M. and Bawa, K.S. (1987). Mating system of a tropical rain forest tree species. *American Journal of Botany*, **74**, 1143–9

O'Malley, D.M., Buckley, D.P., Prance, G.T. and Bawa, K.S. (1988). Genetics of Brazil nut (*Bertholletia excelsa* Humb. Bonpl.:Lecythidaceae). 2. Mating system. *Theoretical and Applied Genetics*, **76**, 929–32

Stiles, F.G. (1975). Ecology, flowering phenology and hummingbird pollination of some Costa Rican *Heliconia* species. *Ecology*, **56**, 285–301

CHAPTER 7

SOME ASPECTS OF POLLINATION BIOLOGY IN CENTRAL AMERICAN FORESTS

George E. Schatz

ABSTRACT

Plant–pollinator interactions are examined within a phylogenetic context, emphasis being given to the higher levels of hierarchies. In describing distinctive characteristics of Central American pollination biology, special attention is given to plant–pollinator assemblages that are endemic, or largely so, to the neotropics. Particular focus is given to plant assemblages pollinated by dynastine scarab beetles, whose importance as pollinators has only recently started to emerge. Examples serve to illustrate the as yet poorly investigated role of floral odours in pollination biology. It is concluded that tools now exist to define precisely floral odours and to address evolutionary questions concerning plant–pollinator interactions, co-speciation and the structure of plant–pollinator assemblages.

INTRODUCTION

One can approach the study of pollination biology from several levels of the ecological and genealogical hierarchies. At organism and population levels, autecological studies investigate the pollination of a single plant species, either through abiotic means, or by its pollinator or coterie of pollinators. Similarly, a single pollinator species and the plant(s) it services may be studied. At higher levels, focus upon taxonomically related (both congeneric and confamilial) and/or functionally similar (= guilds) plant species and/or their pollinators reveals patterns of diversification and co-evolution (here adopted in the most "diffuse" of senses), as well as cases of ecological and phenetic convergence, and contributes to our understanding of diversity by illuminating mechanisms of species packing, niche

69

partitioning and coexistence. For any one community, a complete characterization of pollination biology would encompass a variety of parameters including:

(1)   Extent of biotic versus abiotic pollination;

(2)   Relative frequencies of the major classes of pollinators, and the attractants and rewards, or lack thereof, associated with each class;

(3)   Breeding systems of both plants and pollinators; and,

(4)   Spatial and temporal components of plant–pollinator interactions, or, in summary, all those aspects of the pollination milieu that contribute to the reproductive fitness of both plant and pollinator.

To paint a broad picture of Central American pollination biology, I shall emphasize the higher levels of the hierarchies, presenting individual species accounts only insofar as they provide a basis for synthesis at higher taxonomic, guild and community levels. As a systematist, I feel obliged to frame plant–pollinator interactions within a phylogenetic context. Indeed, to describe what is peculiarly characteristic of Central American pollination biology, I shall highlight plant–pollinator assemblages (particularly those involving beetles as pollinators) endemic, or largely so, to the neotropics, of which, of course, Central America represents only an impoverished subset. Among the endemic plant–pollinator assemblages, one such assemblage, plants pollinated by dynastine scarab beetles, whose importance as pollinators has only recently begun to emerge, will serve to illustrate the as yet poorly investigated role of floral odours in pollination biology. I shall confine my remarks largely to a single life zone, the tropical wet forest, as exemplified by the La Selva Biological Station of the Organization for Tropical Studies, located in the Caribbean lowlands of Costa Rica, as well as make occasional comparisons with Barro Colorado Island (BCI), Panama, a tropical moist forest with more pronounced seasonality.

COMMUNITY-LEVEL CHARACTERIZATION OF A TROPICAL WET FOREST

Not unexpectedly, no complete characterization of pollination for a tropical community yet exists. However, subsets of the rich La Selva flora, estimated to consist of 1500 to 2000 vascular plant species (Hammel and Grayum, 1982), have been analysed for sexual systems (333 woody plants, Bawa, Perry and Beach, 1985) and pollination systems (143 trees, Bawa *et al.*,

1985). I shall attempt to expand on the data base employed by Bawa and his co-workers by further extrapolation from floral morphology of congeners, and the inclusion of herbaceous taxa, to evaluate whether results from subsets of woody plants are indicative of the flora as a whole.

Sexual systems

Of the 333 woody plants Bawa, Perry and Beach (1985) included in the sample, 65.5% are hermaphroditic, 11.4% monoecious, and 23.1% dioecious. The proportion of dioecy is comparable to that reported for woody plants in Malaysian evergreen forest and Costa Rican semi-deciduous forest (Bawa, 1974), but over twice that which has been reported for the entire Barro Colorado Island flora (9%) (Croat, 1979), suggesting that the incidence of dioecy is highest among long-lived perennials. Missing from the sample are both lianas and epiphytes, which together account for over one-third of the total flora at La Selva. Prominent liana families at La Selva include Fabaceae, Bignoniaceae, Dilleniaceae, Malphigiaceae, and Sapindaceae, which consist almost entirely of hermaphroditic species. Epiphyte families include Orchidaceae, the largest family in the flora, Bromeliaceae, Gesneriaceae, Cactaceae, and Piperaceae, nearly all of which are also hermaphroditic. Epiphytic and climbing Cyclanthaceae are all monoecious, as are many Araceae. Additional inclusion of understorey monocot families, e.g. Poaceae, Costaceae, Heliconiaceae, Marantaceae, and Zingiberaceae, further increases the proportion of hermaphroditism. In all likelihood, consideration of dioecy in the La Selva flora as a whole would result in a figure comparable to that reported for BCI.

Among the hermaphroditic species in the sample of Bawa and his co-workers, 28 were further examined for self-incompatibility through controlled pollinations, of which 80% were found to be self-incompatible. Many of the understorey monocot taxa not included in the sample are known to be self-compatible (Kress, 1983). However, the presentation of only a few flowers at a time over an extended period (also a common pattern in epiphyte flowering), necessitates broad spectrum foraging by pollinators, and probably results in these groups being largely outcrossing.

Two of the five species that Bawa and co-workers found to be self-compatible are Annonaceae, a primitive family in which self-compatibility may be widespread (Gottsberger, 1974), and in which beetle pollination is prevalent (Gottsberger, 1970, 1986; Schatz, 1987). The strongly protogynous nature of beetle-pollinated taxa may well have obviated the need for the evolution of self-incompatibility systems (Bawa and Beach, 1981). Young (1986) has also reported self-compatibility in *Dieffenbachia* (Araceae), a beetle-pollinated herb with protogynous monoecious

Table 7.1 Frequency of pollinator classes among a sample of 143 tree species at La Selva Biological Station (Bawa *et al.*, 1985)

Pollinator class	% Tree species
Bat	3.0
Hummingbird	4.3
Small bee	14.0
Medium-sized to large bee	27.5
Beetle	7.3
Butterfly	4.9
Moth	
Sphingid	8.0
Other	7.9
Wasp	4.3
Small diverse insect	15.8
Thrips	0.6
Wind	2.5

inflorescences. The temporal separation of female and male anthesis (dichogamy) is the rule in all monoecious Cyclanthaceae, as well as both hermaphroditic and monoecious Araceae and Arecaceae (which accounted for 45% of the monoecious species in the samples of Bawa and co-workers).

Taking into account the distylous and dichogamous taxa, as well as the percentage of hermaphroditic taxa found to be self-incompatible, the overwhelming implication is that the majority of species at La Selva are outcrossing.

Pollination systems

Table 7.1 enumerates the major classes of pollinators recognized by Bawa *et al.* (1985) and their relative frequencies among a sample of 143 tree species. Consideration of the flora as a whole would likely result in large increases in hummingbird, bee, and small diverse insect groups, with concomitant decreases in bat, moth and wasp pollination. Floral morphologies of canopy flowering liana families referred to earlier suggest pollination by medium to large bees and small diverse insects. Among epiphytes, the majority of Orchidaceae are probably pollinated by bees (van der Pijl and Dodson, 1966), whereas Bromeliaceae and Gesneriaceae are predominantly hummingbird-pollinated (Stiles, 1978, 1980). Further hummingbird pollination of understorey Acanthaceae, Costaceae (in part), and *Heliconia*, supports the finding of a scarcity of hummingbird pollination in the canopy. At Monteverde, Costa Rica, a mid-elevation site, hummingbirds have been

Table 7.2 Taxa at La Selva Biological Station and Barro Colorado Island known or inferred to be pollinated exclusively by Coleoptera

	La Selva	BCI*		La Selva	BCI*
Annonaceae			Cyclanthaceae		
Anaxagorea	2	1	*Asplundia*	8	1
Annona	2	3	*Carludovica*	2	2
Cymbopetalum	2	–	*Cyclanthus*	1	1
Desmopsis	2	1	*Dicranopygium*	2	–
Guatteria	3	2	*Evodianthus*	1	–
Rollinia	1	–	*Ludovia*	1	1
Sapranthus	1	–	*Sphaeradenia*	2	–
Unonopsis	2	1	Magnoliaceae		
Xylopia	2	1	*Magnolia (Talauma)*	2	–
Malmea	–	1	Nymphaeaceae		
Araceae			*Nymphaea*	–	1
Dieffenbachia	7	3	Zamiaceae		
Homalomena	1	1	*Zamia*	1	–
Philodendron	32	13			
Rhodospatha	2	2	Total	96	48
Syngonium	5	1	% of the flora	7	4
Xanthosoma	3	2			
Arecaceae					
Astrocaryum	2	1			
Bactris	4	4			
Cryosphila	1	1			
Desmoncus	1	1			
Elaeis	–	1			
Scheelea	–	1			
Socratea	1	1			

* excludes *Annona glabra* (Annonaceae), *Xylopia frutescens* (Annonaceae), and *Montrichardia arborescens* (Araceae), beetle-pollinated taxa believed to have arrived since the formation of the island

reported to pollinate 16.7% of a sample of 600 species (Feinsinger, 1983). It is noteworthy, however, that one particularly diverse group of hummingbird-pollinated species at mid-elevation, the Ericaceae, is nearly absent at La Selva. Additional diverse understorey taxa missing from the sample of Bawa and co-workers include Melastomataceae, Myrsinaceae, and Solanaceae, all of which exhibit poricidal anther dehiscence indicative of buzz pollination by a variety of bees (Buchman, 1983). Wind pollination would also increase with the addition of Poaceae, although it is probable that some of the deep forest grasses at La Selva are entomophilous.

A further examination of beetle pollination in the flora as a whole reveals a similar proportion to that for just the tree subset (Table 7.2). Among herbaceous taxa, all Cyclanthaceae are probably beetle-pollinated (see below), as well as the majority of Araceae. As inferred from beetle-pollinated taxa at La Selva, twice as many species, and nearly twice the proportion of the flora, are pollinated exclusively by beetles at La Selva in comparison to Barro Colorado Island (Croat, 1978) (Table 7.2). The lower proportion of beetle pollination on BCI can largely be attributed to decreased diversity of Araceae, especially *Philodendron* (32 species at La Selva, 13 at BCI), and Cyclanthaceae (17 at La Selva, five at BCI). Both families, along with Annonaceae, attain their maximum diversity in the wettest neotropical forests.

TAXONOMICALLY RELATED AND/OR FUNCTIONALLY SIMILAR NEOTROPICAL PLANTS AND POLLINATORS

Hummingbirds and their flowers

Thousands of plant species in the neotropics rely exclusively upon over 300 species of endemic hummingbirds (Trochilidae) for pollination. Subfamily Phaethorninae, the so-called hermit hummingbirds with markedly curved bills, occur at low elevations, and exhibit "trapline" foraging behaviour to nectar "rich" flowers; subfamily Trochilinae, with, for the most part, shorter and straighter bills, predominate at higher elevations visiting less specialized "moderate" flowers containing considerably less nectar, which are potentially visited by butterflies and bees as well (Feinsinger, 1976, 1978, 1983; Stiles, 1975, 1977, 1978, 1980). Whereas hermits may forage at a number of different simultaneously flowering rare species, which may or may not place pollen at specific spots along the bill, short-billed hummingbirds may temporarily specialize on and defend a dense patch of flowers (Linhart, 1973). As a consequence of the year-round energy requirements of hummingbirds, hummingbird-pollinated plants tend to be in flower at all times of the year, raising the possibility of phenological displacement of flowering times as a means of reducing interspecific pollen flow (Stiles, 1977, 1978, 1980).

Among those plant taxa endemic to the neotropics that have converged upon the hummingbird syndrome (tubular, often red corolla, with sucrose rich or sucrose dominant nectar) are such diverse genera as *Aphelandra* (Acanthaceae) (200 species), *Centropogon* (Lobeliaceae) (230 species), *Cavendishia* (Ericaceae) (100 species), *Columnea* (Gesneriaceae) (200 species), and *Psitticanthus* (Loranthaceae) (50 species), to name only a few of the more commonly known. The nearly endemic family Heliconiaceae (over 150 species) is exclusively pollinated by hummingbirds in the neotropics, and Stiles (1975) has suggested that the evolution of subfamily Phaethorninae is closely tied to *Heliconia*. Investigations of the South Pacific species of *Heliconia* (Kress, 1985) raise the possibility that the direction of migration of the genus has been eastward from South-east Asia to the neotropics, with the subsequent explosive radiation of *Heliconia* in the neotropics in conjunction with hummingbird pollination.

Neotropical palm groups

Palms are an especially conspicuous part of the vegetation at La Selva, both in their diversity (30 species in 16 genera) and abundance (Chazdon, 1985). Studies of Iriarteoid palms in Venezuela reveal contrasting pollination mechanisms in *Iriartea* and *Socratea* in relation to bee and beetle pollinators respectively (Henderson, 1985), corroborating previous reports from La Selva (Bullock, 1981). The derelomine weevil pollinators of *Socratea* appear to be widespread pollinators of palms, including the endemic Cocosoid genus *Bactris* (over 200 species) (Beach, 1984; Essig, 1971; Mora Urpi and Solis, 1980), as well as the endemic neotropical genus *Cryosophila* (Henderson, 1984). Additional Cocosoid genera endemic to the neotropics likely to be pollinated by beetles are *Scheelea* (36 species) and *Astrocaryum* (44 species) (Bullock, 1981). Within the Geonomoid palms (six genera and 92 species endemic to the neotropics (Moore, 1973), a variety of pollination systems have been reported: bee pollination in *Welffia* (Bullock, 1981); syrphid fly pollination in *Astreogyne* (Schmid,1970); and bat pollination in *Calypterogyne* (Beach, pers. comm.). Nearly all of the Chamaedoreoid palms are neotropical, the genus *Chamaedorea* accounting for 133 of the 143 species (Moore, 1973). Bawa *et al.* (1985) list *Chamaedorea* among wind-pollinated species. However, the strong fragrance of both male and female inflorescences, reminiscent of overripe cheese, and abundant visitation by Thysanoptera, strongly suggest entomophily (Schatz, pers. obs.). The majority of neotropical palms remain to be examined for pollination.

Cyclanthaceae

The endemic neotropical Cyclanthaceae comprise 12 genera and over 200 species of terrestrial and climbing or epiphytic herbs (Hammel, 1986; Harling, 1958). The sole member of the subfamily Cyclanthoideae, *Cyclanthus bipartitus*, is pollinated by dynastine scarab beetles (Beach, 1982). The other subfamily Carludovicoideae is noteworthy for the conservatism of its inflorescence morphology. Strong floral odours and protogynous, nocturnal anthesis correlate with pollination by beetles, contrary to one published report of Meliponine bees as pollinators (Schremmer, 1982). Derelomine weevils and Nitidulidae are abundant visitors, and may use the inflorescence as a mating and oviposition site (Harling, 1958; Schatz, pers. obs.). The role of the conspicuous spaghetti-like staminodia characteristic of subfamily Carludovicoideae has yet to be determined, but in all likelihood, the staminodia serve to increase the surface area involved in odour volatilization, given their turgid state at female anthesis during odour production and beetle arrival, and subsequent loss of turgidity by the onset of male anthesis. The uniformity of inflorescence morphology within subfamily Carludovicoideae may well result in generalized patterns of visitation by pollinators in any one community. Based on observations at La Selva, microhabitat specialization (e.g. *Dicranopygium* congeners with overlapping flowering times, identical floral odours (Schatz and Holman, unpub. data), and weevil visitors (Schatz, pers. obs.), but *D. wedelii* restricted to rocks in streams, and *D. umbrophila* terrestrial in forest understorey), vertical stratification (e.g. *Asplundia* spp.), and phenology (e.g. *Asplundia* spp.) appear to be important factors in reducing interspecific (and intergeneric) pollen flow.

"Lasciviously coloured, scandalously scented blossom after blossom flaunted its genitalia openly, enticing with visual and heretofore unknown olfactory charms any who might be inclined to sample its pleasures."

Tom Robbins, from *Jitterbug Perfume*

Two odour-driven systems

The structural and biotic complexity of tropical wet forest would seem to demand especially strong signals by plants to their pollinators. One need only fly over an unbroken tract of lowland forest to see widely dispersed, spectacular mass-flowering trees in the canopy, or inhale the intoxicating fragrances that permeate the heavy tropical evening air, as nocturnal pollinators who have sequestered themselves from daytime predators come to life, to realize just how strongly tropical rain forest plants must advertise themselves. The dichotomy of diurnal versus nocturnal activity in the tropics in response to year-long, high night-time temperatures, enables two separate worlds to occupy the same space. Among the sample of trees Bawa *et al.* (1985) examined for pollination systems, a combined total of 26.2% were either bat, beetle, or moth-pollinated, i.e. nocturnally pollinated. Floral odours are largely, if not entirely, the means by which these nocturnal flowers direct their pollinators. Correspondingly, nocturnal pollinators must discern from among a bewildering array of airborne volatiles – emitted as point sources, but soon overlapping and mixing – in order to fulfil energy requirements or locate conspecific mates (Kullenberg and Bergstrom, 1975; Schneider, 1969).

Of course, floral odours also function during the daytime. But in all probability, visual cues are equally as important, or more so, in diurnal pollination. However, one diurnal endemic neotropical association, male

Euglossine bees and the plants they pollinate, is entirely dependent upon odour volatiles, and has received considerable attention in the last 15 years (reviews by Williams, 1982; Williams and Whitten, 1983). More recently, an endemic nocturnal assemblage in which odour plays a critical role, dynastine scarab beetles and the plants they pollinate, has been the subject of collaborative research at La Selva (Beach, 1982; Schatz, 1985, 1987; Young, 1986). Following a brief summary of pollination by male euglossines, I shall describe in greater detail various aspects of scarab pollination, comparing and contrasting the two odour-driven systems.

Euglossini/Orchidaceae

The Euglossini comprise five genera and about 180 species of mostly solitary and nest parasitic, medium to large, often colourfully metallic bees restricted to the neotropics (Dressler, 1982). They are unique among the Apidae in that males possess inflated pouches on their hind tibiae into which odour volatiles, initially collected on specialized front tarsal brushes, are deposited. Male euglossines collect "perfumes" from a wide variety of sources including Aracaeae, Bignoniaceae, Euphorbiaceae, Gesneriaceae, Haemadoraceae, Myrsinaceae, and Solanaceae, as well as ooze from tree bark, roots, and perhaps fungus. The bulk of perfume sources, however, are about 625 species of orchids in 55 genera, which rely exclusively upon the male euglossines for pollination (Dodson *et al.*, 1969; Dressler, 1982; Williams, 1982).

Only male euglossines are attracted to and collect odour volatiles. Both male and female euglossines utilize a variety of other plants as nectar and pollen sources, and in so doing can also effect their pollination (e.g. *Costus* (Schemske, 1981); Marantaceae (Kennedy, 1978)). Both sexes exhibit trapline foraging behaviour to either nectar or perfume sources, and are primarily active in the early morning (Ackerman *et al.*, 1982; Janzen, 1971). The ultimate fate of the fragrance compounds collected by male euglossines has yet to be determined, but is thought to be linked to their reproductive biology, the compounds perhaps serving as precursors of male sex pheromones.

Patterns of visitation by male euglossines to perfume sources can often be highly specific with only one or several species present in the community attracted to a given source. Specificity is based upon species-specific combinations of odour compounds (Williams and Dodson, 1972), and has been shown to play an important role as a reproductive isolating mechanism between sympatric species (Hills *et al.*, 1972). Insofar as male euglossines can utilize a variety of perfume sources, the degree of mutual dependence is highly skewed in the plant direction (Ackerman, 1983).

Dynastine scarab beetles as neotropical pollinators

Bates (1886–1890) was perhaps the first author to allude to the possible pollinator role of neotropical dynastines, citing field notes of the collector Champion as having encountered great numbers of *Cyclocephala* deeply embedded in the viscous pollen at the bottom of the spathes of Arums (Araceae). Malme (1907) listed *Cyclocephala* as a visitor to *Victoria cruziana* (Nymphaeaceae), as well as species of *Annona* (Annonaceae). The comprehensive study of the floral biology of *Victoria amazonica* and the behaviour of its *Cyclocephala* visitors by Prance and Arias (1975) established what has become the standard pattern of dynastine scarab pollination. Subsequent studies at La Selva and elsewhere (Beach, 1982, 1984; Bullock, 1981; Cramer *et al.*, 1975; Gibbs *et al.*, 1977; Gottsberger, 1986; Gottsberger and Amaral, 1984; Meeuse and Schneider, 1979/1980; Prance, 1980; Prance and Anderson, 1976; Schatz, 1985, 1987; Valerio, 1984; Young, 1986) reveal the extent of dynastine scarab pollination in the neotropics, and allow one to predict that as many as 900 species rely upon dynastine scarabs for pollination in the neotropics (Table 7.3). The endemic dynastine scarab genus *Cyclocephala* comprises over 220 species (Pike *et al.*, 1975), whereas the related genus *Erioscelis* consists of four species (Endrodi, 1966)

A number of intriguing parallels and contrasts to male euglossine pollination emerge. Scarab-pollinated plants volatilize strong odours to attract pollinators, but do so by elevating flower and inflorescence temperatures through a process referred to as thermogenic respiration (Meeuse, 1975; Nagy *et al.*, 1972). Odour production is timed to coincide with the crepuscular/nocturnal behaviour patterns of dynastine scarabs. Both male and female scarabs are attracted to flowers and inflorescences at the female stage of anthesis (protogyny is common to all scarab-pollinated taxa), where they will remain for usually 24 hours, mating and feeding upon specialized food tissues, until the male stage of anthesis, at which point they depart and move to another female phase inflorescence.

Insofar as flowers and inflorescences fulfil their energy requirements and may be the only aggregation sites for male and female scarabs, the degree of mutual dependence between plant and pollinator is both high and equivalent. In contrast to floral volatiles perhaps serving as precursors for euglossine male sex pheromones, for dynastine scarabs, the floral volatiles themselves have taken on the actual role of the sexual pheromone in the scarab life cycle. As has been suggested for euglossine-pollinated orchids (Dressler, 1981), one can easily imagine a scenario for sympatric speciation in which a mutation resulting in an altered odour profile leads to the attraction of a select subset of visitors, effectively isolating both plant and pollinator.

Table 7.3 Neotropical genera known or predicted to be pollinated by Scarabaeidae: Dynastinae

Annonaceae
Annona section *Annona*
Campicola
Helogenia subsection *Aculeato-papillosae*
Macrantha
Phelloxylon
Pilanona
Psammogenia
Ulocarpus
Cardiopetalum
Cymbopetalum
Duguetia (in part)
Froesiodendron
Fusaea
Malmea (in part)
Porcelia

Araceae
Anthurium (in part)
Caladium
Caladiopsis
Chlorospatha
Dieffenbachia
Homalomena
Monstera (in part)
Montrichardia
Philodendron
Rhodospatha (in part)
Syngonium (in part)
Xanthosoma

Arecaceae
Ammandra
Astrocaryum
Bactris
Palandra
Phytelephas

Cyclanthaceae
Cyclanthus

Magnoliaceae
Magnolia (Talauma) (in part)

Nymphaeaceae
Nymphaea (in part)
Victoria

The scarab-pollinated guild at La Selva is particularly diverse, with perhaps as many as 60 species in four families serviced by over a dozen scarab species, including 14 sympatric *Cyclocephala* (Table 7.4). The hypothesis that patterns of visitation by scarab pollinators would correlate with odour similarities and differences has led to initial screening of fragrance volatiles of scarab-pollinated taxa. Preliminary results (Schatz and Holman, unpub. data) are graphically summarized in Figure 7.1, which depicts all pairwise distances among the odours of 21 scarab-pollinated species at La Selva using Kruskall Multidimensional Scaling (MDS) in the statistical package for personal computers known as SYSTAT.

To some degree, the patterns of visitation thus far observed are nicely explained by odour. The high degree of specificity exhibited between *Philodendron radiatum* and *Cyclocephala sp. nov.* is reflected in the degree to which the odour profile of *P. radiatum* (N-PHI RAD) stands apart from all other scarab-pollinated taxa. Indeed, the odour of *P. radiatum* is composed nearly entirely of compounds unique to it. In addition to odour as a structuring parameter, both phenology (of flowering and adult beetle emergence) and vertical stratification are hypothesized to play important roles in the organization of the scarab/plant assemblage at La Selva.

Table 7.4   Scarab-pollinated plants and flower visiting scarabs at La Selva Biological Station

Scarab-pollinated plants		*Flower visiting scarabs*
Annonaceae		Cyclocephala
Annona	1	*C. amazona*
Cymbopetalum	2 + hybrid	*C. amblyopsis*
		C. atripes
Araceae		*C. brittoni*
Dieffenbachia	7 + hybrids	*C. conspicua*
Homalomena	1	*C. gravis*
Philodendron	32 (all?)	*C. kaszabi*
Rhodospatha	1	*C. ligyrina*
Syngonium	5	*C. mafaffa*
Xanthosoma	3	*C. sexpunctata*
		C. sparsa
Arecaceae		*C. stictica*
Bactris	4	*C. tutilina*
		C. sp. nov.
Cyclanthaceae		
Cyclanthus	1	Erioscelis
		E. columbica
		Mimeoma
		M. acuta

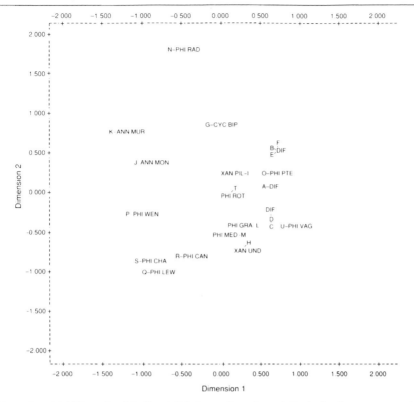

Figure 7.1 Multidimensional Scaling (MDS) in two dimensions of all pairwise distances among the odour profiles of 21 scarab-pollinated species at La Selva Biological Station, Costa Rica

DIRECTIONS FOR FUTURE RESEARCH

Despite our innate behaviour of poking our noses into flowers, and the common wisdom that flowers smell nice to attract their pollinators, there is a dearth of information on floral fragrances and their role in pollination biology. The tools now exist (Williams, 1983) to go well beyond our woefully inadequate verbal descriptions of floral odour, and begin to address evolutionary questions concerning plant–pollinator interactions (Pellmyr and Thien, 1986), co-speciation, and the structure of plant–pollinator assemblages.

REFERENCES

Ackerman, J.D. (1983). Specificity and mutual dependance of the orchid-euglossine bee interaction. *Biological Journal of the Linnean Society*, **20**, 301–4
Ackerman, J.D., Mesler, M.R., Lu, K.L. and Montalvo, A.M. (1982). Food-foraging behavior of

male Euglossini (Hymenoptera: Apidae):Vagabonds or trapliners? *Biotropica*, **14**, 241–8

Bates, H.W. (1886–1890). Insecta. Coleoptera. Dynastinae. In *Biologia Centrali-Americana*., **Vol. 2 (Part 2)**, 299–311

Bawa, K.S. (1974). Breeding systems of tree species of a lowland tropical community. *Evolution*, **28**, 85–92

Bawa, K.S. and Beach, J.H. (1981). Evolution of sexual systems in flowering plants. *Annals of Missouri Botanical Garden*, **68**, 254–74

Bawa, K.S., Bullock, S.H., Perry, D.R., Coville, R.E., and Grayum, M.H. (1985). Reproductive biology of tropical lowland rain forest trees. II. Pollination systems. *American Journal of Botany*, **72**, 346–56

Bawa, K.S., Perry, D.R. and Beach, J.H. (1985). Reproductive biology of tropical lowland rain forest trees. I. Sexual systems and incompatibility mechanisms. *American Journal of Botany*, **72**, 331–45

Beach, J.H. (1982). Beetle pollination of *Cyclanthus bipartitus* (Cyclanthaceae). *American Journal of Botany*, **69**, 1074–81

Beach, J.H. (1984). The reproductive biology of the Peach or "Pejibaye" palm (*Bactris gasipaes*) and a wild congener (*Bactris porschiana*) in the Atlantic lowlands of Costa Rica. *Principes*, **28**, 107–19

Buchmann, S.L. (1983). Buzz pollination in angiosperms. In Jones, C.E. and Little, R.J. (eds) *Handbook of Experimental Pollination Biology*, pp. 73–113. (New York: Van Nostrand Reinhold)

Bullock, S.H. (1981). Notes on the phenology of inflorescences and pollination of some rain forest palms in Costa Rica. *Principes*, **25**, 101–5

Chazdon, R.L. (1985). The palm flora of Finca La Selva. *Principes*, **29**, 74–8

Cramer, J.M., Meeuse, A.D.J. and Teynissen, P.A. (1975). A note on the pollination of nocturnally flowering species of *Nymphaea*. *Acta Botanica Neerlandica*, **24**, 489–90

Croat, T.B. (1978). *Flora of Barro Colorado Island*. (Palo Alto, CA: Stanford California University Press)

Croat, T.B. (1979). The sexuality of the Barro Colorado Island flora (Panama). *Phytologia*, **42**, 319–48

Dodson, C.H., Dressler, R.L., Hiulls, H.G., Adams, R.M. and Williams, N.H. (1969). Biologically active compounds in orchid fragrances. *Science*, **164**, 1243–9

Dressler, R.L. (1981). *The Orchids. Natural History and Classification*. (Cambridge, Massachusetts: Harvard University Press)

Dressler, R.L. (1982). Biology of the orchid bees (Euglossini). *Annual Review of Ecology and Systematics*, **13**, 373–94

Endrodi, S. (1966). Monographie der Dynastinae (Coleoptera: Lamellicornia). I. *Teil. Entomol. Abh. Mus. Tierk*., **33**, 1–460

Essig, B.F. (1971). Observations on pollination of *Bactris*. *Principes*, **5**, 20–4

Feinsinger, P. (1976). Organization of a tropical guild of nectarivorous birds. *Ecological Monographs*, **46**, 257–91

Feinsinger, P. (1978). Ecological interactions between plants and hummingbirds in a successional tropical community. *Ecological Monographs*, **48**, 269–87

Feinsinger, P. (1983). Coevolution and pollination. In Futuyma, D.J. and Slatkin, M. (eds) *Coevolution*, pp. 283–310. (Sunderland, Massachusetts: Sinauer Associates)

Gibbs, P.E., Simir, J. and Diniz Da Cruz, N. (1977). Floral biology of *Talauma ovata* St. Hil. (Magnoliaceae). *Ciencia y Culture*, **29**, 1437–44

Gottsberger, G. (1970). Beitrage zur Biologie von Annonaceen-Bluten. *Oesterreichische Botanische Zeitschrift*, **118**, 237–79

Gottsberger, G. (1974). The structure and function of the primitive angiosperm flower – discussion. *Acta Botanica Neerlandica*, **23**, 461–71

Gottsberger, G. (1986). Some pollination strategies in neotropical savannas and forest. *Plant Systematics and Evolution*, **152**, 29–45

Gottsberger, G. and Amaral, A. (1984). Pollination strategies in Brazilian *Philodendron* species. *Berichte der Deutschen Botanischen Gesellschaft*, **97**, 391–410

Hammel, B.E. (1986). The vascular flora of La Selva Biological Station, Costa Rica Cyclanthaceae. *Selbyana*, **9**, 196–202

Hammel, B.E. and Grayum, M.H. (1982). Preliminary report on the flora project of La Selva field station, Costa Rica. *Annals of Missouri Botanical Garden*, **69**, 420–5

Harling, G. (1958). Monograph of the Cyclanthaceae. *Acta Horti Bergiani*, **18**, 1–428

Henderson, A. (1984). Observations on pollination of *Cryosphila albida* (Palmae). *Principes*, **28**, 120–6

Henderson, A. (1985). Pollination of *Socratea exorrhiza* and *Iriartea ventricosa*. *Principes*, **29**, 64 71

Hills, H.G., Williams, N.H. and Dodson, C.H. (1972). Floral fragrances and isolating mechanisms in the genus *Catasetum* (Orchidaceae). *Biotropica*, **4**, 61–76

Janzen, D.H. (1971). Euglossine bees as long-distance pollinators of tropical plants. *Science*, **171**, 203–5

Kennedy, H. (1978). Systematics and pollination of the "closed-flowered" species of *Calathea* (Marantaceae). *University of California Publications in Botany*, **71**, 1–90

Kress, W.J. (1983). Self-incompatibility systems in Central American *Heliconia* (Heliconiaceae). *Evoution*, **37**, 735–44

Kress, W.J. (1985). Bat pollinatoin of an old world *Heliconia*. *Biotropica*, **17**, 302–8

Kullenberg, B. and Bergstrom, G. (1975). Chemical communication between living organisms. *Endeavour*, **34**, 59–66

Linhart, Y.B. (1973). Ecological and behavioral determinants of pollen dispersal in hummingbird-pollinated *Heliconia*. *American Naturalist*, **107**, 511–23

Malme, G.O.A.N. (1907). Nagra Anteckningar om *Victoria* Lindl., Sarskildt om *Victoria cruziana* D'Orb. *Acta Horti Bergiani*, **4(5)**, 3–16

Meeuse, B.J.D. (1975). Thermogenic respiration in aroids. *Annual Review of Plant Physiology*, **26**, 117–26

Meeuse, B.J.D. and Schneider, E.L. (1979/1980). *Nymphaea* revisited: a preliminary communication. *Israel Journal of Botany*, **28**, 65–79

Moore, H.E. Jr. (1973). The major groups of palms and their distribution. *Gentes Herbarum*, **11(2)**, 27 141

Mora Urpi, J. and Solis, E. (1980). Polinazacion en *Bactris gasipaes* HBK (Palmae). *Revista de Biologia Tropical*, **28**, 153 74

Nagy, K.A., Odell, D.K. and Seymour, R.S. (1972). Temperature regulation by the inflorescence of *Philodendron*. *Science*, **178**, 1195 7

Pellmyr, O. and Thien, L.B. (1986). Insect reproduction and floral fragrances: keys to the evolution of the angiosperms. *Taxon*, **35**, 76–85

Pike, K.S., Rivers, R.L., Ratcliffe, B.C., Oseto, C.Y. and Mayo, Z.B. (1975). *A World Bibliography of the Genus* Cyclocephala (*Coleoptera: Scarabaeidae*). (University of Nebraska, Lincoln: Institute of Agriculture and Natural Resources)

Prance, G.T. (1980). A note on the pollination of *Nymphaea amazonum* Mart. and Zucc. (Nymphaeaceae). *Brittonia*, **32**, 505 7

Prance, G.T. and Anderson, A.B. (1976). Studies of the floral biology of neotropical Nymphaeaceae. 3. *Acta Amazonica*, **6**, 163–70

Prance, G.T. and Arias, J.R. (1975). A study of the floral biology of *Victoria amazonica* (Poepp.) Sowerby (Nymphaeaceae). *Acta Amazonica*, **5**, 109 39

Schatz, G.E. (1985). A new *Cymbopetalum* (Annonaceae) from Costa Rica and Panama, with observations on natural hybridization. *Annals of Missouri Botanical Garden*, **72**, 535–8

Schatz, G.E. (1987). Systematic and ecological studies of Central American Annonaceae. Ph.D. Thesis. University of Wisconsin, Madison.

Schemske, D.W. (1981). Floral convergence and pollinator sharing in two bee-pollinated tropical herbs. *Ecology*, **62**, 946–54

Schmid, R. (1970). Notes on the reproductive biology of *Asterogyne martiana* (Palmae). II. Pollination by syrphid flies. *Principes*, **14**, 39–49

Schneider, D. (1969). Insect olfaction: deciphering systems for chemical messages. *Science*, **163**, 1031–7

Schremmer, F. (1982). Bluhverhalten und Bestaubungsbiologie von *Carludovica palmata* (Cyclanthaceae), ein okologisches Paradoxon. *Plant Systematics and Evolution*, **140**, 95 107

Stiles, F.G. (1975). Ecology, flowering phenology and hummingbird pollination of some Costa Rica *Heliconia* species. *Ecology*, **56**, 285 301

Stiles, F.G. (1977). Coadapted competitors: the flowering seasons of hummingbird-pollinated plants in a tropical forest. *Science*, **198**, 1177–8

Stiles, F.G. (1978). Temporal organization of flowering among the hummingbird food plants of a tropical wet forest. *Biotropica*, **10**, 194–210

Stiles, F.G. (1980). The annual cycle in a tropical wet forest hummingbird community. *Ibis*, **122**, 322–43

Valerio, C.E. (1984). Insect visitors to the inflorescence of the aroid *Dieffenbachia oerstedii* (Araceae) in Costa Rica. *Brenesia*, **22**, 139–46

van der Pijl, L. and Dodson, C.H. (1966). *Orchid Flowers: Their Pollination and Evolution*. (Coral Gables, Florida: University of Miami Press)

Williams, N.H. (1982). The biology of orchids and euglossine bees. In Arditti, J. (ed.) *Orchid Biology: Reviews and Perspectives*. II. pp. 119–71. (Ithaca, New York: Cornell University Press)

Williams, N.H. (1983). Floral fragrances as cues in animal behaviour. In Jones, C.E. and Little, R.J. (eds) *Handbook of Experimental Pollination Biology*, pp. 50–72. (New York: Van Nostrand Reinhold)

Williams, N.H. and Dodson, C.H. (1972). Selective attraction of male euglossine bees to orchid floral fragrances and its importance in long distance pollen flow. *Evolution*, **26**, 84–95

Williams, N.H. and Whitten, W.M. (1983). Orchid floral fragrances and male euglossine bees: methods and advances in the last sesquidecade. *Biological Bulletin*, **164**, 355–95

Young, H.J. (1986). Beetle pollination of *Dieffenbachia longispatha* (Araceae). *American Journal of Botany*, **73**, 931–44

CHAPTER 8

PLANT–POLLINATOR INTERACTIONS IN MALAYSIAN RAIN FORESTS

S. Appanah

ABSTRACT

The lowland dipterocarp forests of Malaysia are reputed to be the most complex terrestrial ecosystems known, and are especially renowned for their extraordinary wealth of plant species. Matching such a diversity are the plant–pollinator interactions. Although such interactions have hardly been investigated, the few systematic investigations have yielded some interesting results. Flowering patterns can be classified into four synusiae based on the presentation of the flowers. This grouping is reflected in the attendant differentiation of the pollinator guilds. A further distinguishing feature of these forests is the heavy supra-annual general flowering of the family Dipterocarpaceae as well as several other families. Among the large array of plant–pollinator interactions observed, certain unique relationships such as the pollination of dipterocarps by thrips provide clues to help explain the tremendous diversity of dipterocarps in this region.

INTRODUCTION

In the tropics, abiotic pollination plays only a marginal role. Instead, a huge variety of animal pollinators, matching the diversity of plants, are engaged in a myriad of interactions with plants. The subject of plant–pollinator interactions fascinates biologists endlessly, and is adorned with some of the most dramatic examples in the story of evolution. The subject has been dealt with great elegance at textbook length by several authors (e.g. Faegri and van der Pijl, 1966; Proctor and Yeo, 1973), and for the tropics, especially the neotropics, by numerous workers (see, for example, review by Baker *et al.*, 1982). By comparison, there are few systematic studies in Malaysia or the

Far East for a comprehensive treatment. Nevertheless, from the few detailed studies, some unique features in plant–pollinator interactions in Malaysia can be discerned to be fundamentally different from the other tropical forests, and these unique interactions can partly explain differences in the evolution of the Malaysian tropics from the rest.

The Malaysian region, which includes Peninsular Malaysia and much of north Borneo, is known to be the least seasonal of all the tropical regions. The climax phase rain forests, called the lowland dipterocarp forests, are renowned for their diversity of flora and fauna. In terms of tree species diversity, these humid forests are very rich and are among the most diverse in the world (Whitmore, 1975). What distinguishes such diversity further is the presence among several genera of a huge number of apparently sympatric species (Ashton, 1969, 1988). This is trenchantly expressed in the family Dipterocarpaceae. This family of emergent trees is highly species rich, with some 400 species in west Malesia. It is not unusual to see packed together as many as 40 species within one community.

Such a complex forest poses numerous questions to an ecologist:

(1)   What are the pollinators of the huge diversity of trees, and, related to it, how do so many closely related tree species coexist without competition for pollinators?

(2)   What is the nature of plant–pollinator interactions, and how do they influence pollen flow in this complex plant community?

(3)   Are the flowering patterns observed here different from the rest of the tropics, and, if so, are they related to the unique aseasonality in the region?

In addressing some of these questions, I shall first touch upon some aspects of pollination processes in the marginal vegetation, and then focus mainly on the lowland dipterocarp forests. Here, the different floral presentations and pollinators within the community are differentiated. Next, the various flowering phenologies are introduced, greater emphasis being paid to the unique general flowering observed in the region. A summary of the floral attractants and rewards is presented. Finally, two Malaysian case studies are discussed in terms of the plant–pollinator interactions, and how they influence pollen dispersal. I conclude with a discussion of how certain pollinators like thrips may have allowed the diversification and maintenance of the extraordinary diversity of dipterocarps in this region.

POLLINATION AND SPECIES DIVERSIFICATION STUDIES IN MALAYSIA

Some of the earliest observations were merely cursory and undertaken in botanical gardens (Burkill, 1919). Detailed studies were restricted to 'curious pollination' syndromes, for example, *Amorphophallus* and *Rafflesia* (van der Pijl, 1937; Faegri and van der Pijl, 1966), or to the foraging of *Xylocopa* in secondary vegetation (van der Pijl, 1939). Corner (1952) described interactions of *Ficus* and fig wasp, while in the early 1970s work was undertaken on bat pollination and foraging behaviour (Start, 1974; Start and Marshall, 1976; Gould, 1977, 1978).

While these studies represented exciting beginnings for pollination ecology in the tropics, the gene flow patterns to elucidate speciation and diversification of plants in the tropics remained obscure. It was in fact Corner (1954) who made the first bold step in that direction. From his observations, he concluded that the resulting infrequency of neighbouring conspecific individuals in the closed species-rich forest ecosystem, and the intermittent and non-synchronized massive flowering of individual trees drawing the attention of whole swarms of pollinators, would severely reduce cross-pollination and promote tendencies for individuals to inbreed, and for mutants to propagate colonies as new species.

Later, Fedorov (1966) similarly echoed the same ideas and invoked 'automatic genetic processes' such as genetic drift to explain the presence of closely related sympatric species. Fedorov's interpretation was disputed by Ashton (1969), who argued for the existence of an unspecialized pollination system consisting of numerous pollen vectors, which enables frequent out-crossing, especially within clumps of conspecifics. He further suggested that allopatric adaptive speciation is the main mode for diversification in tropical forest trees. In order to determine whether inbreeding or outbreeding is the prevalent mode, a major research programme to study the breeding systems and pollination mechanisms of Malaysian rain forest trees was undertaken by Ashton and his colleagues (Appanah, 1979; Ashton, 1976; Ashton *et al.*, 1977; Chan, 1977; Gan, 1976; Ha, 1978; Kaur, 1977; Yap, 1976). This contribution is largely based on these studies conducted in Malaysia. Parallel studies were initiated by Bawa and his associates in Central America (reviewed by Schatz in this volume).

POLLINATION IN MANGROVE SWAMP AND SECONDARY FORESTS

In mangrove swamp forests, large stretches of the coast are dominated by a few species of trees such as *Rhizophora*, *Sonneratia*, *Bruguiera* and

Avicennia. The flowering pattern commonly observed is annual and extended. Some species such as *Rhizophora* set only a few, large, and conspicuous flowers per day, and are visited by the wide-ranging cave-dwelling bat, *Eonycteris*. Other common pollinators are the bees, *Apis* and *Trigona*.

The secondary forests, which fringe the primary forests, are dominated by pioneers and late-seral vegetation. Pioneers such as *Macaranga*, *Mallotus*, *Callicarpa* and *Melastoma* flower and are visited annually or continuously by insects such as hoppers, trigonid bees, honey bees and the wide-ranging carpenter bees. Several other species such as in *Musa* and *Oroxylum*, and the late-seral species, for example, in *Durio*, *Parkia* and *Artocarpus* are commonly found here. They generally flower annually, producing few flowers per day and are visited by the bats. Sun birds (Nectarinidae) are also common in this habitat, visiting flowers of *Erythrina*, etc.

FLOWERING SYNUSIAE IN LOWLAND DIPTEROCARP FORESTS

Vegetation is stratified in the lowland dipterocarp forests. Based on the position in which flowers are presented, Ashton *et al.* (1977) divided the forest trees into two synusiae: those that presented their flowers above the canopy, and those that held the flowers within the foliage. It is presently possible, however, to recognize two more synusiae. To the third synusiae belong the cauliflorous trees, which hold their flowers along the trunk (trunciflory), and or around the main branches (ramiflory). A fourth synusiae is formed by the shrubs and herbs. Here, too, belongs the interesting group of some of the unusually large geophilous flowers. Pollinators can be differentiated among the various synusia.

Flowers held above the canopy

Many emergent and canopy species present their flowers above the canopy. Most of these species have hermaphrodite flowers, which are usually large, brightly coloured, conspicuous, and thus visible from long distances. The trees bear a very large number of flowers which further enhances their conspicuousness. Although deciduousness is rare in the Malaysian forests, the few examples are emergent trees which drop their leaves before flowering and this feature increases their visibility (e.g. *Dyera costulata*, *Bombax valetonii*, *Scaphium* species, *Koompassia* species, *Ficus* species and *Parkia* species). The flowering is generally seasonal or intermittent, with high conspecific synchrony. Olfactory lures are usually present, but seem to be effective only over short distances among diurnal flowers. The flowers possess large amounts of nectar and/or pollen; large-sized and wide-ranging vertebrates and bees are their most common visitors. A contrast in this synusiae is provided by the family Dipterocarpaceae. Except for a few species such as *Shorea ovalis*, *Neobalanocarpus heimii* and *Dryobalanops* species, the majority of the species belonging to *Shorea*, *Hopea*, *Vatica* and *Dipterocarpus* have flowers with reduced floral rewards, strong odours and crepuscular anthesis.

Flowers held within the canopy

To this synusia belong many members of the understorey, and dioecy is a common feature. These trees usually flower for extended periods annually or frequently, and present a small number of flowers at a time. The tiny flowers are white to greenish and have small amounts of nectar and/or pollen. Their pollinators and flower visitors are restricted to the wide array of unspecialized insects commonly found in the immediate vicinity of the trees.

Cauliflorous flowers

A few emergents, some canopy, and many understorey species exhibit cauliflory, and dioecy is a common feature here too. The flowers held on the trunk and/or main branches are generally larger (or small but clumped), coloured from whitish to pink, and are strongly olfactorial. The flowering can be annual to irregular, is synchronous and lasts only for a few days. The pollinators include beetles and trigonid bees from the understorey, and the above-canopy forager *Apis dorsata*. Even bats and birds that are usually confined to the upper canopy have been observed to descend to forage on ramiflorous flowers of *Durio malaccensis* and *Ganua* species (see later).

Forest floor flowers, shrubs and herbs

A wide variety of flowers are lumped within this synusia, including herbs, shrubs and parasitic plants. The flowers are often very specialized (e.g. thermogenic flowers of *Arum*), sizes range from small to the largest flowers in the world (e.g. *Rafflesia*), and colours from white to pink, purple and often earth brown. Usually, only a few flowers are presented at a given time. Flowers have strong odours. The pollinators are usually flies, beetles, tiny bees, wasps, traplining butterflies and moths. Also present here are flowers that emit dung and carrion odour (e.g. *Arum*, *Rafflesia*, *Amorphophallus*), which draw the dung and carrion flies that are normally resident there.

Differences among synusiae

The reasons for differences in the flowers and pollinators among the four synusiae can be adduced. Visibility is highly restricted in the understorey, and a flower with strong visual lures in this stratum is not likely to draw pollinators from a wide range. Although some of the flowers smell strongly, odour has a restricted effect in a windless microclimate. Indeed, it is not surprising that the majority of the flower visitors are exclusively insects, drawn from the ever present pool of short-ranging, low-energetic, polylectic forms. Predominant among them are apparently imprecise trapliners (see Janzen, 1971) such as trigonid bees, solitary wasps and butterflies which may be drawn by both visual and olfactory lures over short ranges. The lowest strata also include unspecialized beetles, midges, flies and thrips which seem to be drawn more by olfactory lures, though final alighting can be mediated visually.

The poor floral rewards in this understorey synusiae seem to reflect the low energy needs of the visitors too, as well as the low numbers of visitors being lured. The flowering, however, is usually for extended periods, with wide synchrony, and individual flowers may remain receptive for several days, raising the chances of pollination by these haphazard visitors. These characteristics are clearly evidenced in the understorey species *Xerospermum intermedium* (Appanah, 1982).

The dominant pollinators in the upper stratum are *Apis dorsata* with a very wide foraging range, and many trapliners such as carpenter bees, birds and bats. These pollinators are known to have acute vision, and the bright flowers can lure a potentially greater number of pollinators from a wider area. Nevertheless, many of the site-specific understorey pollinators do visit, probably opportunistically, some of these flowers too. The flowering within species is highly synchronous and brief. Differing from this norm is a whole group of dipterocarps of the genera *Shorea*, *Hopea* and *Dipterocarpus* which have developed a unique syndrome: they present numerous but often tiny flowers with reduced floral rewards and nocturnal anthesis. They form the breeding grounds for thrips and other such tiny flower-feeding insects which then act as their pollinators.

The cauliflorous synusia is interesting in that the flowers have the same characteristics as the synusia with flowers held above the canopy. In the former, though, the flowers are instead brought down and packed along the branches and/or trunk. A number of their pollinators descend from the upper stratum too, probably drawn below by olfactory lures. For example, *Apis dorsata*, which is blind in the red range of the visual spectrum (Burkhardt, pers. comm.), does not usually forage below the canopy in the closed forest, where the visual light available is richer in the reds. Yet this bee can penetrate below the canopy and forage downwards on the trunciflorous flowers of *Baccaurea* species. Bats are the likely foragers on the cauliflorous *Durio malaccensis* flowers. *Ganua* species, with its pinkish flowers on the main branches, is highly visible in the understorey and draws birds found in that stratum. In conclusion, it must be recognized that these observations lack the quantitative data that Bawa and his co-workers (Bawa, Perry and Beach, 1985; Bawa *et al.*, 1985) gathered in their ecosystem level studies of the neotropical rain forests.

FLOWERING PATTERNS

Flowering in the Malaysian primary forests is highly variable, but can be categorized into several patterns as follows:

(1)   Flowering almost continuously, setting a few flowers per day. This characteristic is common in a small but significant number of understorey genera (e.g. Rubiaceae – *Lasianthus, Randia, Ixora*);

(2)   Annual flowering, for extended periods, with synchrony in population, but with periodic heavy flowering years among many understorey species (e.g. *Xerospermum intermedium*); and,

(3)   The supra-annual general flowering of Malaysian climax forests, which at long intervals of 2–8 years, involves a majority of Dipterocarpaceae, and several other diverse families such as the Leguminosae, Polygalaceae and Bombacaceae in the canopy, and Myristicaceae and Euphorbiaceae in the understorey. Flowering is heavy, and can extend throughout west Malesia (Wood, 1956; Medway, 1972; Ashton, 1969; Appanah, 1985). Although this is the dominant pattern, at least some of the species, especially the non-dipterocarps (e.g. Leguminosae – *Koompassia, Sindora, Milletia*) are known to flower less gregariously and at shorter intervals (2–3 years).

General flowering in lowland dipterocarp forests

The general flowering in the lowland dipterocarp forests is unique to this region and has not been reported elsewhere. During one such flowering in Pasoh Forest Reserve in 1981, as many as 161 tree species belonging to 41 families were observed to flower within the span of 4 months (Appanah, 1985). This represented 33.6% of the 478 species and 75% of the families of trees >10 cm diameter at breast height (dbh) recorded in a 10-ha survey of the forest (Ashton, 1971). The flowering was much heavier in 1976 (pers. obs.). No other forests have been reported to approach such a diversity of tree species and abundance of trees flowering in a single habitat, and for so brief a period. In the neotropics, gregarious flowering is generally an intraspecific phenomenon, with much less synchronization among species, and flowering is more evenly distributed on a regular basis (Frankie *et al.*, 1974).

Sources of pollinators during a general flowering

The general flowering, being so intense, can pose an immense demand for pollinators. Furthermore, the irregularity and long intervals between general flowerings make it unlikely that long-lived pollinators can build up their populations sufficiently rapidly to meet the demands. Observations suggest that, upon advance of the general flowering, there seems to be a local increase in the numbers of bees, wasps, butterflies and other ubiquitous flower visitors. In addition, some of the wide-ranging pollinators such as carpenter bees (*Xylocopa* species), sun birds and bats, whose usual domicile is the secondary forests, begin to migrate into the mature phase forest to forage on many above-canopy flowers.

However, the source of pollinators for the dipterocarps, *Shorea* section *Mutica*, is in sharp contrast to other known pollination systems. The millions of flower buds of the series of closely related *Shorea* species form the breeding ground for tiny flower thrips (Thysanoptera; predominantly *Thrips* species, *Thrips hawiiensis* (Morgan) and *Megalurothrips* species) (Appanah and Chan, 1981). The adults of *Thrips* species emerge from the buds and flowers within 8 days of egg deposition and begin to feed on the petal tissue and pollen of these nocturnal flowers. As the corolla are shed in the morning, they displace the thrips. When the new flowers bloom in the evening, the flowers emanate a strong sweet smell which lures the thrips pollinators back to the new flowers. Considering that a single tree may produce as many as 4 million flowers, and the thrips have such a short life-cycle and a fecundity of about 30 eggs per female, these thrips are able to build up their populations very rapidly and meet the pollinatory needs of millions of these flowers.

Competition for pollinators

Several tree genera that share similar pollinator guilds exhibit sequential flowering, suggesting the existence of competition for pollinators (Appanah, 1985; Ashton *et al.*, 1988). For example, several unrelated species, *Xanthophyllum discolor*, *X. rufum*, *Connarus* species, *Sindora velutina*, *S. coriacea*, *Mesua ferrea*, *Callicarpa maingayi*, *Milletia atropurpurea* and *Tetracera* species pollinated by *Xylocopa* bees, have staggered blooming periods (Appanah, 1985). Interspecific sequential flowering also occurs among the thrips-pollinated *Shorea* species (Chan and Appanah, 1980), and among species in other sections of *Shorea* such as *Richetioides* and *Anthoshorea* (Wood, 1956). Finally, it seems that many *Apis dorsata* pollinated trees such as *Pentaspadon, Koompassia, Scaphium* and several others also bloom sequentially during the general flowering.

FLORAL ADAPTATIONS

Attractants for pollinators

There are few systematic studies on the floral attractants among Malaysian rain forest plants. Thus, it is only possible to make a general commentary, and to highlight known case studies. The review by Faegri and van der Pijl (1966) covers all floral attractants, and the substance of the observations made for neotropical plants (Baker and Baker, 1983; Frankie and Haber, 1983; Kevan, 1983) also apply in general terms to the Malaysian situation.

Floral colour

Floral colour has been generally treated in the earlier section on flowering synusiae. Some distinction in the colouration of the flowers between the various synusiae can be perceived:

(1)   Usually bright colours are seen among the flowers held above the canopy;

(2)   The colouration is generally very dull among the flowers held within the canopy;

(3)   The flowers of the cauliflorous synusia are less bright than those of (1) above; and,

(4)   The flowers of the forest floor synusia are highly variable.

One unique colour here is the earth-brown colour of the parasitic *Rafflesia* flowers found on the forest floor. On this subject of colour resemblance, another interesting observation is the similarity in appearance between the red-and-green parakeets (*Loriculus*) and the red flowers of *Erythrina*. The birds "disappear" as soon as they alight on the flowers (Faegri and van der Pijl, 1966), an adaptation which probably provides protection during feeding when the birds are most vulnerable.

Floral scents

As indicated in the section on flowering synusiae, it is apparent that floral odours increase as one descends to the ground, completely in contrast to floral colours. Odour is greater among many cauliflorous flowers (e.g. Annonaceae) and is greatly heightened among the geophilous flowers. Notable exceptions in the upper canopy are the nocturnal flowers of Dipterocarpaceae and Bombacaceae, which have sickly-sweet and sour smells respectively. These smells pervade the forests when the flowers are in anthesis late in the evening.

The generalized correlations between pollinators and floral odours (Faegri and van der Pijl, 1966) apply in Malaysia too. Bee- and butterfly-pollinated flowers are mildly sweet scented; beetle flowers have strong aminoid smells, somewhat like the thrips-pollinated dipterocarp flowers; and the carrion and dung beetles and flies are drawn to flowers that mimic such smells.

The floral odours of the mass flowering dipterocarps are enigmatic. During the mass flowering, as many as four or five species of *Shorea* and *Dipterocarpus* may be in anthesis simultaneously, all within proximity of each other. The human nose is unable to distinguish these odours. Evidence is also accruing to suggest that, apart from thrips, which seem omnipresent among all the flowers, individual sections/species of dipterocarps may have a distinct set of pollinators probably unique to them. Of those known, the pollinators of *Shorea* section *Mutica* are thrips, that of *Shorea* subsection *Shorea* are hoppers, and tiny beetles have been seen on *Dipterocarpus*. This observation presents a conundrum: do these insects differentiate between the odours, and is there preferential selection of one insect over the rest?

FLORAL REWARDS FOR POLLINATORS

Nectar

There exist a few case studies on nectar production and chemistry of Malaysian rain forest plants. A study on the nectar production of the androdioecious understorey tree *Xerospermum intermedium* revealed that nectar flows in temporally rhythmic alternating pulses between the two sexes (Appanah, 1982). The activity pattern of the insect visitors, trigonid bees and butterflies, to the two sexes echoed rather closely this rhythm of nectar pulses. This temporal variation in nectar production creates the impetus for nectar foragers to shift between the two sexes, effecting pollen transfer. Such a pattern may be widespread among dioecious species, and has also been detected among neotropical trees, including hermaphrodites (Frankie and Haber, 1983).

Biotic deterrents are also beginning to receive attention in the neotropics (Frankie *et al.*, 1976). In Malaysia, I have observed a swarm of *Trigona canifrons* invade the crown of *Neobalanocarpus heimii* and drive off all other bees. This kind of aggression in fact promotes out-crossing by forcing the evicted bees to forage on other trees. The classic case of biotic deterrence is that of ant guards in the extra-floral nectary of *Thunbergia grandiflora* which discourage *Xylocopa* pollinators from biting the flower and stealing the nectar (Faegri and van der Pijl, 1966).

POLLINATOR–PLANT INTERACTIONS AND POLLEN DISPERSAL

Flowers and their pollinators, and the diversity of interactions between them, can greatly influence the pattern of pollen dispersal between trees. Two Malaysian case studies illustrate this point. In the androdioecious understorey tree, *Xerospermum intermedium*, the anthers of the 'hermaphrodite' flower are not normally dehiscent, the male trees flower more frequently and the sex ratio is male biased (Appanah, 1982). Male trees also bear more flowers than the 'hermaphrodite' trees. The 'hermaphrodite' flower is receptive for 3 days. The nectars differ qualitatively between the two sexes, and they are secreted in alternating pulses. The flower visitors represented the short-ranging pool of opportunistic insects, mostly *Trigona* and some Danaid butterflies. The pollen loads of *Trigona* also included pollen from as many as seven other species. The butterflies carried pollen of fewer species but they rarely visited the 'hermaphrodite' trees. It seems that bees transfer pollen over short distances, probably among clumps, while the butterflies are engaged in occasional long distance transfer of pollen.

The six sympatric, sequentially flowering, thrips-pollinated *Shorea* species show a high level of outbreeding (Chan, 1981). Inter-tree pollen transfer is obviously mediated by thrips (Appanah and Chan, 1981). I surmise this transfer to result from the displaced thrips taking off from the carpet of corollas on the forest floor, and making their way back haphazardly to the canopy. The oblique spiral trajectory of the falling corollas, combined with the returning upward flight of adult thrips will promote cross-pollination within the breeding clumps, whereas a small gust of wind might displace the thrips in the air and this can provide a low degree long distance out-crossing. The female *Thrips* species, with a life cycle of 8 days, can probably multiply for two or three generations and substantially increase their populations before the first species in the series, *S. macroptera* Dyer, blooms. Interestingly, this species has been shown to have multiple seedlings (Foxworthy, 1932), which is tentatively attributed to adventive embryony (Kaur *et al.*, 1978).

SPECIES DIVERSITY AND POLLINATORS

Among other factors, in an outbreeding population, it can be speculated that the availability of pollinators will impose an upper limit on how many species (and individuals) can occur together, especially if they are closely allied species flowering simultaneously and competing for similar pollinators. The dipterocarps, with their high species richness, are excellent material to examine how so many conspecific individuals and numerous related species can exist apparently allopatrically.

Peninsular Malaysia has nine dipterocarp genera, but the species richness is not fairly distributed. The genera can be divided into two groups: those that are species rich (*Vatica* – 21 spp., *Dipterocarpus* – 31 spp., *Hopea* – 32 spp., *Shorea* – 57 spp.); and those that are not (*Neobalanocarpus* – one spp., *Cotylelobium* – two spp., *Dryobalanops* – two spp., *Parashorea* – three spp., *Anisoptera* – six spp.) (Appanah, 1987).

Floral biology and pollinators, for those that can be confirmed, can also be distinguished into two clear groups. All the genera with low species richness have flowers with diurnal anthesis, they open fully, exposing well developed, yellow anthers, with or without nectar, and are visited by bees, predominantly the wide-ranging *Apis dorsata*. The trees flower more frequently, even annually. In contrast, the flowers in all the species-rich genera have nocturnal anthesis, do not open fully, have poorly developed anthers (excepting *Dipterocarpus*), and lack nectar. This group usually flowers during the general flowering. Their pollinators and/or visitors are tiny, and are from the pool of flower-feeding and breeding insects such as thrips, beetles, bugs and hoppers. The only exception seen so far is *S. ovalis* whose floral characters belong to the less species-rich group, and is pollinated by bees.

High species diversity and abundance of populations seem to be the rule among the dipterocarps pollinated by insects like thrips (Appanah, 1987). These dipterocarps occur sympatrically in large consectional series with tightly packed populations. If outbreeding is common among dipterocarps (Chan, 1981), a limiting source of pollinators, such as bees with their slow capacity to increase numbers with increment of food supply, will restrict the number of species and density of individuals that can be pollinated. The trees, when they exceed this density, will experience a decline in out-crossing and fruit set. In contrast, the fecund pollinators like thrips, with their capacity to increase their populations rapidly in response to the availability of a massive source of flowers, may have removed the restraint on the number of species and abundance of individuals that can utilize their pollinatory services without the deleterious effects of competition. Insects like thrips seem to have allowed the dipterocarps to take a quantum leap in their diversification.

If indeed it is true that each section/genus of dipterocarps has its own distinctive pollinators, then stigma contamination would be much reduced. We may thus hypothesize that: a) the pollinators, by being able to multiply rapidly at short notice, may have permitted the large packing of sympatrically growing consectional species; and b) the variety of these tiny, fast-breeding insects that seem to be separately available for each of the various sections of dipterocarps may have permitted the concurrent flowering of dipterocarp species in different sections without competition for pollinators.

High species diversity is not unique to dipterocarps. Other families such as Euphorbiaceae, Annonaceae and Myristicaceae also have species-rich genera. These species are not all associated with fast-breeding, small insects for pollination. For example, the genus *Eugenia* (Myrtaceae) has some 150 species in Peninsular Malaysia. The common pollinators of the genus are butterflies (Danaidae) (pers. obs.). It is hard to speculate what role these pollinators may have in the maintenance of the diversity of *Eugenia*. Perhaps other explanations such as apomixis may be more probable in this case.

CONCLUDING REMARKS

From our present understanding of pollination processes in the lowland dipterocarp forests, we can derive the following conclusions:

(1) The position in which flowers are presented can be divided into four synusiae, viz. flowers held above the canopy, flowers held below the canopy, cauliflorous flowers, and the forest floor flowers. Their respective pollinators can be similarly differentiated.

(2) The flowering patterns, although highly variable, can be classified into three categories, viz. continuously flowering species, those flowering periodically and annually, and the supra-annually flowering species.

(3) Complex pollinator–plant interactions exist which can promote xenogamy, despite the apparent difficulties posed by the vegetation.

(4) Unique pollinators like thrips are able to multiply rapidly at short notice to meet the unusually heavy demand for pollinatory services during a heavy supra-annual general flowering. This partly explains the existence of large consectional diversity of dipterocarps in a single community.

These conclusions are rather tentative, based as they are on few detailed studies. Further studies of a more rigorous and quantitative basis are urgently needed.

REFERENCES

Appanah, S. (1979). *The ecology of insect pollination of some tropical rain forest trees*. Ph.D. Thesis. University of Malaya, Kuala Lumpur.

Appanah, S. (1982). Pollination of androdioecious *Xerospermum intermedium* Radlk. (Sapindaceae) in a rain forest. *Biological Journal of the Linnean Society*, **18**, 11–34

Appanah, S. (1985). General flowering in the climax rain forests of South-east Asia. *Journal of Tropical Ecology*, **1**, 225–40

Appanah, S. (1987). Insect pollination and the diversity of dipterocarps. In Kostermans, A.J.G.H. (ed.) *Proceedings of the Third Round Table Conference on Dipterocarps*. Samarinda, East Kalimantan, April 1985, pp. 277–91. (Jakarta: Unesco)

Appanah, S. and Chan, H.T. (1981). Thrips: the pollinators of some dipterocarps. *Malaysian Forester*, **44**, 234–52

Ashton, P.S. (1969). Speciation among tropical forest trees: some deductions in the light of recent evidence. *Biological Journal of the Linnean Society*, **1**, 155–96

Ashton, P.S. (1971). International Biological Programme (IBF) – Malayan Project. Pasoh Forest Vegetation Survey. Second Report. Mimeographed. Botany Department, University of Malaya, Kuala Lumpur.

Ashton, P.S. (1976). An approach to the study of breeding systems, population structure and taxonomy of tropical trees. In Burley, J. and Styles, B.T. (eds) *Tropical Trees: Variation, Breeding and Conservation*, pp. 35–42. Linnean Society Symposium Series No.2 (London: Academic Press)

Ashton, P.S. (1982). Dipterocarpaceae. *Flora Malesiana*, I, **9**, 237–552

Ashton, P.S. (1988). Dipterocarp biology as a window to the understanding of tropical forest structure. *Annual Review of Ecology and Systematics*, **19**, 347–70

Ashton, P.S., Soepadmo, E. and Yap, S.K. (1977). Current research into the breeding systems of rain forest trees and its implications. In Bruenig, E.F. (ed.) *Transactions of the International MAB-IUFRO Workshop on Tropical Rain Forest Ecosystems and Resources*. Hamburg-Reinbek. 12–17 May 1977, pp. 187–92. Special Report No. 1. Chair of World Forestry, Hamburg-Reinbek.

Ashton, P.S., Givnish, T.J. and Appanah, S. (1988). Staggered flowering in the Dipterocarpaceae: New insights into floral induction and the evolution of mast fruiting in the aseasonal tropics. *American Naturalist*, **132**, 44–60

Baker, H.G. (1978). Chemical aspects of the pollination of woody plants in the tropics. In Tomlinson, P.B. and Zimmermann, M.H. (eds) *Tropical Trees as Living Systems*, pp. 57–82. (Cambridge: Cambridge University Press)

Baker, H.G. and Baker, I. (1975). Studies on nectar-constitution and pollinator-plant coevolution. In Gilbert, L.E. and Raven, P.H. (eds) *Animal and Plant Coevolution*, pp. 100–40 (Austin, Texas: University of Texas Press)

Baker, H.G. and Baker, I. (1983). Floral nectar sugar constituents in relation to pollinator type. In Jones, C.E. and Little, R.J. (eds) *Handbook of Experimental Pollination Biology*, pp. 117–41 (Amsterdam: Van Nostrand Reinhold)

Baker, H.G., Bawa, K.S., Frankie, G.W. and Opler, P.A. (1982). Reproductive biology of plants in tropical forests. In Golley, F.B. (ed.) *Tropical Rain Forest Ecosystems: Structure and Function*, pp. 183–215. Ecosystems of the World Volume 14A. (Amsterdam: Elsevier)

Bawa, K.S., Bullock, S.H., Perry, D.R., Coville, R.E. and Grayum, M.H. (1985). Reproductive biology of tropical rain forest trees. II. Pollination systems. *American Journal of Botany*, **73**, 346–56

Bawa, K.S., Perry, D.R. and Beach, J.H. (1985). Reproductive biology of tropical rain forest trees. I. Sexual systems and incompatibility mechanisms. *American Journal of Botany*, **73**, 331–45

Burkill, I.H. (1919). Some notes on the pollination of flowers in the botanic gardens. Singapore and other parts of the Malay Peninsula (1918–1921). *Gardens Bulletin of the Straits Settlements*, **2**, 165–76

Chan, H.T. (1977). The reproductive biology of some Malaysian dipterocarps. Ph.D. Thesis.

University of Aberdeen, Aberdeen.

Chan, H.T. (1981). Reproductive biology of some Malaysian dipterocarps. III. Breeding systems. *Malaysian Forester*, **44**, 28–36

Chan, H.T. and Appanah, S. (1980). Reproductive biology of some Malaysian diperocarps. I. Flowering biology. *Malaysian Forester*, **43**, 132–43

Corner, E.J.H. (1952). *Wayside Trees of Malaya.*, **Vol. 1.** (Singapore: Government Printers)

Corner, E.J.H. (1954). The evolution of tropical forests. In Huxley, J., Hardy, A.C. and Ford, E.C. (eds) *Evolution as a Process*, pp. 34–46. (London: Allen and Unwin)

Faegri, K. and van der Pijl, L. (1966). *The Principles of Pollination Ecology*. (London: Pergamon Press)

Fedorov, A.A. (1966). The structure of the tropical rain forest and speciation in the humid tropics. *Journal of Ecology*, **54**, 1–11

Foxworthy, F.W. (1932). *Dipterocarpaceae of the Malay Peninsula*. Malayan Forest Record No. 10.

Frankie, G.W. and Haber, W.A. (1983). Why bees move among mass–flowering Neotropical trees. In Jones, C.E. and Little, R.J. (eds) *Handbook of Experimental Pollination Biology*, pp. 360–72 (Amsterdam: Van Nostrand Reinhold)

Frankie, G.W., Baker, H.G. and Opler, P.A. (1974). Comparative phenological studies of trees in tropical wet and dry forests in the lowlands of Costa Rica. *Journal of Ecology*, **62**, 881–919

Frankie, G.W., Opler, P.A. and Bawa, K.S. (1976). Foraging of solitary bees: Implications for outcrossing of a neotropical forest tree species. *Journal of Ecology*, **64**, 1049–57

Gan, Y.Y. (1976). Population and phylogenetic studies on species of Malaysian rain forest trees. Ph.D. Thesis. University of Aberdeen, Aberdeen.

Gilbert, L.E. (1972). Pollen feeding and reproductive biology of *Heliconius* butterflies. *Proceedings of the National Academy of Sciences*, **69**, 1403–7

Gould, E. (1977). Foraging behavior of *Pteropus vampirus* on the flowers of *Durio zibethinus*. *J. Malaysian Nature Journal*, **30**, 53–7

Gould, E. (1978). Foraging behavior of Malaysian nectar-feeding bats. *Biotropica*, **10**, 184–92

Ha, C.O. (1978). Embryological and cytological aspects of the reproductive biology of some understorey rain forest trees. Ph.D. Thesis. University of Malaya, Kuala Lumpur.

Janzen, D.H. (1971). Euglossine bees as long distance pollinators of tropical plants. *Science*, **171**, 203–5

Kaur, A. (1977). Embryological and cytological studies on some members of the Dipterocarpaceae. Ph.D. Thesis. University of Aberdeen, Aberdeen.

Kaur, A., Ha, C.O., Jong, K., Sands, V., Chan, H.T., Soepadmo, E. and Ashton, P.S. (1978). Apomixis may be widespread among trees of the climax rain forest. *Nature*, **271**, (5644), 440–2

Kevan, P.G. (1983). Floral colours through the insect eye: What they are and what they mean. In Jones, C.E., and Little, R.J. (eds) *Handbook of Experimental Pollination Biology*, pp. 3–30. (Amsterdam: Van Nostrand Reinhold)

Medway, Lord. (1972). Phenology of tropical rain forest in Malaya. *Biological Journal of the Linnean Society*, **4**, 117–46

Opler, P.A. (1981). Nectar production in a tropical ecosystem. In Bentley, B.L. and Elias, T.S. (eds) *Biology of Nectaries*, pp. 30–79. (New York: Columbia University Press)

Percival, M.S. (1961). Types of nectar in Angiosperms. *New Phytologist*, **60**, 235–81

Proctor, M.C.F. and Yeo, P. (1973). *The Pollination of Flowers*, (Collins: London)

Start, A. (1974). The feeding biology in relation of food sources of nectarivorous bats (Chiroptera: Macroglossinae) in Malaysia. Ph.D. Thesis. University of Aberdeen, Aberdeen.

Start, A. and Marshall, A.G. (1976). Nectarivorous bats as pollinators of trees in West Malaysia. In Burley, J. and Styles, B.T. (eds) *Tropical Trees: Variation, Breeding and Conservation*, pp. 141–50 (London: Academic Press)

van der Pijl, L. (1937). Biological and physiological observations on the inflorescence of *Amorphophallus*. *Recueil de Travaux Botaniques Neerlandais*, **34**, 157–67

van der Pijl, L. (1939). Over de meedraden van enkele Melastomataceae. *De trop. nat.*, **28**, 169–72

van der Pijl, L. (1956). Remarks on pollination by bats in the genera *Freycinetia, Duabanga*, and *Haplophragma*, and on chiropterophily in general. *Acta Botanica Neerlandica*, **6**, 291–315

Whitmore, T.C. (1975). *Tropical Rain Forests of the Far East*. (Oxford: Clarendon Press)

Wood, G.H.S. (1956). The dipterocarp flowering season in North Borneo, 1955. *Malayan Forester*, **19**, 193–201

Yap, S.K. (1976). The reproductive biology of some understorey fruit tree species in the lowland dipterocarp forest of West Malaysia. Ph.D. Thesis. University of Malaya, Kuala Lumpur.

CHAPTER 9

PHENOLOGY AND FLORAL MORPHOLOGY IN RELATION TO POLLINATION OF SOME SRI LANKAN DIPTEROCARPS

S. Dayanandan, D.N.C. Attygalla, A.W.W.L. Abeygunasekera, I.A.U.N. Gunatilleke and C.V.S. Gunatilleke

ABSTRACT

Phenological observations of several lowland dipterocarp species made between November 1985 and April 1987 in the Sinharaha Biosphere Reserve in Sri Lanka indicate that leaf flushing was continuous throughout the year in Shorea congestiflora *(Thw.) Ashton and* S. trapezifolia *(Thw.) Ashton and restricted to certain periods in others. In* S. affinis *(Thw.) Ashton, flushing preceded flowering, while in* S. cordofolia *(Thw.) Ashton,* S. disticha *(Thw.) Ashton and* S. worthingtonii *Ashton flushing took place 4 5 months before flowering, and in* S. megistophylla *Ashton and* Vateria copallifera *(Retz,) Alston it immediately followed flowering. Although the time and duration of flushing varied considerably and overlapped between species, their blooming peaks were sequential, except in* S. megistophylla, S. worthingtonii *and* S. cordifolia. *Populations of the three species, and of* S. affinis *and* S. disticha *bloomed between March and May before the onset of the south-west monsoons, those of* S. congestiflora *and* S. stipularis *in October and November during the north-east monsoons, and those of* S. trapezifolia *in April, June, November and December of 1986. In contrast to these* Shorea *spp.,* V. copallifera *bloomed continuously from December 1985 to April 1986. The average blooming period of each inflorescence, individual tree and population of these species was brief and ranged between 2 and 7 days, 5 and 15 days and 10 and 18 days respectively. In* V. copallifera, *however, the corresponding values were 3 and 14 days, 45 and 50 days and 80 and 150 days respectively. Inflorescences of these* Shorea *spp. are axillary and terminal, but those of* V. copallifera *are only axillary. Flowers of the section* Doona, *to which all but one of the study species of* Shorea *belong, resemble each other structurally, but differ in their sizes,* S. megistophylla *having the largest flowers and* S. affinis *and* S. congestiflora *the smallest. Flowers of* V. copallifera *differ from these* Shorea *species in having equal sized externally fulvous sepals, 40–60 stamens and stellate hairs on the ovary. In both genera, anther dehiscence accompanies anthesis and flower longevity varies from 8 to 12 hours. Among a host of flower visitors to these species,* Apis dorsata *and* A. indica *appeared to be the most effective pollinators, particularly because of their large body size, abundance of individuals, and method of pollen collection. All species studied seem to be self-incompatible.*

INTRODUCTION

The subfamily Dipterocarpoideae, of the tropical timber family Dipterocarpaceae, is confined to South and South-east Asia. Its greatest diversity is in the humid forests, and the entire family is represented by 13 genera and some 470 species (Ashton, 1982), but, in South Asia, it has 10 genera and 99 species (FAO, 1985). Of the latter, 44 are recorded in Sri Lanka, where all but one (*Vatica chinensis* L.) are endemic to the island. Among the higher taxa, the genus *Stemonoporus* and section *Doona* of the genus *Shorea* are also endemic. Sri Lanka, with 98% of its dipterocarp species endemic, records the highest endemicity of the family in South and South-east Asia (Jacobs, 1981). In Sri Lanka, the highest diversity of species is found among its endemic taxa, viz. 10 in *Shorea* section *Doona* and 15 or 26, as recognized by Ashton (1980) and Kostermans (1981) respectively, in *Stemonoporus*; each of the other taxa is represented by one each in *Shorea* section *Anthoshoreae* and in *Vateria*, two in *Cotylelobium*, three in *Vatica*, and four each in *Shorea* section *Shorea*, *Dipterocarpus* and *Hopea*.

Not only do most species of Dipterocarpaceae have good timber value, some also yield products useful to villagers (Gunatilleke, 1988). However, these species are never cultivated by the local people. Our studies on pollination biology are a part of a long-term project to examine the population biology of these species, with the intention of gathering sufficient knowledge as a first step in their domestication in village home gardens and in the buffer zones of biosphere reserves. Studies on pollination biology of dipterocarps in Sri Lanka have been initiated for six of the 10 *Shorea* species of section *Doona*, [*Shorea affinis* (Thw.) Ashton, *S. cordifolia* (Thw.) Ashton, *S. megistophylla* Ashton, *S. congestiflora* (Thw.) Ashton, *S. disticha* (Thw.) Ashton, and *S. trapezifolia* (Thw.) Ashton, *Shorea stipularis* Thw. of section *Anthoshoreae* and *Vateria copallifera* (Retz.) Alston.]

The particular features examined in this study and presented here are:

Table 9.2 Floral measurements (to the nearest mm) of some members of the Dipterocarpaceae at Sinharaja rain forest

Species	Flower parts Length (mm) Pedicel M (R)	Sepals Short M (R)	Sepals Long M (R)	Petals M (R)	Androecium Fil.	Androecium Ant.	Androecium Con.	Ova.	Gynoecium Style M (R)	Sample size No. of flowers measured
Section Doona										
S. trapezifolia	5 (4–7)	3 (3–4)	5 (4–5)	6 (5–7)	1	1	1	1	5 (4–6)	25
S. affinis	2 (3–6)	3 (2–3)	4 (4–5)	7 (6–9)	1	2	1	1	5 (4–6)	50
S. congestiflora	2 (1–2)	3 (2–3)	4 (4–5)	7 (6–8)	2	1	1	1	3 (2–5)	25
	2 (1–3)	3 (2–4)	4 (3–5)	7 (6–8)	2	1	1	1	4 (3–5)	25
S. cordifolia	6 (5–7)	3 (3–4)	5 (4–6)	9 (7–10)	2	2	1	1	5 (4–5)	50
	7 (5–9)	3 (3–4)	6 (5–7)	10 (7–11)	2	2	1	1	5 (4–5)	25
S. disticha	6 (4–8)	6 (5–7)	9 (8–10)	12 (12–13)	3	2	1	1	9 (8–10)	25
S. megistophylla	7 (6–8)	7 (6–7)	9 (8–10)	12 (11–14)	3	4	2	2	10 (9–11)	30
	8 (6–11)	8 (7–9)	10 (9–11)	14 (12–16)	4	4	2	2	9 (8–12)	30
	9 (7–11)	7 (6–9)	11 (10–12)	15 (14–16)	4	3	2	2	10 (8–11)	30
Section Anthoshorea										
S. stipularis	3 (3–4)	4 (3–5)	5 (4–6)	17 (15–19)	1	1	2	1	3 (2–3)	30
	3 (3–4)	5 (5–6)	5 (5–6)	15 (12–17)	1	1	2	1	2 (2–3)	30
V. copallifera	16 (12–20)	16 (13–18)		17 (12–21)	N/A	N/A	N/A	3	9 (8–10)	30

Data in each row represent details obtained from an individual tree. Thus in *S. trapezifolia*, *S. affinis*, *S. disticha* and *V. copallifera* one tree per species, in *S. congestiflora*, *S. cordifolia* and *S. stipularis* two trees per species, and in *S. megistophylla* three trees per species, were sampled. Fil. = Filament, Ant. = Anther, Con. = Connective, Ova. = Ovary, M = Mean, R = Range, N/A = Not available to observe quite clearly the foraging behaviour of the visiting animals. In the case of those visitors that were slow moving and less sensitive to touch, their foraging pattern was observed using a handlens as well.

The activity of some animals was also noted by determining the frequency of their visitation. Thus, the total number of landings by a given species at 15-minute intervals, on 25–30 selected flowers in *Shorea* spp. and 50 flowers in *V. copallifera* that were clearly visible, were monitored. This procedure enabled us to compare the foraging patterns of at least some of the different visitor species.

Pollination experiments and fruit-set

The following pollination treatments were carried out:

(1)   Flowers non-emasculated and bagged;

(2)   Flowers emasculated and bagged;

(3)   Flowers non-emasculated, selfed, and bagged;

(4)   Flowers emasculated, selfed, and bagged;

(5)   Flowers non-emasculated, crossed, and bagged;

(6)   Flowers emasculated, crossed, and bagged;

(7)   Flowers non-emasculated and left open; and,

(8)   Flowers emasculated and left open.

In the case of emasculated treatments, the undehisced stamens were removed in the mature bud stage, i.e. between 14:00–18:00 h on the day before the flowers opened. Pollen used in hand pollinations was obtained from previously bagged flowers. Pollinations were done between 08:00–11:00 h for *Shorea* spp. and 08:00–12:00 h for *V. copallifera* when the flowers were open and pollen readily available. For selfing, pollen obtained from different flowers of the same tree was used. Large amounts of pollen were easily obtained by tapping the flowers over a glass slide which was then gently touched to the stigma to effect pollen transfer.

After hand pollinations were completed, the flowering branches were bagged using brown paper pollination bags (Pollen Tector, Carpenter Co., USA). Because a few flowers open each day on a given inflorescence, hand pollinations were carried out on successive days on the same inflorescence.

On the last day of pollination, those buds that were not treated were picked to prevent contamination of the flowers already treated. The bags were removed one day after the last set of flowers was pollinated. If twigs are bagged for too long, the leaves become etiolated and flowers are shed. Therefore, it was important to remove the bags as quickly as possible.

For each species, at least two trees were selected for these pollination experiments. On individual trees, the number of flowers that could be manipulated per treatment varied as given in Table 9.3 and was limited by the number of flowers that could be reached from the canopy platforms.

RESULTS

Vegetative and reproductive phenology

There is no single pattern for the *Shorea* spp. with respect to the relationship among flushing, flowering (blooming or anthesis) and fruiting (Table 9.4). Three main groups may be identified, based on whether flushing takes place more or less continuously throughout the year, or is restricted to a short period before or after flowering. In the case of *V. copallifera*, each flowering twig produced new leaves upon fruit maturity or soon after flowering on those twigs without fruits

The sequence of flowering of each species (Figure 9.1) shows that all species bloomed more than once during the sampling period of 18 months (November 1985 to May 1987), except *S. megistophylla* which flowered only once. In species with more than one flowering episode per year, individual trees did not bloom more than once, with the exception of 13 trees of *S. trapezifolia*. *S. disticha* and *S. stipularis* each had only a few individuals flowering during the period studied, but the remaining species had good flowering episodes during this period.

Two flowering seasons appear to prevail among the Shoreas being studied, viz. March–June and September–December. In general, the most intense flowering period of different species did not overlap, except in the case of *S. cordifolia* and *S. megistophylla*. In *V. copallifera*, the pattern of flowering was different in that each tree continued to bloom over a period of 2.5 months, with two bursts of flowering (Figure 9.1). The first (November–January) involved less than 10% of the crown and occurred in only three individuals. The second episode was more intense, commenced in January, lasted 115 days and was exhibited by 17 of the 19 individuals in the population. The forest population of this species did not come into bloom during the entire study period, while plants in the disturbed forest bloomed twice. Only one tree (reference number 2VC 10) in the forest population had a few inflorescences (less than 1% of the crown in flower) and no fruits were set.

Table 9.3 Sample size of flowers used in pollination experiments of *S. megistopylla, S. cordifolia, S. congestiflora, S. trapezifolia,* and *V. copallifera* at Sinharaja

		Species, pollination period and number of flowers manipulated				
Treatments		*S. meg.* 31 Mar./4 Apr. 1986	*S. cord.* 18/27 Mar. 1986	*S. cong.* 6/12 Nov. 1986	*S. trap.* 20/23 Nov. 1985	*V. copal.* Jan./Mar. 1986
Bagged only						
E	I	174	82	86	106	116
	II	43	98	103	252	87
	III	–	–	58	–	173
NE	I	384	103	100	478	103
	II	243	186	151	149	106
	III	–	–	253	–	113
Selfed and bagged						
E	I	214	70	71	112	99
	II	427	224	115	245	106
	III	–	–	107	–	104
NE	I	224	65	127	141	201
	II	179	144	95	250	111
	III	–	–	188	–	144
Crossed and bagged						
E	I	233	106	108	48–54	133
	II	427	179	111	155–252	112
	III	–	–	106	–	102
NE	I	178	73	134	29–72	152
	II	175	112	119	87–155	288
	III	–	–	176	–	101
Open pollinated						
E	I	168	93	127	122	99
	II	72	127	102	238	113
	III	–	–	66	–	103
NE	I	1223	201	380	366	209
	II	1101	278	347	490	112
	III	–	–	250	–	115

E and NE = emasculated and non-emasculated treatments
I, II and III the number of trees sampled per species

110

Table 9.4 Flushing period of some dipterocarp species in relation to the flowering season

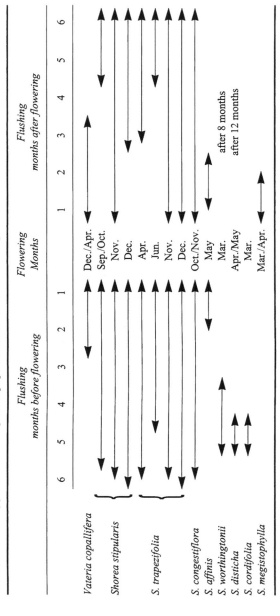

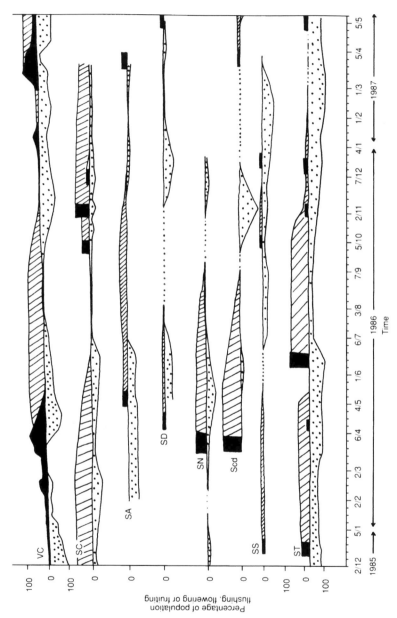

Figure 9.1 Duration and intensity (indicated by height of each graph from 0 line) of flowering (black), fruiting (striped) and flushing (dotted) observed in *S. trapezifolia* (ST), *S. stipularis* (SS), *S. cordifolia* (Scd), *S. megistophylla* (SM), *S. disticha* (SD), *S. affinis* (SA), *S. congestiflora* (SC) and *V. copallifera* (VC). at Sinharaja. Scale on vertical axis for each species is similar to that given for *V. copallifera* or *S. trapezifolia*

Floral morphology

Morphologically, the inflorescences and flowers of all the *Shorea* species section *Doona* are similar, except that they show size variation among species (Tables 9.1 and 9.2). Inflorescences of *Shorea* spp. are compound racemes or panicles, borne at the ends of branches at a density of one inflorescence per node. *S. cordifolia* is an exception, with one to four panicles per node; generally in *S. cordifolia* there is one (rarely two) axillary inflorescence, but, at the terminal end, there are often more than one and frequently up to four (Figure 9.2). The lengths of that part of the branch bearing inflorescences and of first order inflorescence axes, and the number of flowers per inflorescence, are given for each species in Table 9.1.

Individual flowers of *Shorea* species section *Doona* are pedicelled, bearing three long and two short, light greenish white sepals. The corolla consists of five white petals which are fused together at the base but separate along their lengths. Distally, the petals are rounded. The androecium consists of 15 stamens, whose filaments are white, broad and loosely connate at the base, but narrow and free at the point of attachment to the orange anther lobes. Filaments are dorsifixed to the anthers and extend beyond them into a small spoon shaped or spathulate yellow orange connective (Figure 9.3). The pollen grains are slightly sticky and bright yellow orange. The gynoecium consists of a white superior ovary, a slender style, and an unbranched simple stigma. The ovary has three carpels, each with one loculus wherein a single central axis bears two pear shaped pendulous ovules. Thus, each ovary has three loculi andsix ovules.

Among the shoreas of section *Doona*, the largest inflorescences and flowers were found in *S. megistophylla*. On the whole, while *S. congestiflora* had the smallest inflorescences, *S. trapezifolia* had the smallest flowers (Tables 9.1 and 9.2). In all six species of section *Doona* studied, filaments and anther lengths varied between 1–4 mm, and the size of the connective in five of the species was less than 1 mm. In *S. megistophylla*, it was 2 mm. The ovary length did not vary between species, but in general, the style progressively increased in length in species with larger flowers (Table 9.2).

In *S. stipularis* of section *Anthoshoreae*, there are only axillary inflorescences which arise at a density of one per node on the inflorescence bearing part of the twig. Like flowers of *Shorea* section *Doona*, these are pendulous and pedicelled with three long and two short, light green sepals. The petals are quite dissimilar to those of section *Doona* in having a basal urceolate portion with a narrow opening, at which point the free distal parts radiate out. The latter are also twisted and are perpendicular to the basal part. The androecium comprises 15 stamens which are much smaller and more delicate than those of section *Doona*. They occupy a very small volume within the urceolate base. There is no conspicuous colour difference between the filaments, anthers and connectives,

113

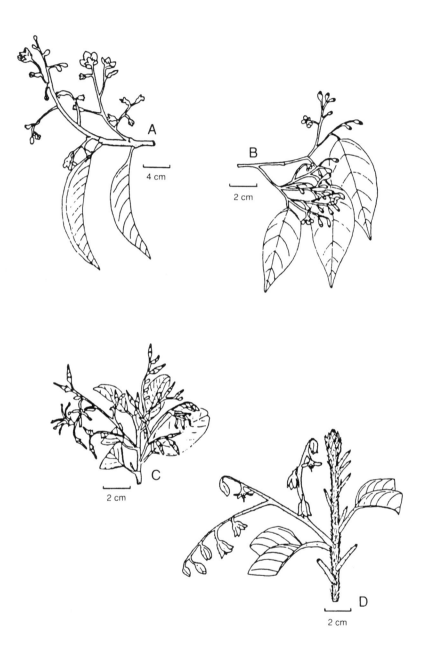

Figure 9.2 Comparison of the morphological details of inflorescences of *S. megistophylla* (A), *S. cordifolia* (B), *S. stipularis* (C) and *V. copallifera* (D) studied at Sinharaja

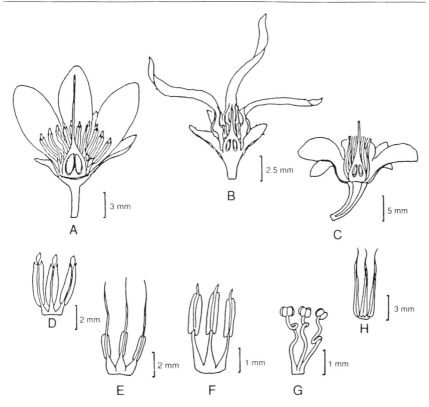

Figure 9.3 Details of half flowers and stamens of *S. megistophylla* (A and D), *S. stipularis* (B), *V. copallifera* (C and H), *Parashorea* (E), *Neobalanocarpus* (F) and *S. ovalis* (G), to compare the differences among them

all of which are creamish white. The connective is acicular in shape and always as long as the filament and anther together; the style is inconspicuous and barely extends beyond the stamens (Figure 9.3).

In *V. copallifera*, inflorescences are compound racemes or panicles. Unlike *Shorea* species section *Doona*, which have both terminal and axillary inflorescences, *V. copallifera* has only axillary inflorescences. An inflorescence-bearing twig has at its terminal end an aggregation of stipules from whose axils future leaves are produced. Leaves may or may not be borne within the inflorescence-bearing portion of the twig, but always in the region immediately below it. Inflorescences are only rarely present in the region of mature leaves (Figure 9.2).

The flowers of *V. copallifera* are pendulous, borne on short pedicels, have five sepals and five petals and differ from those of *Shorea* species section *Doona* in the following:

(1)   Presence of two conspicuous, tomentose, acute bracteoles which fall off at maturity;

(2)   Equal sized sepals which are quite fulvous on the outside, and glabrous within;

(3)   Petals are much more fleshy, creamy white or ivory, with subacute free ends;

(4)   Stamens exceed 15 and are not constant in number, either within or between trees. The mean number of stamens per flower of four different individuals ranges between 40–72;

(5)   The connective extends as an acicular structure beyond the anthers (Figure 9.3); and,

(6)   The outer wall of the ovary has stellate hairs, whereas that of the shoreas is glabrous.

Anthesis, flower longevity and fruit development

In *Shorea* species section *Doona*, anthesis of flowers commences between 05:30–06:00 h and flowers are fully open by 09:00–10:00 h. At the time of anthesis, a pleasant fragrance emanates and can be detected several metres away. Anther dehiscence accompanies anthesis, and by 09:30–10:00 h anthers are fully split and most of the pollen dispersed. By 13:00 h, the corollas, together with the stamens, begin to shed and the process peaks between 15:00–16:00 h. By late evening, only buds and young fruits remain on the inflorescence. Duration of fruit development among the different *Shorea* spp. ranged between *c.* 95 days for *S. megistophylla* to 160–180 days for *S. affinis*, with a range of 112–140 days for the remaining *Shorea* spp. (Table 9.5).

Anthesis in *V. copallifera* begins before midnight and requires about 6–8 hours, possibly because of its hard sepals. Usually, by 06:00 h, all flowers are fully open. Anther dehiscence is concurrent with anthesis. However, before 08:00 h the slightly sticky pollen remains attached to the anther lobes and is not easily shed. Maximum pollen is released between 08:00–09:00 h. By 15:00 h of the same day, anthers of open flowers begin to abcise from the rest of the flower and fall away, several at a time. The petals on the other hand fall within the next 24 hours, either separately or two or three together.

Unlike those of *Shorea* species section *Doona*, the sepals of *V. copallifera* do not enlarge after pollination and do not twist to cover the

Table 9.5 Duration of phenological events in some species of Dipterocarpaceae at Sinharaja in Sri Lanka

Species	Inflorescence: initiation to anthesis	Blooming of each inflorescence	Blooming of each tree	Blooming of whole population	Fruit development up to ripe/ mature stage	Number of flowers blooming each day per inflorescence Mean (Range)
Section Doona						
S. megistophylla	14	2–7	10	18	95	3 (1–5)
S. disticha	8	5–10	8	15	N/A	2 (1–3)
S. trapezifolia	13–20	3–6	5–6	12	112–133	2 (1–5)
S. cordifolia	16	6	N/A	19	112–140	N/A
S. congestiflora	40–50	5	15	18	140	7 (3–11)
S. affinis	N/A	N/A	N/A	N/A	160–180	N/A
Section Anthoshoreae						
S. stipularis	N/A	10	N/A	N/A	140	1 (1–3)
V. copallifera	30	3–14	45–50	80–150	150	2 (1–4)

Data gathered from one individual per species where a single value is given and three individuals per species where a range is given. Data in the last column are from 25 inflorescences per species. N/A = Data not available ovary and style, apparently affording no protection to the developing fruit. However, they play a vital role in the protection of the corolla, androecium and gynoecium in the developing bud. Flower predators were never observed to feed on the sepals, except those that have the ability to pierce it.

Colour changes associated with fruit development are from brownish or pinkish white to brown. Of all the Sri Lankan dipterocarps, the largest fruits were observed in *V. copallifera* (Table 9.6). Among the shoreas the period of floral bud development, from first signs of inflorescence emergence up to blooming, took 8 days in *S. disticha* and 40–50 days in *S. congestiflora*. The remaining *Shorea* species ranged between 13–20 days, and *V. copallifera* required 30 days for maturation of inflorescences and buds (Table 9.5).

The duration of the blooming periods for each inflorescence, individual, and population is given in Table 9.5. Each inflorescence in *Shorea* spp. bloomed for 2–10 days, each individual 5–15 days and each population for 10–18 days. In *V. copallifera*, the corresponding periods were 3–14 days, 45–50 days and 60 days respectively. The number of flowers blooming per day per inflorescence in the species studied varied between 1–7 flowers (Table 9.5).

Colour changes of flowering and fruiting crowns

During bud stage and flowering, the crowns of *Shorea* species section *Doona* can easily be located from a distance because they are fringed white. In flowering crowns, leaves droop vertically downwards, while the inflorescences are held upwards. Once flowering is complete and the tree is

Table 9.6 Mature fruit data of some dipterocarp species in Sri Lanka

Species	Nut length (mm) Mean (Range)	Wing length (mm) Mean (Range)	Nut and wing weight (g) Mean (Range)
Section *Doona*			
Shorea megistophylla	22 (18–25)	53 (43–56)	3.7 (3.3–4.9)
S. cordifolia	21 (17–24)	40 (36–45)	1.7 (1.2–2.2)
S. trapezifolia	19 (17–21)	50 (34–49)	1.8 (1.5–2.1)
S. congestiflora	19 (16–22)	42 (35–49)	0.9 (0.6–1.1)
V. copallifera			
Tree 1	78 (53–96)	Not	103 (54–199)
2	73 (61–90)	applicable	107 (50–169)
3	96 (66–102)		136 (78–209)

Corresponding data for *S. disticha*, *S. affinis* and *S. stipularis* were not available during the period of study in young fruit, the crown takes on a spectacular lilac colour in the case of *S. megistophylla* and a brownish-pink colour in other species. In *S. stipularis*, the axillary inflorescences arise among the foliage leaves which do not bend downwards to fully expose these inflorescences. Instead, they are intermingled with the leaves that are held upright. Nevertheless, because inflorescences are borne on the uppermost nodes, the crown takes on a white colour that is easily recognizable from a distance. In *V. copallifera*, the crown colour during inflorescence development, flowering and early fruiting is more or less the same and is creamy yellow to light pink. In this species, too, the inflorescences are held above the leaves and quite well exposed (Figure 9.2).

Flower visitors and pollination

Although detailed studies of flower visitors and pollinators were carried out only for *S. megistophylla*, the type of visitors, their activity, and foraging pattern appears to be similar for the other *Shorea* species in section *Doona*. Twenty-one species of insect visitors were sighted at the flowers of *S. megistophylla* (Table 9.7). Based upon their activity at the flowers, they may be grouped into the following categories:

(1) Pollen and nectar gatherers or feeders. These insects did not cause extensive damage to the flowers. All bee species gathered pollen and possibly fed on nectar, while a beetle of the family Elateridae, flies of family Calliphoridae, and possibly some thrips and ants, only fed on pollen and did not collect or store it on their bodies;

(2) Stamen and corolla feeders. Beetles of the families Chrysomelidae and Scarabacidae were observed biting away the stamens and corolla; and

(3) Predators of insects. Spiders and wasps were frequently observed to prey on bees that visited the inflorescences.

Apart from these insects, two mammals, the giant squirrel and western purple-faced leaf monkey, were observed to feed on the inflorescences of *S. megistophylla* before 07:00 h and at about 17:00 h. Two lizards and a shell-less tree snail were also observed on one of the *S. megistophylla* trees, at a canopy height of 29 m. While the former is a likely predator of insects that visit the flowers, the latter may be feeding on young inflorescences (the tree was not in young flush at the time).

In *V. copallifera*, 17 species of insects were observed at inflorescences. One *Calotes* sp. was also observed. The behaviour of the insects at the

Table 9.7 Insect species visiting *S. megistophylla* and the duration of their visitation times, on 3 consecutive days of the blooming period

Insect species	Day and time of visitation	Day 1 (6.4.85)	Day 2 (7.4.86)	Day 3 (8.4.86)

Hymenoptera
1. *Apis dorsata*
2. *Apis Indica*
3. *Trigona* sp. ?
4. *Drauns opis* sp. ?
5. Like 4 but different
6. *Pachyhalictus* sp.
7. *Nomia* sp.

Coleoptera
8. Elateridae, Alcianoxanthus sp.
9. Chrysomellidae sp.
10. Scarabacidae, Rutalinae, Anomala sp.

Diptera
11. Calliphoridae sp. 1
12. Calliphoridae sp. 2

Other Species
13. Ant sp. 1
14. Ant sp. 2
15. Ant sp. 3
16. Thrips

17. *Vespa cinata* (Wasp sp. 1)
18. Eumenidae Rhynchlum sp. (Wasp sp. 2)
19. Beetle sp. (Small black)
20. Butterfly like insect
21. Alosquito spp.. Several others

flowers was in most instances similar to that observed at *Shorea*. However, Thysanoptera and larvae of several lepidopteran species that did not visit *Shorea* were predators of flowers and consequently destroyed them. The times of visitation of the species and their behaviour patterns are given in Tables 9.7 and 9.8.

Our observations from canopy platforms on insect activity in dipterocarp flowers at Sinharaja reveal that the pollen smeared abdomens of the larger *Apis* spp. always touch the stigma of flowers when visiting them. As such, they appear to be the most effective pollinators. In contrast, the small social

Table 9.8 Insect species visiting flowers of *V. copallifera* and the time of their visitations

	Time of visiting	
Insects visiting flowers	*Adult*	*Larvae*
Hymenoptera		
Family-Apidae		
Apis dorsata	08:00 – 12:00	–
A. indica	08:00 – 14:00	–
Family-Helictidae		
Pachyhalictus kalutarae	08:00 – 16:00	–
Nomia sp.	08:00 – 14:00	–
Family-Megachilidae		
Chalicodoma ardens	08:00 – 14:00	–
Diptera		
Family-Calliphoridae		
Celliphora 'Blow fly'	08:00 – 12:00	–
Family-Unknown		
Species 1	08:00 – 12:00	–
Coleoptera		
Family-Unknown		
Beetle	02:00 – 15:00	–
Family-Lamperidae		
Photinus sp.	Night–before 06:00	–
Lepidoptera		
Family-Lymantriidae		
Euproctis scintillans	Night and early morning	All day
Dasychira mendosa	"	"
Thiacidas vilis	"	"
Leucoma submarginata	"	"
Family-Tortricidae		
Species 1	"	
Family-Geometridae		
Thalera sp.	"	"
Family-Gelechiidae		
Species 1	"	"
Thysanoptera	All the time	All the time

bees such as *Trigona*, Braun's *Apis* and *Pachyhalictus* are less effective pollinators. Not only were they short-distance foragers, but, being of relatively small size, they always alight on the petals or stamens without actually touching the stigma.

Only negligible numbers of thrips were present on the *Shorea* species of section *Doona*, but, in *V. copallifera* they were abundant, particularly in young buds and young fruits. During the first flowering episode of November–January 1986, almost all the developing inflorescences of three individuals demarcated for pollination studies were completely destroyed by thrips. However, during the second episode of flowering in January–April 1986, thrip damage was much less, and affected only one of the three trees. Thrip damage manifested itself as radially expanding black spots on the outside of flower parts. In time, the affected tissues became brittle and were easily broken off. Growth of inflorescences bearing such damaged buds also ceased and the inflorescences were abscised shortly afterwards.

Floral features related to pollination

Among the floral features related to pollination, those of particular importance are:

(1)   Inflorescence axes are held upright and away from the leaves, so that they are well presented to the pollinating agents;

(2)   Flowers are relatively large and conspicuous. They are fragrant at the time of anthesis, thus providing both visual and sensory cues to pollinators;

(3)   The pendulous flowers make it easy for bees to collect pollen onto their bodies. The style protruding beyond the stamens invariably touches the insect's abdomen when pollen is being gathered, particularly by the large bees;

(4)   The pendulous flowers are also well protected from rain and the possibility of pollen being washed out is minimized; and,

(5)   The number of flowers that open per day per inflorescence is limited to 1–7. This effect is compensated by a single inflorescence blooming for 2–10 days, thus maintaining a higher level of pollinator activity throughout the flowering episode. Limiting the number of open flowers per day per inflorescence could also be advantageous if bad weather prevails on a given day and flowers fail to pollinate.

Pollination experiments, fruit-set and fruit losses

In the absence of a) pollen tube germination studies and b) clearly discernible external morphological changes following pollination and during early stages of fruit development, it was not possible to record with certainty the extent of pollination success, i.e. whether successful fertilization took place or not. Consequently, "fruit" losses observed soon after pollination could be due to both pre- and post-zygotic abortions. However, results of the pollination experiments carried out suggest the following:

(1)   In each species, fruit success among the four treatments using non-emasculated flowers was less variable within the first 10 days after pollination than that between 25–30 days after pollination (Fig 9.4 and Table 9.9). At this latter stage, fruit success in *S. megistophylla* and *S. cordifolia* decreased in the order of crossed, selfed, open and bagged treatments. In *S. congestiflora*, the results of the selfed and open treatments and in *S. trapezifolia* those between selfed and bagged treatments were nearly equal. At 25–30 days after pollination, the crossed treatment gave the highest proportion of fruits in all species. At mature fruit stage, when fruits were ripe and ready to be dispersed, fruit success in the non-emasculated treatments in all four species of *Shorea* showed a similar trend and decreased in the order of crossed, open, selfed and bagged treatments, except in *S. trapezifolia* where the order between the selfed and bagged treatments was reversed.

(2)   Comparison of results of the non-emasculated treatments at the ripe fruit stage in the four *Shorea* species (Table 9.9) shows that mean crossing success is 5% and 9% in *S. trapezifolia* and *S. megistophylla*, and 18% and 21% in *S. congestiflora* and *S. cordifolia* respectively. Selfing success in all species was much less than the crossed treatments: 0.5% each in *S. trapezifolia* and *S. cordifolia*, 1% in *S. megistophylla* and 2.8% in *S. congestiflora*. Thus, the degree of cross- and self-compatibility varies from species to species.

The differences between results of the crossed and selfed treatments suggest that each species has some mechanism to limit or restrict the extent of successful self-pollination and that there is selection to preferentially reject fruits resulting from self-pollination and self-fertilization. Fruit set in non-emasculated and bagged treatments, even though low, suggests that reproduction is possible without the intervention of an insect pollinator.

(3)   Fruit-set from crossed treatments of both emasculated and non-emasculated flowers was similar in all species examined except in *S.*

trapezifolia (Fig 9.4). In contrast, fruit-set of open and self-pollinated flowers appear to be affected by emasculation.

Failure to set mature fruits in the emasculated and open pollinated treatments indicates that the absence of stamens may adversely affect the pollination process. If wind was important in bringing about successful pollinations, then some percentage of fruits would have been formed in the emasculated and open pollinated treatment.

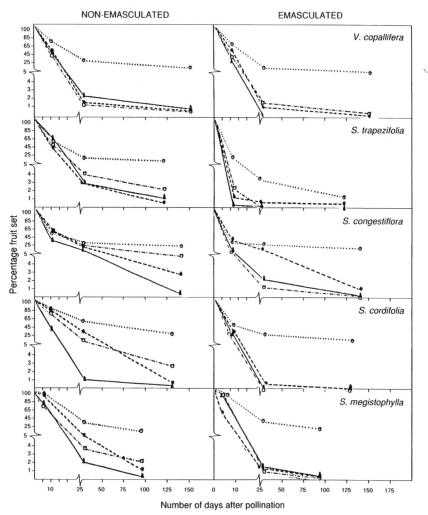

Figure 9.4 Proportion of fruits retained by the respective species of Dipterocarpaceae after different pollination treatments (O···O = artificially crossed; □–•–□ = open pollinated; ●——● = artificially selfed and Δ–Δ = bagged) using non-emasculated and emasculated flowers

Table 9.9 Results of pollination studies carried out on five species, *S. megistophylla*, *S. cordifolia*, *S. congestiflora*, *S. trapezifolia* and *Vateria copallifera*, of Dipterocarpaceae at Sinharaja rain forest

Treatments			Percentage fruiting success														
			S. meg.			*S. cord.*			*S. cong.*			*S. trap.*			*V. copal.*		
			A	D	R	B	D	R	B	D	R	B	C	R	B	D	R
Bagged	E	I	85	(0.6)	0.5	2	(0)	0	0	(0)	0	0	(0)	0	17	(0)	0
		II	91	(2)	0	44	(0)	0	3	(0)	0	0.4	(0)	0	28	(0)	0
		III	–	–	–	–	–	–	12	(7)	0	–	–	–	26	(0)	0
	NE	I	73	(0.8)	0.9	2	(0)	0	4	(0)	0	33	(4)	2	35	(0)	0
		II	76	(3)	0	68	(2)	0.5	46	(12)	2	74	(1)	0	42	(4)	2
		III	–	–	–	–	–	–	40	(18)	0.4	–	–	–	41	(2)	0
Selfed	E	I	74	(0.5)	0.4	13	(0)	0	1	(0)	0	2	(1)	0.9	57	(0)	0
		II	28	(2)	0	36	(1)	0	30	(4)	0	0	(0)	0	51	(2)	0.9
		III	–	–	–	–	–	–	59	(15)	4	–	–	–	37	(0)	0
	NE	I	97	(3)	2	57	(0)	0	4	(0)	0	40	(3)	1	20	(1)	0
		II	88	(8)	0	90	(54)	1	54	(11)	0.9	21	(2)	0	64	(1)	0.9
		III	–	–	–	–	–	–	83	(25)	7	–	–	–	–	–	–
Crossed	E	I	98	(40)	16	40	(22)	6.6	1	(1)	1	17–21*	(2–6)*	2	64	(5)	3
		II	80	(20)	0.8	46	(19)	12	23	(11)	5	0.9	(0)	0	59	(10)	7
		III	–	–	–	–	–	–	55	(44)	31	–	–	–	56	(6)	5
	NE	I	96	(39)	15	70	(45)	14	16	(2)	0	48–61*	(13–14)*	6–7*	67	(5)	5
		II	86	(16)	2	81	(56)	28	37	(26)	22	34	(6)	4	60	(4)	2
		III	–	–	–	–	–	–	80	(43)	33	–	–	–	62	(16)	12
Open	E	I	86	(0.6)	0	6	(0)	0	0	(0)	0	2	(0)	0	37	(2)	1
		II	**	**	**	43	(0)	0	6	(0)	0	2	(0)	0	30	(2)	2
		III	–	–	–	–	–	–	9	(2)	0	–	–	–	33	(0)	0
	NE	I	60	(5)	3	63	(5)	3	14	(2)	0.3	34	(5)	3	18	(0)	0
		II	80	(2)	0.9	82	(8)	2	77	(18)	6	43	(2)	1	40	(1)	0.8
		III	–	–	–	–	–	–	56	(20)	8	–	–	–	–	–	–

A=1–5, B=6–10, C=20–25 and D=25–30 days after pollination and R=ripe fruit stage, when fruit success data were gathered. E and NE indicates emasculated and non emasculated treatments. I, II and III represents number of trees studied. * Range of fruiting success using 2 different pollen parents.
**=Data not available as this branch was accidentally broken during this study

The lack of fruit-set in emasculated and bagged treatments in three species suggests the absence of apomixis. Low (0.3%) fruit-set in the corresponding treatment in *S. megistophylla* could be due either to apomixis or to contamination by xenogamous pollen.

Although three individuals for each of the study species were selected to carry out pollination experiments, one individual in each of three species had to be abandoned due to the following reasons. In *S. cordifolia*, the ovaries of one tree developed galls. In *S. megistophylla*, pollination bags were taken off too late. Consequently, leaves of enclosed branches were etiolated and fruits were prematurely shed. In *S. trapezifolia*, too, fruits were prematurely shed for no obvious reason on one of the study trees. In the fourth species, all three individuals could be compared (Table 9.9). Except in one individual of *S. congestiflora*, where crossing success was 0%, all others showed some degree of crossing. The level of crossing as well as selfing varied between different individuals of a given species.

DISCUSSION

Time, duration and sequence of flowering

In Sri Lanka, the flowering of dipterocarp species (Ashton, 1980; Kostermans, 1981, 1983, 1984, and our own observations) takes place throughout the year, with a strong peak in April and a lesser one in November–December. The least number of species come into flower in July (Table 9.10). In Pasoh Forest Reserve (Appanah and Chan, 1981) and Kepong (Ng, 1977) in Malaysia, dipterocarps are reported to flower from March to June with a peak in April. However, a few trees in Kepong also flowered in October, November and December in some years (Ng, 1977). Dipterocarps in seasonal evergreen forests flower from November to March in the Indo-Burmese region and Philippines (Ashton, 1982). Thus, flowering time of some dipterocarps in Sri Lanka coincides with that in aseasonal evergreen forests of Malaysia, while that of others corresponds to seasonal evergreen forests in the Asian tropics.

The Sri Lankan *Shorea* spp. of sections *Doona* and *Anthoshoreae* and *V. copallifera* studied at Sinharaja bloomed in consecutive years (November 1985–April 1987), indicating that they flower annually. In this respect, they resemble dipterocarps of the seasonal tropics and are quite unlike species of the aseasonal tropics which bloom supra-annually (Ashton, 1980). Duration of blooming for each population of the Sri Lankan *Shorea* spp. section *Doona* studied ranged between 10–18 days, while that of six *Shorea* spp. section *Mutica* in Pasoh (Malaysia) ranged between 15–25 days (Appanah and Chan, 1981). In contrast, the population of *V. copallifera* bloomed for

Table 9.10   Time of flowering of each Sri Lankan dipterocarp species as given by Ashton (1980), Kostermans (1981, 1983, 1984) and our own field observations

Species		J	F	M	A	M	J	J	A	S	O	N	D
Cotylelobium lewisianum (Trimen ex Hook.f.) Ashton	?	X	.										
C. scabriusculum (Thw.) Brandis	IF				X	X							
Dipterocarpus glandulosus Thw.	IF			X	X								
D. hispidus Thw.	IF				X								
D. insignis Thw.	IF				X								
D. zeylanicus Thw.	R		X										
Hopea brevipetiolaris (Thw.) Ashton	R				X								
H. cordifolia (Thw. Trimen	R				X								
H. discolor Thw.	IF				X								
H. jucunda Thw.	IF				X								
Section *Shorea*													
Shorea dyeri Thw.	IF	X		X									
S. lissophylla Thw.	IF			X	X	X	X						
S. oblongifolia Thw.	IF			X	X	X	X			X			
S. pallescens Ashton								X					
Section *Anthoshoreae*													
S. stipularis Thw.	?			X	X						X	X	X
Section *Doona*													
S. affinis (Thw.) Ashton	R		X	X	X	X	X						
S. congestiflora (Thw.) Ashton	IF			X					X	X	X	X	X
S. cordifolia (Thw.) Ashton	IF		X	X	X	X	X						
S. disticha (Thw.) Ashton				X	X	X							
S. gardneri (Thw.) Ashton	IF	X	X	X				X				X	X
S. megistophylla Ashton	IF		X	X	X	X							
S. ovalifolia (Thw.) Ashton	IF			X	X								X
S. trapezifolia Thw.	R			X	X	X				X		X	X
S. worthingtonii Ashton	IF			X	X								
S. zeylanica	R			X	X	X		X					
Stemonoporus acuminatus (Thw.) Beddome	IF	◄						X					►
St. affinis Thw.	IF	◄						X		X	X		►
St. canaliculatus Thw.	IF					X	X						
St. ceylanicus (Wight) Alston	?					X							
St. cordifolius (Thw.) Alston	?					X		X	X		X	X	X
St. elegans Thw.	?					X		X	X			X	
St. lanceolatus Thw.	?			X				X					
St. lancefolius (Thw.) Ashton	IF	◄		X				X		X			►
St. nitidus Thw.	IF			X						X			
St oblongifolius Thw.	IF	◄					X	X				X	►
St. petiolaris Thw.	?		X	X									
St. reticulatus Thw.	IF	◄					X	X	X			X	X►
St. revolutus Trimen ex Hook.f.	IF	◄						X					►
St. rigidus Thw.	?												X
St. gardneri		X		X	X							X	
Vateria copallifera (Retz.) Alston	R	X	X	X	X								X
Vatica affinis Thw.	IF			X									
V. chinensis L.	IF	◄					X	X					►
V. obscura Trimen	?					X	X	X					
No. of spp. flowering each month		12	14	22	36	19	16	12	10	11	9	14	15

R = Flowering regularly (annually). IF = Flowering infrequently (supra-annually)

127

80–150 days. *Shorea robusta* Gaertn. f. of the savanna forests of northern India and Nepal is reported to bloom for about 35 days (Ashton, unpub.).

As in *Shorea* spp. of sections *Richetioides* and *Mutica* in Andulau Forest Reserve (Brunei) and Pasoh Forest Reserve (Malaysia), the *Shorea* species of section *Doona* also flowered sequentially. However, unlike those of the former group and like those of the latter, the period of anthesis in most species of *Doona* did overlap, although the major flowering periods of each species failed to coincide, except for *S. cordifolia* and *S. megistophylla*. Synchrony in anthesis of different species could permit hybridization between them, but among the individuals studied, only one depicted features representative of *S. megistophylla* and *S. cordifolia*, suggesting that it is a product of hybridization. One other species, *S. worthingtonii* Ashton (which was not included in our present study), also bloomed about the same time as *S. megistophylla* and *S. cordifolia*. In this species, however, many different morphologically distinct leaf forms, which showed continuous variation, were observed and could well be the result of hybridization with either or both of the two species that bloom at the same time.

The general consensus that flowering phenology of species in disturbed or selectively logged forests does not necessarily reflect that of undisturbed sites was supported by the different flowering patterns of *V. copallifera* observed in disturbed and undisturbed forest populations. In contrast, casual observations revealed that *Shorea* spp. in section *Doona* bloomed concurrently in the disturbed and undisturbed forest populations. This is being substantiated by accurate and systematic studies in the 20 ha block of undisturbed forest at Sinharaja demarcated for this purpose.

Floral biology and pollination

Differences exist between the floral biology of six closely allied *Shorea* species of section *Mutica*, the Red Meranti group, studied at Pasoh Forest Reserve (Malaysia) by Appanah and Chan (1981), and the Sri Lankan species of section *Doona* investigated at Sinharaja. In *Shorea* section *Mutica*, anthesis is between 17:00–18:30 h. The flowers remain open the whole night, abscising their corollas the following morning. In *Shorea* section *Doona*, anthesis occurs between 06:00–06:30 h and most corollas are shed between 14:00–18:00 h; by late evening, inflorescences do not bear any open flowers.

In both sections, a strong fragrance is emanated at anthesis time. Similar information is also available for *Shorea ovalis* (Korth.) Bl., where anthesis occurs between 15:30–16:00 h and anther dehiscence precedes anthesis; its almost scentless flowers remain open through the night and corollas are shed the following morning. In *Neobalanocarpus heimii* (King) Ashton, anthesis and anther dehiscence occur early in the morning and corollas are shed the same afternoon, thus showing a close similarity to the corresponding phenomena in section *Doona*.

On the basis of species diversity in each genus of the Dipterocarpaceae, Appanah (1987) recognized two broad groups among the Malaysian taxa in this family. He further pointed to a combination of floral features exhibited in genera of each group that contribute to distinguishing two types of pollinators which service them. Genera flowering regularly, with diurnal anthesis, bearing fully open flowers with long, bright yellow, conspicuous anthers and mainly bee pollinated, occur in the less species-rich group I. Genera with mass supra-annual flowering with nocturnal anthesis, bearing partially open nodding flowers with relatively small, white inconspicuous anthers enclosed by the urceolate base of the corolla, are pollinated by tiny flower-feeding insects and are found in the species-rich group II. Based on floral biology of the dipterocarp species studied at Sinharaja, *Shorea* species of section *Doona* and *V. copallifera* are typical of group I. However, these species have nodding flowers and belong to the second most species-rich taxon (section *Doona* with 10 species) of the family in the island. The genus *Stemonoporus* with 15 (or 26 according to Kostermans 1981) species ranks first. Floral features of *S. stipularis* of *Anthosphoreae* justify its classification in group II, although its diurnal anthesis and species paucity of the section (which is represented only by one or two species as recognized by Kostermans, 1983) is atypical of this group.

Flower visitors of dipterocarps are very varied and include members of almost all the major insect groups. Appanah (1987) recognizes two groups of pollinators among the dipterocarps, viz. the slow breeding, large, active, wide-ranging bee pollinators, and the fast breeding, tiny, somewhat passive pollinators such as flower thrips, hoppers, bugs, beetles and flies. *Shorea ovalis*, *Shorea curtisii* Dyer ex King, two species of *Dryobalonops* and *Neobalanocarpus heimii* have been observed to be pollinated by Meliponid bees (Ashton, 1982; Appanah, 1985). The last mentioned species was also observed to be visited by *Apis dorsata*, its principal pollinator, and *Apis cerana*. It is also reported that peak activity time was between 09:00–10:00 h, while visitation time extended between 06:30–15:00 h (Chan, 1977; Appanah, 1979, 1981, as cited in Ashton, 1988). *Neobalanocarpus heimii* resembled *Shorea* spp. of section *Doona* in all features except the activity time of large bees which showed a peak between 07:00–08:30/09:00 h and the time when all insect activity ceased, after 12:00 h. Appanah (1985) argues that many more allied species are present among dipterocarps pollinated by thrips and other tiny insects, as shown by a number of species in the following sections: eight in *Richetioides*, seven in *Mutica*, six in *Anthosphoreae*, eight in *Shorea*. In Sri Lanka, there are 10 *Shorea* species of section *Doona* which are closely allied. Pollinators of

129

seven of them were found to be large bees – *Apis dorsata* and *Apis indica*; *V. copallifera* however is also bee pollinated, but the genus is not represented by any other species in Sri Lanka. This suggests that pollinators may not be a suitable feature to differentiate the Dipterocarpaceae taxa in Sri Lanka into the two groups identified by Appanah for Malaysia.

It is indeed quite interesting to note that thrips in six to eight species of each of the Malaysian *Shorea* sections mentioned earlier are in fact their effective pollinators. Possibly these species, as well as the Sri Lankan *Shorea* species of section *Doona* examined, have evolved a mechanism to overcome damage by thrips. The Malaysian *Shorea* species have gone a step further, making use of this very same group for their own benefit as pollinators. On the other hand, pollination by thrips could even be considered primitive in this group of *Shorea* species.

Breeding systems

The breeding systems for five *Shorea* spp. of section *Mutica*, two in *Richetioides*, two in *Pachycarpa* and one in *Ovales* studied by Chan (1981) in Peninsular Malaysia, when compared with those of section *Doona* in Sri Lanka (Table 9.11), reveal the following:

(1)   Fruit-set within 7–10 days after pollination in all treatments, on the whole, was much higher (to 98%) in section *Doona* than those of Malaysia (to 53%), suggesting that fruit from unsuccessful pollinations are rejected much sooner in Malaysian species;

(2)   Fruit-set within 30 days after pollination in selfed treatments was 0–54% in section *Doona*, 16% in section *Ovales*, 0–3% and 0% in sections *Mutica* and *Pachycarpa* respectively. Chan (1981) reported that species of the latter two sections exhibit self-incompatibility while *S. ovalis* was self-compatible. In section *Doona*, one of the two/three study individuals was self-incompatible and the other(s) self-compatible. In the absence of embryological studies, however, it is not possible to say whether apomixis is partly or even altogether responsible for this apparent self-compatibility. *S. ovalis* has been confirmed to be apomictic (Kaur, 1977);

(3)   The proportion of fruit-set within 30 days after pollination in emasculated and crossed treatments was similar among the *Shorea* spp. of sections *Mutica*, *Pachycarpa* and *Doona* and was higher than in *S. ovalis*; and,

Table 9.11 Comparison of percentage fruit-set, 7–10 days (outside parentheses) and 30 days (within parentheses) after pollination, in breeding studies on some Malaysian (*) (Chan, 1981) and Sri Lankan (**) dipterocarps

Sections	Percentage fruit success				
	Emasculated and bagged	*Non-emasculated and bagged only*	*Non-emasculated selfed*	*Emasculated crossed*	*Non-emasculated open*
*Mutica**	0 (0)	0–1 (0)	4–20 (0–3)	35–53 (15–42)	0.1–4 (0–0.9)
*Pachycarpa**	0 (0)	0–0.6 (0)	15 (0)	45 (38)	0.7–2 (0–1)
*Richetioides**	N/A	0.2 (0)	N/A	N/A	0–1 (0–0.9)
*Ovales**	0 (0)	22 (16)	30 (16)	34 (18)	21 (16)
*Doona***	0–91 (0–7)	2–78 (0–18)	4–90 (0–54)	1–98 (0–44)	0–82 (0–20)

Data given converted to the nearest whole number for values greater than one. N/A = Data not available

(4) The proportion of fruit-set within 30 days after pollination in non-emasculated and open pollinated treatments was extremely low (0–1%) in the respective species of *Mutica, Pachycarpa* and *Richetioides* as compared to those of *S. ovalis* (16%) and *Doona* (0–20%). From available results of Malaysian studies on sections *Mutica, Pachycarpa* and *Richetioides*, it appears that the production of mature fruit from open pollination would be low (< 1%) as compared to 0.3–8% in section *Doona*.

The observation made by Chan (1981) that most fruit losses occur within the first 30 days after pollination is in agreement with the Sri Lankan data.

Comparison of fruit-set among the Sri Lankan *Shorea* spp. showed that fruit remaining at maturity in the open pollinated treatment or under natural conditions was highest (0.3–8%) for *S. congestiflora*. This may be attributed to the fact that *S. congestiflora* is the first of the *Shorea* spp. to bloom during the November–December flowering season. Consequently, competition from other mass blooming *Shorea* spp. for pollinators is at a minimum.

The phytosociological studies carried out in 100 non-contiguous plots in five different locations of the 8800 ha Sinharaja MAB Reserve (Gunatilleke and Gunatilleke, 1985) revealed that the density distribution of individuals over 30 cm girth at breast height (gbh) is highest in *S. trapezifolia* followed by *S. congestiflora, S. megistophylla* and *S. cordifolia* (13, 10, 7 and 5 individuals per ha respectively). The first two of these species were recorded in all five sites studied (300–1000 m), depicting their wide ecological amplitude.

The reproductive biology of the few dipterocarps studied in Sri Lanka shows that some are similar to Malaysian taxa while others show more resemblance to dipterocarps from South Asia. Furthermore, the presence of both archaic (*Stemonoporus* and *Vateria*) and advanced (*Shorea*) taxa, all but one endemic and restricted to a very small area of about 14 800 km^2, could contribute much valuable information to our understanding of the evolutionary biology of this important timber family. Two of the Sri Lankan *Shorea* species (*S. trapezifolia* and *S. congestiflora*) exhibiting annual flowering, heavy fruit-set, high densities, a wide spatial distribution and faster growth rates (unpublished data) appear to be outstanding candidates for future silviculture, germplasm conservation and continued research.

ACKNOWLEDGEMENTS

We wish to thank Professors P.S. Ashton and K.S. Bawa, the co-principal investigators of the Harvard/Peradeniya/Massachusetts collaborative research project, for their inspiration, encouragement and advice during this study. We also appreciate the expert advice of Dr. K.V. Krombein and other research entomologists of the Systematic Entomology Laboratory, US Department of Agriculture, for identification of insects collected during the project. This investigation was supported by US-AID Research Grant DPE-5542-G-SS-4073-00, through Harvard University and locally administered by the Natural Resources, Energy and Science Authority of Sri Lanka. Finally, we thank the US Agency for International Development, the Asian Network for Biological Sciences in Singapore and the International Foundation for Science, Sweden, for sponsoring the participation of Drs. I.A.U.N. Gunatilleke and C.V.S. Gunatilleke at the Bangi workshop.

REFERENCES

Appanah, S. (1985). General flowering in the climax rain forests of South-east Asia. *Journal of Tropical Ecology*, **1**, 225–40

Appanah, S. (1987). Insect pollinators and the diversity of dipterocarps. In Kostermans, A.J.G.H. (ed.) *Proceedings of the Third Round Table Conference on Dipterocarps*. Samarinda, Indonesia. 16–20 April 1985, pp. 277–91 (Jakarta: Unesco)

Appanah, S. and Chan, H.T. (1981). Thrips: the pollinators of some dipterocarps. *Malaysian Forester*, **44**, 234–52

Appanah, S. and Chan, H.T. (1982). Methods of studying the reproductive biology of some Malaysian primary forest trees. *Malaysian Forester*, **45**, 10–20

Ashton, P.S. (1980). Dipterocarpaceae. In Dassanayake, M.D. and Fosberg, F.R. (eds) *A Revised Handbook to the Flora of Ceylon*. **Volume 1**, pp. 343–423 (New Delhi: Amerind Publishing Company)

Ashton, P.S. (1982). Dipterocarpaceae. *Flora Malesiana*, Series 1, **92**, 237–552

Ashton, P.S. (1988). Dipterocarp biology as a window to the understanding of tropical forest structure. *Annual Review of Ecology and Systematics*, **19**, 347–70

Chan, H.T. (1981). Reproductive biology of some Malaysian dipterocarps. III. Breeding systems. *Malaysian Forester*, **44**, 28–36

FAO. (1985). Dipterocarps of South Asia. RAPA Monograph 1985/84. FAO Regional Office for Asia and the Pacific, Bangkok.

Gunatilleke, C.V.S. (1988). Trees of the Sinharaja forest and their non-timber products. In Ng, F.S.P. (ed.) *Trees and Mycorrhiza*, pp. 251–60 Forest Research Institute Malaysia, Kepong

Gunatilleke, C.V.S. and Gunatilleke, I.A.U.N. (1985). Phytosociology of Sinharaja – contribution to rain forest conservation in Sri Lanka. *Biological Conservation*, **31**, 24–40

Jacobs, M. (1981). Dipterocarpaceae: the taxonomic and distributional framework. *Malaysian Forester*, **44**, 168–89

Kaur, A. (1977). Embryological and cytological studies on some members of the Dipterocarpaceae. Ph.D. Thesis. University of Aberdeen, Aberdeen.

Kostermans, A.J.G.H. (1981). *Stemonoporus* Thw. (Dipterocarpaceae): a monograph (parts 1 and 2). *Bulletin Museum National d'Histoire Naturelle, Paris, 4 sér., 3. Section B, Adansonia*, **3**, 321–58 and **4**, 373–405

Kostermans, A.J.G.H. (1983). The Ceylonese species of *Shorea* Roxb. (Dipterocarpaceae). *Botanische Jahrbuecher fuer Systematik Pflanzeneschichte und Pflanzengeographie*, **104**, 183–201

Kostermans, A.J.G.H. (1984). Monograph of the genus *Doona* Thwaites (Dipterocarpaceae). *Botanische Jahrbuecher fuer Systematik Pflanzeneschichte und Pflanzengeographie*, **104**, 425–54

Ng, F.S.P. (1977). Gregarious flowering of dipterocarps in Kepong, 1976. *Malaysian Forester*, **40(3)**, 126–37

CHAPTER 10

BEETLE POLLINATION IN TROPICAL FORESTS OF AUSTRALIA

Anthony K. Irvine and Joseph E. Armstrong

ABSTRACT

Pollinating beetles can be separated into two main feeding types, pollen/nectar and herbivorous feeders. Some beetle families that contribute to these two types of feeding behaviour are listed, together with some plant families which have species pollinated by each feeding type. Examples of beetle pollination in rain forests of north Queensland are described. Beetles pollinate plants throughout the vertical profile of forests. Plants described in the pollination studies are Alphitonia petriei *(Rhamnaceae),* Balanophora fungosa *(Balanaphoraceae),* Diospyros pentamera *(Ebenaceae),* Eupomatia laurina *(Eupomatiaceae),* Flindersia brayleyana *(Flindersiaceae), and* Myristica insipida *(Myristicaceae). An estimate of the frequency of beetle pollination in tropical forests of north Queensland is compared with estimates in the American tropics. Trends suggest beetle pollination may be more significant in Australian tropical forests and this may be due to differences in insect groups of each region. In both regions, the level of beetle pollination appears to be underestimated and poorly studied. The importance of beetle pollination to forest ecology and implications for management are discussed. Many useful plants are beetle-pollinated and knowledge of beetle–plant interactions can contribute to the conservation and development of these genetic resources.*

INTRODUCTION

Beetle pollination has been a subject of intense controversy, mainly on two accounts. The first deals with the origin of insect pollination in angiosperms. Diels (1916) suggested beetles were the first pollinators of angiosperms based on Hamilton's (1897) account of pollination of *Eupomatia laurina* R.

135

Br. by herbivorous beetles. This account was criticized by Grinfel'd (1975) on grounds that herbivory in beetles was regarded more recently evolved than pollen feeding. Gottsberger (1974, 1977) observed pollen-feeding beetles on flowers of an ancient genus *Drimys* and described beetle-pollination syndromes supporting the concept of beetles as the first pollinators. Pollination by herbivorous beetles was regarded as a later specialized development. Grinfel'd (1975) suggested beetles were among several other ancient pollen-eating insect groups that began visiting early angiosperms. Recently, Bernhardt and Thien (1987) supported this view.

The second concern was whether beetles effectively pollinate flowers. Despite published examples of beetle pollination (Hamilton, 1897; Grant, 1950), claims were made that beetles generally do not participate in pollination (Percival, 1965; Faegri and van der Pijl, 1976; Armstrong, 1979) especially not in cross-pollination (Meeuse, 1978). Recent work indicates that beetles are effective pollinators and that beetle pollination is common and widespread (Thien, 1980; Beach, 1982; Henderson, 1984; House, 1985). The importance of the process in tropical forest ecology is highlighted by the fact that many economically useful plants such as pejibaye palm (Beach, 1984), nutmeg (Armstrong and Drummond, 1986), *Magnolia* (Thien, 1974), custard apple (Gazit *et al.*, 1982) and relatives such as soursop, etc. as well as oil palm are pollinated by beetles. Despite this observation, beetle pollination is not well studied.

Our aim is to make it clear that there are two main types of beetle-pollination behaviour, and list some families of beetles and plants associated with each. A second aim is to demonstrate that beetle pollination occurs throughout the vertical profile of rain forests. Third, we estimate the frequency of beetle pollination in some forests of north Queensland and compare this estimate with estimates for some American tropical forests.

TERMINOLOGY

Dioecious plants bear male and female flowers on separate plants. Monoecious plants have male and female flowers on the same plant. Hermaphroditic plants have male and female sex organs within the one flower. Protandrous flowers have male organs maturing before female organs. Protogynous flowers have female organs maturing before the male organs. Homogamous flowers have both male and female organs maturing at the same time. "Micro-beetles" refer to very small beetles up to 3 mm in length. A lens (x10) is needed to observe their behaviour on flowers visited. Small beetles are 4–8 mm long. The Cairns region refers broadly to the rain forest area around Cairns (north Queensland) between latitudes 15–19°S. This rain forest area extends 60 km inland at its widest point.

METHODS

Using species lists for several tropical communities in north Queensland, trees, shrubs, vines, epiphytes and herbs were rated as pollinated by beetles from observations by Irvine, case histories, and extrapolation from related species with similar flower morphologies. A species was rated as beetle pollinated if beetles were prominent among pollinating visitors. The degree of importance of beetles in each case was not assessed. In American studies, estimates of the "most probable pollinator" were based on observations and inferred data (Frankie, in Bawa, 1980; Bawa *et al.*, 1985) and on known pollinators in Venezuela (Sobrevila and Arroyo, 1982). Crome and Irvine (1986) showed that, in a tree that normally would have been rated bat and bird pollinated, less conspicuous insects fertilized flowers to the same degree as birds. Hence the difficulty in all these assessments is that observed conspicuous pollinators may mean non-scoring of other significant pollinators. Consequently, we scored taxa as both vertebrate and insect pollinated if we knew that both groups participated in the pollination. We tried to assess equivalent communities in each tropical area, but more field data are needed before comparisons between amorphous entities such as Australian and American tropics have meaning.

TYPES OF BEETLE POLLINATION

Beetle pollination can be separated into two main types, based on the manner of feeding of the beetles. In one, beetles are mostly pollen and/or nectar feeders and do little or no damage to flowers visited. In the other, beetles are herbivorous and feed on flower parts as well as pollen. In both types, beetle sizes range from micro-beetles to large beetles around 3 cm long. Flowers pollinated by herbivorous beetles are usually robust and range from single flowers to large inflorescences characterized by mostly fleshy development of supporting rachii, scales, modified hairs, bracts, and numerous flower organs. Often, ovules are protected by thick fibrous tissue containing tannins, raphides and silica bodies (Uhl and Moore, 1973). Flowers that are pollinated by pollen/nectar feeding beetles can be of any shape or size, as very small beetles can gain access to most flower designs. Flowers attracting beetles usually have a strong odour (see Pellmyr and Thien, 1986 and references therein) at some stage of maturity. These odours tend to be sweet or sweet-fruity in flowers visited by pollen/nectar-feeding beetles, and more musky, ripe-fruity, or aminoid in flowers visited by herbivorous beetles.

Most work has highlighted herbivorous-beetle pollination, and relatively few cases of the pollen/nectar-beetle type have been reported until recently

Table 10.1   Some beetle families that provide species for beetle pollination. The list is probably more complete for herbivore-beetle families. Numbers refer to reference source below

Pollen/nectar beetles		Herbivore beetles	
Alleculidae	1.	Allocorhynidae	2.
Anthicidae	3.	Cerambycidae	4.
Bruchidae	1.	Curculionidae	5.
Buprestidae	6.	Languriidae	2.
Byturidae	7.	Nitidulidae	8.
Cantharidae	8.	Scarabaeidae	9.
Chrysomelidae	10.	Chrysomelidae?	?
Cerambycidae	1.	Staphylinidae?	10.
Cleridae	1.		
Coccinellidae	1.		
Curculionidae	8.		
Cryptophagidae	11.		
Dermestidae	10.		
Elateridae	12.		
Helodidae	12.		
Lagriidae	8.		
Languriidae	11.		
Lycidae	8.		
Meloidae	1.		
Melyridae	1.		
Mordellidae	1.		
Nitidulidae	1.		
Oedmeridae	1.		
Phalacridae	11.		
Pythidae	12.		
Rhipiphoridae	11.		
Scarabaeidae	8.		
Scraptiidae	11.		
Staphylinidae	11.		
Trixagidae	11.		

Sources: 1. (Grinfel'd, 1975) 2. (Tang, 1987) 3. (Armstrong and Drummond, 1986) 4. (Norman and Clayton, 1986) 5. (Hamilton, 1897) 6. (Hawkeswood, 1978) 7. (Pellmyr, 1984) 8. (Irvine, pers. obs.) 9. (Beach, 1982) 10. (Thien, 1980) 11. (Britton, 1973) 12. (Armstrong, 1979) ? (Information not clear)

(Grinfel'd, 1975; Pellmyr, 1984; House, 1985; Armstrong and Drummond, 1986; Armstrong and Irvine, 1989*a*). This is surprising considering the diverse array of beetle families that have species that are mainly pollen/nectar feeders and the diverse array of plant families visited by these beetles. On the other hand, only a few families of beetles provide herbivorous-pollinating species (Tables 10.1 and 10.2)

Table 10.2   Some plant families with species pollinated by pollen/nectar beetles and herbivorous beetles

Pollination by pollen/nectar beetles		Pollination by herbivorous beetles	
Araceae	1.	Annonaceae	2.
Asteraceae	3.	Aizoaceae	2.
Cunoniaceae	4.	Araceae	5.
Ebenaceae	6.	Arecaceae	7.
Elaeocarpaceae	4.	Balanophoraceae	4.
Flindersiaceae	4.	Calycanthaceae	8.
Lauraceae	6.	Caryophyllaceae	2.
Magnoliaceae	9.	Clusiaceae	2.
Myristicaceae	10.	Cyclanthaceae	11.
Myrtaceae	12.	Degeneriaceae	13.
Orchidaceae	14.	Dilleniaceae	2.
Polygonaceae	15.	Eupomatiaceae	16.
Rosaceae	2.	Magnoliaceae	9.
Rutaceae	4.	Nymphaeaceae	17.
Sterculiaceae	4.	Orchidaceae	14.
Theaceae	4.	Polemoniaceae	2.
Winteraceae	13.	Rosaceae	2.
Aizoaceae?	2.	Winteraceae	13.
Cactaceae?	2.	Asteraceae?	2.
Caryophyllaceae?	2.	Cactaceae?	2.
Clusiaceae?	2.	Cistaceae?	2.
Meliaceae?	?.	Chloranthaceae	18.
Polemoniaceae?	2.	Paeoniaceae	2.
Proteaceae?	2.	Proteaceae?	2.
Sapindaceae?	?.	Theaceae?	2.
		Tiliaceae?	2.

Sources: 1. (Monteith, 1986) 2. (Gottsberger, 1977) 3. (Britton, 1973) 4. (Irvine, pers. obs.) 5. (Bawa and Beach, in Bawa *et al.*, 1985) 6. (House, 1985) 7. (Henderson, 1984) 8. (Grant, 1950) 9. (Thien, 1974) 10. (Armstrong and Irvine, 1989*a*) 11. (Beach, 1982) 12. (Hawkeswood, 1978) 13. (Thien, 1980) 14. (Armstrong, 1979) 15. (Grinfel'd, 1975) 16. (Hamilton, 1897) 17. (Schneider, 1979) 18. (Endress, 1987) ? (Information not clear)

BEETLE POLLINATION IN NORTH QUEENSLAND RAIN FORESTS

The following case studies describe beetle pollination in a herb, in understorey, subcanopy, canopy and pioneer trees in rain forest.

Pollen/nectar beetle pollination – micro-beetles

Myristica insipida R.Br. (Myristicaceae), is a dioecious subcanopy or canopy tree 12–25 m tall, in lowland rain forest across northern Australia, north of 20°S latitude. In the Cairns region, it predominates in forests with fertile soils, from sea level to around 1000 m altitude. In some wet lowland forests, it occurs on poorer soils at lower densities. Flowers occur in small clusters in leaf axils or behind leaves in male and female trees. They are inconspicuous, 5–6 mm long, 2–3 mm wide, cream green, and urn-shaped. Central sexual parts are easily accessible to very small visitors. Female flowers do not secrete nectar, although the stigmatic surface can appear moist. Both sexes produce a similar sweet floral fragrance, which appears to be the primary attractant to flower visitors. Flowers open at night, but pollinators are diurnal.

Four groups of insects have been recorded visiting flowers, a minute parasitic fly (whose larvae destroy both male and female flowers), one ant species, thrips, and micro-beetles (Armstrong and Irvine, 1989a). Thrips dwell in the flowers, feeding on the internal wall of the receptacle and pollen. Their movements within the female flower by-pass the central stigma. The ant's foraging habits make it an unlikely pollinator of a dioecious tree. Eight species of micro-beetles (including four weevils) were seen eating pollen in an upland rain forest. Two of the weevil species were observed probing stigmas of female flowers in a lowland forest. Beetles occur in higher numbers on male trees, which bear more flowers than female trees and provide pollen as a reward (Armstrong and Irvine, 1989b). Time spent on female flowers is about one-third to one-sixth of the time spent at male flowers (Armstrong and Irvine, 1989a). This observation has led Armstrong and Drummond (1986) to suggest that female flowers of *Myristica* may mimic male flowers, although beetles are attracted to the moist stigmatic surface and do not immediately leave female flowers. The beetles are quite skilful fliers and move readily from tree to tree.

Pollen/nectar beetle pollination – small to medium-sized beetles

Diospyros pentamera (Woolls and F. Muell.) Woolls and F. Muell. ex Heirn (Ebenaceae) is a dioecious, subcanopy or canopy tree 10–25 m tall, in rain forests of northern New South Wales, south Queensland and north Queensland, where it is more common in upland rain forest areas. It is mostly found on fertile soils in moist to drier rain forest. Flowers are grouped in small clusters in leaf axils. They are inconspicuous, small (4–5 mm wide), with the overall colour tending to be white with brownish pink patches on the petals, and bowl shaped. Sexual organs are accessible to small insects, and female flowers secrete nectar. Both male and female flowers exude a similar strong fruity odour, which appears to be the main attractant to flower visitors.

A large number of beetles, flies and wasps (mostly 3–6 mm long), visit flowers of both sexes, but insect numbers are highest at male trees. Visits of beetles and flies to female trees increase as flower shedding increases. Beetles carried the highest pollen loads and moved more regularly between male and female trees than other insects, and are considered the main pollinators (House, 1985).

Flindersia brayleyana F. Muell. (Flindersiaceae) is a hermaphroditic, canopy tree 30–50 m tall, occurring in rain forests around the Cairns region. It is present in some lowland forests, but mainly occurs between 600 and 1000 m altitude, in moist to wet rain forests on fertile volcanic soils. The species is mass flowering with numerous terminal panicles of bright white flowers 5–7 mm wide, each with a bright orange central disc surrounding the ovary. En masse flowers are conspicuous. They are quite shallow and sexual organs are very accessible to small and medium-sized insects. Anthers mature before the stigma (protandrous). Flowers secrete nectar with a heavy sweet smell. Visual effects and odour probably attract flower visitors.

A vast number of insects, mostly 2–12 mm long, visit the flowers, with beetles and flies dominating. Lepidopterans, wasps and bees are less prominent. Of these, larger lepidoptera are less likely to contact the anthers. Seasonal conditions can result in the pollination system being dominated by beetles. Trees flower in November to early January, a hot period, with the degree of wetness/dryness being determined by storms. If storms are frequent, waves of scarab beetles *Phyllotocus apicalis* Macleay (Scarabaeidae: Melolonthinae) emerge daily and swarm over flowers, competing for feeding sites and mates during early morning. This social interaction results in movement to other trees. In the afternoon, the feeding and mating frenzy ceases, and nearly every flower is occupied quietly by an individual beetle. *Chauliognathus flavipennis* Macleay (Cantharidae) is another noticeable beetle visitor. Both beetles have long slender legs ideally suited for walking over rough outlines of inflorescences. In drier seasons, beetles and flies probably contribute equally to pollination. Flowers tend to be too small for vertebrate pollinators, and such pollinators have not been observed.

Alphitonia petriei Braid and C. T. White (Rhamnaceae) is a hermaphroditic pioneer tree 5–25 m tall, which regenerates in disturbed areas in wet rain forests of the Cairns region, from sea level to 1200 m altitude on most soil types, except on drier volcanic soils. It is mass flowering, and inflorescences arise in leaf axils and extend beyond the leaves to form a conspicuous array. Individual flowers are small, 5–6 mm wide, dull creamy-green, shallow and protandrous, with sex organs readily accessible. They produce copious nectar and have a heavy sweet fragrant smell. Both visual and odour effects are probably attractants for flower visitors.

Flowers attract an array of visitors of similar size to those that visit *F. brayleyana*. Flies tend to be more prominent numerically than beetles. Wasps and bees also become noticeable. Trees flower in the warm dry season (September to early November,) and the most prominent beetle visitors are *C. flavipennis*, *Metriorrhynchus rhipidius* Macleay (Lycidae), and *Lagria grandis* Gyllh. (Lagriidae). In disturbed areas, the trees spring up in dense stands and pollinators have little difficulty moving between trees. Beetles make a substantial contribution to the pollination of this species.

Herbivorous beetle pollination – micro-beetles

Eupomatia laurina R. Br. (Eupomatiaceae) is a hermaphroditic understorey tree 3–10 m tall, occurring from southern New South Wales to the Cairns area, where it ranges from sea level to 1200 m altitude. It is confined to rain forest in tropical lowlands and mid uplands, but, above 900 m, extends into gullies of tall open forests bordering cloud forests over a wide range of soil types. Flowers are born singly or in small groups in leaf axils and behind the leaves. Flowers are relatively large fleshy structures 2–3 cm wide, inconspicuous, dull white with numerous staminodes and stamens fused together basally to form a complete unit, the synandrium. This unit is shed 14–30 hours after flower opening. Petals are absent. The flower is shallow, but movement of staminodes produces a chamber. Stigmas are receptive about 12 hours before anthers dehisce (protogynous). They are readily accessible if a floral visitor arrives within 2–3 hours of flower opening, as this is the time staminodes take to reclose the flower chamber which opened at dawn. Anthers are easily accessible. Flowers exude a strong fruity-musky odour, particularly at the time of opening. This odour is the main attractant for floral visitors.

Despite the size of flowers, only a few insect species have been recorded as visitors. Two species of bees (one in cloud and one in lowland forests) visit flowers when anthers dehisce, but, at this time, stigmas are strongly sealed from access by closure of staminodes. Small fly larvae, possibly drosophilids, have been found in low numbers in floral parts, but adults have not been observed visiting anthers. In the lowlands, predatory green ants may wait on the edge of flowers for weevil pollinators to arrive, but they do not enter flowers. The only pollinator recorded is a weevil, *Elleschodes hamiltonii* T. Blackburn, which may consist of two or more species. *Elleschodes* occurs throughout the range of *E. laurina* and has its life cycle tied to it.

Weevils enter the flower, feed on staminoides and lay eggs in staminode bases. In the early evening, weevils emerge from the flower, feed on pollen as anthers dehisce and then fly away. During the night, the synandrium abscises and sheds. At dawn, weevils arrive at newly opening flowers on other individual plants as there is usually a 1 or 2 day pause between the opening of flowers on the same plant (Endress, 1984). This synchronization facilitates out-crossing (Endress, 1984) and may involve flights of over 200 m to find the nearest flowering neighbours. Weevil eggs hatch on the fallen synandrium, the larvae feed on it, and pupate in the soil. In the tropics, they emerge as adults some 14–16 days after the original flower opened, and can commence a second cycle on flowers that are still available. No ovules are attacked by the pollinator. Both the *Elleschodes* weevils and the tree *Eupomatia laurina* appear completely dependent upon each other for sexual reproduction (Armstrong and Irvine, 1990).

Herbivorous beetle pollination – small beetles

Balanophora fungosa Forster and G. Forster is a monoecious understorey herb growing in clumps 8–20 cm tall. It is parasitic on roots of rain forest trees and lacks chlorophyll. The species occurs in rain forest in Queensland, north of 27°S latitude. Around Cairns, it ranges from sea level to 1000 m altitude on a wide range of soil types. It emerges seasonally above the forest floor between the months of April and September. Individual flowering spikes form penis-like structures about 8 cm tall. Male flowers form a ring around the base of the female spike, which contains thousands of microscopic apetalous flowers. When anthers dehisce, the basal ring turns fluffy white. When the female flowers are receptive, the colour of the spike is dull fawn or flesh pink turning brown. The flowers appear to be homogamous or slightly protandrous. Female flowers exude a strong mousey odour when receptive and produce small pools of nectar during early morning.

A numerous array of pollen and nectar vectors visit the flowers, such as ants, flies, springtails, a noctuid moth, and rats. Rats sniff and chew male flowers bringing pollen into contact with facial hair and appears to be attracted to plants by the mousey odour (Cox and Irvine, unpublished). The plant form is very suitable for ant pollination. Govindappa and Shivamurthy (1975) recorded only bees visiting *B. abbreviata* Blume (also monoecious) in south India, but bees or wasps were not seen visiting *B. fungosa*. Among herbivorous pollinators, two beetle species, about 5 mm long, of *Lasiodactylus* (Nitidulidae), one tipulid fly, and one pyralid moth utilize basal bracts as breeding chambers. Adult beetles, which feed mainly on pollen, can be found in these bracts together with larvae. Adults move between spikes in the one clump and probably between clumps in search of pollen and oviposition sites. The developing larvae of the beetles, the tipulid fly, and the pyralid moth feed on pollen and the central supporting stalk, tunneling between female flowers, breaking them up into islets of tissue which peel off or shed at the fruit dispersal stage. Odour seems to be the main attractant to flower visitors. The beetles have their sexual reproduction linked with the life cycle of *B. fungosa* and are likely to play a prominent role in pollination.

143

FREQUENCY OF BEETLE POLLINATION IN TROPICAL FORESTS

In this first attempt to estimate forest community pollination in north Queensland, the results (Table 10.3) show that insect pollination dominates all communities, and is mostly above 80%. Of these insect-pollinated species, between 45 and 75% in each community are unknown with regard to beetle pollination. Beetle contribution to insect pollination is estimated as above 17% in most communities. The lowest score (3.5%) occurs in a wind-exposed coastal vine-thicket, whereas the highest score (24.9%) occurs in cloud rain forest. The highest individual community score was 28.7% for cloud rain forest on granite bedrock. The results show an estimate of beetle pollination in a vegetation gradient between a mid-upland, moist, complex notophyll vine forest (defined in Webb, 1978), its disturbed edge (consisting mainly of secondary species), and an open forest community. The results were similar, 18.3% for the vine forest, 21.0% for the disturbed edge and 17.0% for the open forest. Table 10.3 also shows the percentage of species in each community definitely known not to be pollinated by beetles, including insect-pollinated species such as *Ficus*, plus wind-pollinated species.

Table 10.4 compares beetle pollination in this study with some American tropical forests. Results indicate a higher degree of beetle pollination in wet and dry vine forests of north Queensland than in Costa Rica. Cloud forest results, in north Queensland and Venezuela, cannot really be compared as workers in Venezuela may have only studied bird and bee systems, though lack of one or two scores for beetles in 38 species could imply lower beetle pollination in Venezuela.

DISCUSSION

The Costa Rican study (Bawa *et al.*, 1985) noted beetle pollination only in the understorey of lowland rain forest, whereas our work shows beetle pollination occurs in all levels of the forest, from herbs to canopy and pioneer trees. Plant families such as Myrtaceae, Elaeocarpaceae, Lauraceae and Flindersiaceae contribute many members to canopies of north Queensland rain forests, and all have species which experience beetle pollination. Perhaps prominence of Myrtaceae in Australian forests could account for the high level of beetle pollination estimated. Work in American tropics and subtropics (Gottsberger, 1970, 1977, 1985, 1986; Norman and Clayton, 1986) describes beetle pollination for Annonaceae. The Costa Rican wet forest study recorded a Lauraceae and all Annonaceae species as also being beetle pollinated. With further study, results in north

Table 10.3 Beetle pollination in north Queensland tropical forests

Forest type	Range of n	W	V	I	−B	B
Coastal dunes						
SNVT †	143	8.4	1.4	86.7	11.2	3.5
Woodland heath *	140	17.1	15.0	81.4	17.9	12.1
Lowland						
Wet lowland CMVF †	191–245^	0.3	10.6	96.2	6.0	18.7
Moist/wet " " ‡ §	181	1.1	16.0	95.0	5.5	18.8
Upland						
Moist upland CNVF †	93–143^	1.6	7.9	93.0	7.4	17.7
Upland – transition						
Moist upland CNVF §	93	3.2	10.6	90.3	9.7	18.3
Margin upland CNVF §	38	7.9	5.3	76.3	7.9	21.0
Upland open forest §	47	6.4	17.0	89.4	8.5	17.0
Upland cloud						
Cloud CNVF and SNVF ‡ §	164–200^	2.9	16.5	94.8	6.1	24.9
Dry tropical						
Dry tropical Vine †	100	4.0	12.0	93.0	7.0	11.0
Dry woodland †	36	19.4	12.0	80.6	19.4	19.4

Key: *n* = Number of species. ^ = Range of plant species number in examples of the community, i.e. 3 communities for Wet Lowland Forest, 2 communities for Moist Upland Forest, 2 communities for Cloud Forest. W = Wind pollinated. V = Vertebrates prominent in pollination. I = Insects prominent in pollination; includes vertebrate-pollinated taxa where insects are also prominent pollinators. −B = Definitely not beetle pollinated; includes insect-pollinated and wind-pollinated taxa. B = Beetles prominent in pollination. SNVT = Simple Notophyll Vine Thicket. CMVF = Complex Mesophyll Vine Forest. CNVF = Complex Notophyll Vine Forest. SNVF = Simple Notophyll Vine Forest. See Webb (1978) for community terms. Sources of Species Lists: * Clarkson (1986), † (Stocker *et al.*, 1981), ‡ (Unwin and Stocker, unpublished), § (Irvine, unpublished)

Queensland are likely to be higher, as of the 20–30 species of Lauraceae and Annonaceae in the wet-forest communities, only a few Lauraceae were scored for beetle pollination and no Annonaceae species were scored, due to the lack of observations on these families in Australia. Other life forms in Queensland forests, such as lianas, vines and epiphytes in the Annonaceae, Araceae and Orchidaceae are also beetle pollinated (Table 10.2). Hence, it is

Table 10.4 Comparison of beetle pollination in north Queensland with similar American tropical communities, (percentages)

Site/country	Tropical wet lowland rain forest	Dry tropical vine forest	Cloud forest
N. Qld. Australia			
All life forms			
!n = 214	18.7	—	—
n = 100	—	11.0	—
+n = 182	—	—	24.9
Trees only			
!n = 156	24.1	—	—
n = 52	—	21.1	—
(Irvine, this study)			
La Selva, Costa Rica			
Trees only			
n = 143	7.3	—	—
(Bawa *et al.*, 1985)			
Costa Rica			
Trees only			
n = 122	—	13.9	—
(Frankie, in Bawa, 1980)			
Altos de Pioe, Venezuela			
All life forms			
n = 38	—	—	0.0
(Sobrevila and Arroyo, 1982)			

Note: !n = the mean of three lowland wet rain forests, n for each being 191, 206, 245. Individual percentages were respectively 22.5, 18.4, 15.2. +n = the mean of two cloud forests, n for each = 164, 200. Individual percentages were respectively 28.7, 21.0. For trees only, in each lowland wet rain forest, n = 169, 151, 152. Individual percentages were respectively 24.9, 23.8, 23.7

likely there is no angiosperm life form in the vine forests of north Queensland that does not have a beetle-pollinated species.

The only example of pollination by pollen/nectar-feeding beetles in the wet Costa Rican forest appears to be a Lauraceae species; all other cases involve pollination by herbivorous beetles. This observation raises the question: Are pollen/nectar-feeding beetles uncommon in Central American forests? Many pollination studies in the Americas describe the activity of masses of small and large bees. We have species of these bees in Australia, but, from all accounts, they are less numerous. Are these bees so dominant in Costa Rica and Venezuela that they have inhibited or displaced fly and beetle pollinators noticeable in north Queensland, or is the contribution of the latter overlooked? It appears that bees have been extraordinarily successful in the American tropics, whereas, in the Australian tropics, flies and beetles are the important pollinators.

The difference in pollinator prominence is likely to reflect differences in the types of flowers noticeable in north Queensland and Costa Rican forests. For example, in Costa Rica, the sampling concerned 143 flowers whose longevity of flowers in a vast majority of species does not exceed 12 hours (Bawa *et al.*, 1985). In contrast, the majority of Australian species have flowers with a longer life (e.g. in a wet lowland forest, at least 70% of 191 species, have a flower life of 24 hours or longer). Based on differences in pollinators and shorter flower life, it is probable that average flower size is larger in Costa Rican than in north Queensland forests. In Australian tropical rain forests, small to literally minute flowers are very common in families such as Anacardiaceae, Celastraceae, Euphorbiaceae, Lauraceae, Meliaceae, Myrsinaceae, Rutaceae, Sapindaceae, Sapotaceae, Symplocaceae and Vitaceae. The minute flowers (less than 3 mm wide) in these families have an arrangement and presentation of sex organs that only minute insects such as micro-beetles, flies, and thrips could utilize.

An analysis of 24 tropical forest communities assessed for this study, indicates that a mean of 18.7% (range 3.5–28.7%) of the species experience beetle pollination. As claimed in the American studies, our results probably underestimate the frequency of beetle pollination, as the process is poorly studied. It can thus be seen from Costa Rican studies and this study, that beetle pollination plays a prominent role in the ecology of tropical forests. Virtually nothing is known about the effects of disturbance or wide-scale clearing upon the diverse array of pollinating beetles. However, we suspect that the single species of beetle pollinator of commercial nutmeg observed in southern India (Armstrong and Drummond, 1986) may be a faunal paucity caused by extensive tropical forest clearing for agriculture.

Knowledge of beetle behaviour also becomes important when beetle-pollinated plants are introduced into cultivation. Orchard designs and conditions may not favour the behaviour or life cycle of the pollinator, and it may be important to have a healthy natural forest nearby to ensure fruit production. Current fruit yields may also be increased by introductions of natural pollinators. One area of potential for biochemical research is the isolation and synthesis of the chemical nature of floral odours that attract beetles. Isolation of these compounds would not only provide information on insect behaviour, and possibly plant/insect evolution, but could provide a means of managing insect populations in agricultural and artificial environments.

ACKNOWLEDGEMENTS

We thank colleagues Frank Crome, Andy Gillison, Andrew Graham, Rod Griffin, Graham Harrington, and Geoff Tracey, Susan House (ANU, Canberra), Kamal Bawa (Univ. Massachusetts) for comments. We thank entomologists Tom Weir, Elwood Zimmerman (Coleoptera), Ted Edwards (Lepidoptera), Don Colless (Diptera), Ray McInnes (Curator), ANIC, Div. Entomology, CSIRO, Canberra for identifications and help. This work resulted from research partly funded by a NSF US-Australia Co-operative Science Programme Grant, INT-8513473 to J. E. Armstrong.

REFERENCES

Armstrong, J.A. (1979). Biotic pollination mechanisms in the Australian flora – a review. *New Zealand Journal of Botany*, **171**, 467–508

Armstrong, J.E. and Drummond, B.A. (1986). Floral biology of *Myristica fragrans* Houtt. (Myristicaceae), the nutmeg of commerce. *Biotropica*, **18**, 32–8

Armstrong, J.E. and Irvine, A.K. (1989*a*). Floral biology of *Myristica insipida* R. Br. (Myristicaceae), a distinctive beetle pollination syndrome. *American Journal of Botany*, **76**, 86–94

Armstrong, J.E. and Irvine, A.K. (1989*b*). Flowering, sex ratios, pollen–ovule ratios, fruit-set, and reproductive effort of a dioecious tree, *Myristica insipida* (Myristicaceae), in two different rain forest communities. *American Journal of Botany*, **76**, 74–85

Armstrong, J.E. and Irvine, A.K. (1990). Functions of staminodia in the beetle-pollinated flowers of *Eupomata laurina*. *Biotropica*, **22** (3) (in press).

Bawa, K.S. (1980). Evolution of dioecy in flowering plants. *Annual Review of Ecology and Systematics*, **11**, 15–39

Bawa, K.S., Bullock, S.H., Perry, D.R., Coville, R.E. and Grayum, M.H. (1985). Reproductive biology of tropical lowland rain forest trees. II. Pollination systems. *American Journal of Botany*, **72**, 346–56

Beach, J.H. (1982). Beetle pollination of *Cyclanthus bipartitus* (Cyclanthaceae). *American Journal of Botany*, **69**, 1074–81

Beach, J.H. (1984). The reproductive biology of the peach or "pejibaye" palm (*Bactris gasipaes*) and a wild congener (*B. porschiana*) in the Atlantic lowlands of Costa Rica. *Principes*, **28**, 107–19

Bernhardt, P. and Thien, L.B. (1987). Self-isolation and insect pollination in the primitave angiosperms: new evaluations of older hypotheses. *Plant Systematics and Evolution*, **156**, 159–76

Britton, E.B. (1973). Coleoptera. In CSIRO Division of Entomology (ed.) *Insects of Australia*, pp. 495–621. (Melbourne: Melbourne University Press)

Clarkson, J.R. (1986). Check list – McIvor River sand dunes. In Radke, A. (ed.) *The Vegetation of the McIvor River – Cape Flattery Sand Dunes*, pp. 1–32. Tolga, Queensland.

Crome, F.H.J. and Irvine, A.K. (1986). "Two bob each way": The pollination and breeding system of the Australian rain forest tree *Syzygium cormiflorum* (Myrtaceae). *Biotropica*, **18**, 115–25

Diels, L. (1916). Käferblumen bei den Ranales und ihre Bedeutung für die Phylogenie der Angiospermen. *Berichte der Deutschen Botanischen Gesellschaft*, **34**, 758–74

Endress, P.K. (1984). The flowering process in the Eupomatiaceae (Magnoliales). *Botanische Jahrbuecher für Systematik Pflanzengeschichte und Pflanzengeographie*, **104**, 297–319

Endress, P.K. (1987). The Chloranthaceae: reproductive structures and phylogenetic position. *Botanische Jahrbuecher für Systematik Pflanzengeschichte und Pflanzengeographie*, **109**, 153–226

Faegri, K. and van der Pijl, L. (1976). *The Principles of Pollination Ecology*. Second revised edition. (Oxford: Pergamon Press)

Gazit, S., Galon, I. and Podoler, H. (1982). The role of nitidulid beetles in natural pollination of *Annona* in Israel. *Journal of Americal Society of Horticultural Science*, **107**, 849–52

Gottsberger, G. (1970). Beiträge zur Biologie von Annonaceen-Blüten. *Oesterreichisch Botanische*

Zeitschrift, **118**, 237–79

Gottsberger, G. (1974). The structure and function of the primitive angiosperm flower — a discussion. *Acta Botanica Neerlandica*, **23**, 461–71

Gottsberger, G. (1977). Some aspects of beetle pollination in the evolution of flowering plants. *Plant Systematics and Evolution*, Supplementum 1, 211–26

Gottsberger, G. (1985). Pollination and dispersal in the Annonaceae. *Annonaceae Newsletter*, (Utrecht) 1, 6–7

Gottsberger, G. (1986). Some pollination strategies in neotropical savannas and forests. *Plant Systematics and Evolution*, **152**, 29–45

Govindappa, D.A. and Shivamurthy, G.R. (1975). The pollination mechanism in *Balanophora abbreviata* Blume. *Annals of Botany*, **39**, 977–8

Grant, V. (1950). The pollination of *Calycanthus occidentalis*. *American Journal of Botany*, **37**, 294–7

Grinfel'd, E.K. (1975). Anthophily in beetles (Coleoptera) and criticism of the cantharophilous hypothesis. *Entomologicheskoe Obozrenie*, **54**, 18–22

Hamilton, A.G. (1897). On the fertilisation of *Eupomatia laurina*, R.Br. *Proceedings of Linnean Society of New South Wales*, **22**, 48–56

Hawkeswood, T.J. (1978). Observations on some Buprestidae (Coleoptera) from the Blue Mountains, N.S.W. *Australian Zoologist*, **19**, 257–75

Henderson, A. (1984). Observations on pollination of *Cryosophila albida*. *Principes*, **28**, 120–6

House, S.M. (1985). Relationships between breeding and spatial pattern in some dioecious tropical rain forest trees. Ph.D. Thesis. Australian National University, Canberra.

Meeuse, A.D.J. (1978). Nectarial secretion, floral evolution and the pollination syndrome in early angiosperms. *Proceedings of the Koninklijke Nederlandse Akademie van Wetenschappen*, Series C, **81**, 300–26

Monteith, G.S. (1986). Some curious insect-plant associations in Queensland. *Queensland Naturalist*, **26**, 105–16

Norman, E.M. and Clayton, D. (1986). Reproductive biology of two Florida pawpaws: *Asimina obovata* and *A. pygmaea* (Annonaceae). *Bulletin of the Torrey Botanical Club*, **113**, 16–22

Pellmyr, O. (1984). The pollination ecology of *Actaea spicata*. *Nordic Journal of Botany*, **4**, 443–56

Pellmyr, O. and Thien, L.B. (1986). Insect reproduction and floral fragrances: keys to the evolution of the angiosperms? *Taxon*, **35**, 76–85

Percival, M.S. (1965). *Floral Biology*. (Oxford: Pergamon Press)

Schneider, E.L. (1979). Pollination biology of the Nymphaeaceae. In *Proceedings IVth International Symposium on Pollination*. Maryland Agricultural Experimental Station. Special Miscellaneous Publication 1, pp. 419–29. Maryland Agricultural Experimental Station, College Park.

Sobrevila, C., and Arroyo, M.T. (1982). Breeding systems in a montane tropical cloud forest in Venezuela. *Plant Systematics and Evolution*, **240**, 19–37

Stocker, G.C., Tracey, J.G., Clarkson, J.R., Lavarack, P.S., Irvine, A.K., Jackes, B.R., Bunt, J.S. and Walker, D. (1981). Guide Field Trip 5, Northern Queensland. *XIII International Botanical Congress*.

Tang W. (1987). Insect pollination in the cycad *Zamia pumila* (Zamiaceae). *American Journal of Botany*, **74**, 90–9

Thien, L.B. (1974). Floral Biology of Magnolia. *American Journal of Botany*, **61**, 1037–45

Thien, L.B. (1980). Patterns of pollination in the primitive angiosperms. *Biotropica*, **12**, 1–13

Uhl, N.W. and Moore, H.E. Jr. (1973). The protection of pollen and ovules in palms. *Principes*, **17**, 111–49

Webb, L.J. (1978). A general classification of Australian rain forests. *Australian Plants*, **9**, 349–63

CHAPTER 11

POLLINATION AND REPRODUCTIVE BIOLOGY OF AN UNDERSTORY NEOTROPICAL AROID

Helen J. Young

ABSTRACT

Dieffenbachia *(Araceae) is a rhizomatous herb of neotropical rain forests, pollinated by nine species of large scarab beetles in the genera* Cyclocephala *and* Erioscelis. *The pollination biology of* D. longispatha *was studied at the La Selva Biological Station in the Atlantic lowlands of Costa Rica. The protogynous, monoecious inflorescence increases in temperature on the evening when stigmas become receptive. Beetles arrive at dusk and spend the next 24 hours eating protein-rich staminodia surrounding the stigmas and mating. As the anthers release pollen on the following evening, beetles crawl over the male flowers, become covered in pollen, and fly away. Over 8000 beetles were marked during three flowering seasons, and about 25% were recaptured. Information about the floral constancy, flight distances, and behaviour of each beetle species suggests that these beetles are advanced and reliable pollinators. Their long distance flights between inflorescences result in extensive pollen flow and large neighbourhood areas of* Dieffenbachia. *Male reproductive success (the number of seeds sired) is estimated for each inflorescence from a variety of morphological and ecological parameters. Male and female success are unrelated to the numbers of male and female flowers possessed by each inflorescence. Male success is positively related to the number of visiting beetles for 2 years; female success is positively related to the beetle number in 1982, but negatively related to beetle number in 1983. Factors affecting inflorescence size and sex ratio are discussed.*

INTRODUCTION

Recent studies have left no doubt that beetles are important pollinators in the tropical forest understorey (Beach, 1982; Essig, 1971; Gibbs *et al.*, 1977; Gottsberger and Amaral, 1984; Mora Urpi and Solis, 1980; Schatz, 1985; Valerio, 1984; Irvine and Armstrong, this volume). The biology of cantharophily is still in its youth, with studies concentrating on documenting the floral biology of beetle-pollinated plants and the mechanics of pollination. Yet, pollinators play a major role in the evolution of the plants they visit, influencing the movement of genes (via pollen) within and between populations, and determining the reproductive success of those plants.

The size of a genetic neighbourhood is used as an index of the degree of genetic substructure of populations. Estimates of neighbourhood sizes (number of individuals involved in random mating) can be made from information on the distance genes travel in one generation, via both pollen and seeds (Wright, 1943). Species made up of numerous small genetic neighbourhoods have the potential for adaptation to local habitats, because of low migration of genes between populations. In contrast, large genetic populations represent large numbers of individuals involved in panmixis, retarding adaptation to microhabitats within the population. The size of genetic neighbourhoods will influence the degree of inbreeding, the level of heterozygosity, and the importance of random genetic drift in a population. Neighbourhood sizes of tropical plants are largely unknown.

By carrying pollen from one plant to another, pollinators determine the reproductive success of the plants they pollinate. Reproductive biology of plants has tended to equate female reproductive success with total fitness, disregarding the role of pollen donation as a component of fitness. Hermaphroditic species contribute both male and female gametes to the next generation, and floral features such as petal size and colour and nectar quantity and quality are likely to be under selection pressures for their role in both female and male reproductive success (Bell, 1985; Queller, 1983; Stanton *et al.*, 1986). Again, little is known about the role of pollen donation in the total reproductive success of plants, with tropical plants being no exception.

The purpose of this contribution is to address several questions:

(1)   Who are the pollinators of *Dieffenbachia longispatha* and how is pollination effected?

(2)   What is the neighbourhood size (an estimate of the number of ramets that mate randomly) of *D. longispatha*? and,

(3)　　What ecological and morphological factors of *D. longispatha* affect
　　　　the success of each ramet to act as maternal and paternal parents?

Dieffenbachia longispatha Engler and Krause (Araceae) is a monoecious,
perennial herb of neotropical rain forests, occurring in both disturbed and
primary forests. The growth form consists of ramets connected by thick
rhizomes, each ramet bearing numerous cauline leaves. Inflorescences are
produced in the axis of the youngest leaf. The inflorescence consists of a
spadix bearing female flowers at the base and male flowers at the tip, and a
fleshy spathe that surrounds the spadix except during flowering. The
spadices are protogynous, the female flowers being receptive 24 h before the
male flowers. Reproductive ramets produce two to seven inflorescences in a
flowering season. Inflorescences on a ramet generally do not overlap in
flowering so there is little potential for geitonogamy. Beetle numbers within
inflorescences were almost three times higher in 1983 than in 1982 (Young,
1986a; $\bar{x} = 8.7$ in 1983, $\bar{x} = 3.2$ in 1982), providing a natural experiment to
test the effect of beetle number on number of fruits produced.

METHODS

This study was conducted from 1982 to 1984 at the La Selva Biological
Station in the Atlantic lowland rain forests of Costa Rica. Some 1017 ramets
of *Dieffenbachia longispatha* in an area of 70 000 m² were marked and
mapped. The flowering season extended from March to September, during
which period reproductive ramets were monitored daily. For a sample of
inflorescences, the relationships between the length of the spadix devoted to
male flowers and the number of male flowers (N = 25), and between the length
of spadix with female flowers and the number of female flowers (N = 84),
were determined. As each inflorescence opened (when no sexual parts were
receptive and before visitation by insects), I measured the length of the spadix
devoted to male flowers and to female flowers. On the following morning, the
number, identity and sex of visiting scarab beetles were determined. Beetles
were marked with a unique series of small notches cut into the elytra and
replaced within the inflorescence. This marking technique allowed me to
document the flight distances between visits to consecutive inflorescences.

As part of a community study of the scarab beetle-pollinated plants at La
Selva, George Schatz and I documented visitation to approximately 60
species of plants. Recapture of marked beetles permitted us to estimate the
floral constancy of each species to *D. longispatha*.

Mature infructescences were collected for inflorescences that flowered in
1982 and 1983 and the numbers of fruits and undeveloped ovules were
determined for each fruit that matured during the 9 months following

pollination. Therefore, for each inflorescence produced, I know its spatial location within the population, the dates that it was female and male, the estimated number of male and female flowers, the number and identity of visiting beetles, and the number of fruits produced. The beetle recapture data provide a frequency distribution of beetle movement distances between visits to inflorescences.

To calculate the neighbourhood size and area, I used a modification of Wright's (1946) equation (Beattie and Culver, 1979; Wright, 1969) which incorporates the leptokurtic nature of beetle flight distances.

RESULTS

Natural history

The phenology of individual inflorescences of *D. longispatha* involves 3 days (see Young, 1986*a* for details). Inflorescences open in the evening, and, for the first 24 hours, no sexual parts are receptive and no insects visit the inflorescence. In the evening of the second day, the stigmas become receptive (at about 17:30 h), the spadix increases in temperature up to 4°C and insects arrive, alighting on the spadix. Table 11.1 lists the insect visitors to inflorescences of *D. longispatha*. By examination of pollen loads on arriving insects, I have determined that only the scarab beetles are pollinators; the Diptera, Hemiptera, Dermaptera, Thysanoptera, and nitidulid beetles do not carry pollen and should be referred to only as visitors.

The beetles are rewarded with fleshy food in the form of protein-rich staminodia surrounding the stigmas (Young, 1986*a*). Most beetles remain

Table 11.1 Insect visitors to inflorescences of *Dieffenbachia longispatha*

Coleoptera	Diptera
	Drosophilidae
Scarabaeidae	Richardiidae
Cyclocephala amblyopsis Bates	
C. gravis Bates	Hemiptera
C. sexpunctata Bates	Miridae
C. tutilina Burm.	
C. conspicua Sharp	Dermaptera
C. kaszabi Endrödi	
C. ligyrina Bates	Thysanoptera
C. atripes Bates	
C. ampliata Bates	
Erioscelis columbica Endrödi	
Nitidulidae	
Staphylinidae	

within the inflorescence for 24 hours, eating staminodia and mating. On the evening of the third day, the anthers release pollen, the spathe begins to tighten around the female flowers and the beetles crawl up the spadix, becoming covered with pollen, and fly off into the darkness. The staminodia serve the important function of feeding the beetles during their 24-hour visit, encouraging them to remain within the inflorescence until the anthers dehisce. When I removed the staminodia from 36 inflorescences before beetle visitation, beetles visited the inflorescences, but all beetles departed from the inflorescences within 12 hours, before pollen was released.

Neighbourhood size

The number of inflorescences open on any given day is low; the number of inflorescences in female phase is even lower due to the temporally changing sex expression (Figure 11.1). Most days have four or fewer inflorescences in female phase in an area of 70 000 m^2. For beetles to visit these inflorescences, they frequently must travel long distances.

Beetles fly an average of 83 m between consecutive visits to *D. longispatha* (range 1–680 m); however, most of these flights represent movements to the nearest neighbour with an inflorescence in female phase (Young, 1986a). The flight distributions of *Cyclocephala gravis* and *C. amblyopsis* exhibit significant leptokurtosis (Young, 1988), therefore calculations of neighbourhood size assuming a normal distribution of pollen will result in underestimates of N_e. The equation used to calculate N_e (Beattie and Culver, 1979; Wright, 1969) is:

$$N_e = \frac{2^{2\alpha}(\Gamma(2\alpha+1)\Gamma(\alpha))}{2\Gamma(3\alpha)} \pi \sigma^2 d$$

where d = the density of reproductive ramets, σ^2 = the variance of pollen dispersal distance (pollinator movement distances), Γ = gamma function, and α is estimated from:

$$\gamma+3 = \frac{\Gamma(\alpha)\,\Gamma(5\alpha)}{(\Gamma(3\alpha))^2}$$

where γ = the kurtosis value of the beetle flight distributions.

Neighbourhood size and area (effective neighbourhood size divided by the density) were calculated for each year separately because reproductive ramet density differs between years (Table 11.2). The resulting estimate of number of randomly mating ramets (227–611) is at least as high, and much higher in some cases, than has been previously reported for other plant species (Table 11.3). The area of random mating (9–18 ha) is much larger than that reported for temperate plant species.

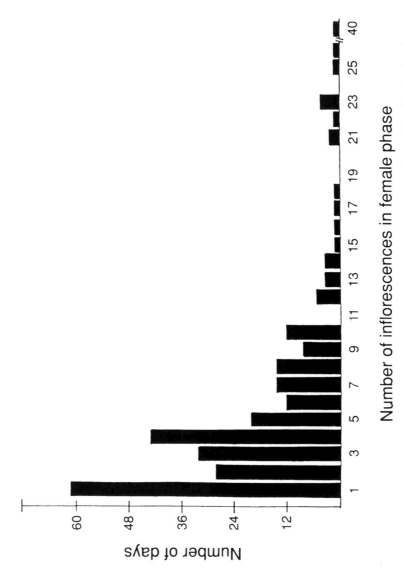

Figure 11.1 Frequency distribution of the number of inflorescences in female phase by day, for 1982–1984 combined

Table 11.2 Neighbourhood size and neighbourhood area of *Dieffenbachia longispatha* for 3 years

Year	Density of reproductive ramets (per m^2)	Kurtosis	N_e (number of ramets)	Area (m^2)
1982	0.0059	1.54	556	94 227
1983	0.0026	1.97	227	87 308
1984	0.0034	8.39	611	179 848

N_e is based on pollen dispersal distances alone, inferred from pollinator flight distances. This calculation incorporates values of kurtosis of flight distances of all beetle species combined, calculated for each year separately

Table 11.3 Estimates of genetic neigbourhood size (N_e) for several herbs and trees

	N_e	References
Herbs		
Borrichia frutescens	20–30	Antlfinger (1982)
Lupinus texensis	42	Schaal (1980)
Viola spp.	102–470	Beattie and Culver (1979)
Senecio spp. (bees)	8–24	Schmitt (1980)
Senecio spp. (butterflies)	990–6150	Schmitt (1980)
Carduus nutans (pollen)	126–378	Smyth and Hamrick (1987)
Carduus nutans (pollen and seed)	1281–3844	Smyth and Hamrick (1987)
Trees		
Ulmus americana	253	Wright (1953)
Fraxinus pennsylvanica	16	Wright (1953)
Cedrus atlantica	208	Wright (1953)
Pinus radiata	1–3200	Bannister (1965)

Male and female reproductive success

Female reproductive success is indexed by the number of fruits matured on each inflorescence. Male success for each inflorescence is estimated from several parameters: the proportion of all beetles in the population that visit a particular inflorescence, the relative floral constancy of the visiting beetles to *D. longispatha*, the number of potential mates (inflorescences in female phase) available, the distances to those inflorescences, the probabilities that individuals of each beetle species will move to those female inflorescences, and the number of fruits produced by those inflorescences. A value was calculated from these ecological parameters to estimate the number of seeds sired by each inflorescence (Young, 1986*b*). The floral constancy of the three major pollinators is estimated as the

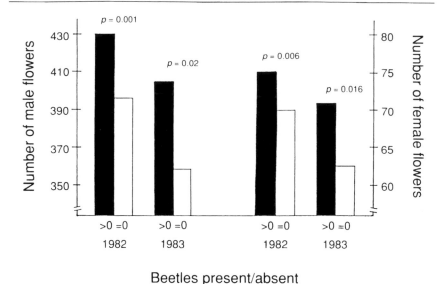

Figure 11.2 Number of male and female flowers possessed by inflorescences that were and were not visited by beetles for 1982 and 1983. In all cases, unvisited inflorescences possessed significantly fewer flowers than visited inflorescences. 1982: N = 72 unvisited infl., N = 168 visited infl.; 1983: N = 10 unvisited infl., N = 262 visited infl.

proportion of all recapture events (recaptures of beetles in all the beetle pollinated plants monitored) that represent movements between inflorescences of *D. longispatha* (0.825 for *Cyclocephala amblyopsis*, 0.962 for *C. gravis*, and 0.941 for *Erioscelis columbica*).

For both years, inflorescences that received visitation from beetles had significantly more male flowers and more female flowers than inflorescences that were not visited (Figure 11.2). Therefore, visited inflorescences were larger than unvisited inflorescences. This result suggests that inflorescences may need to be larger than a minimum size to receive visits by pollinators.

In an analysis of beetle numbers from inflorescences that were visited, the regression of beetle number on flower number is significantly positive for inflorescences flowering in 1983, but the relationship is not significant for 1982 (Figure 11.3). In 1983, large inflorescences were visited by more beetles than smaller inflorescences.

For visited inflorescences, estimated male success (the number of seeds sired by each inflorescence) is significantly positively related to the number of visiting beetles in both years (Figure 11.4). Female success is significantly positively related to beetle number in 1982, but significantly negatively related to beetle number in 1983 (Figure 11.4). However, in

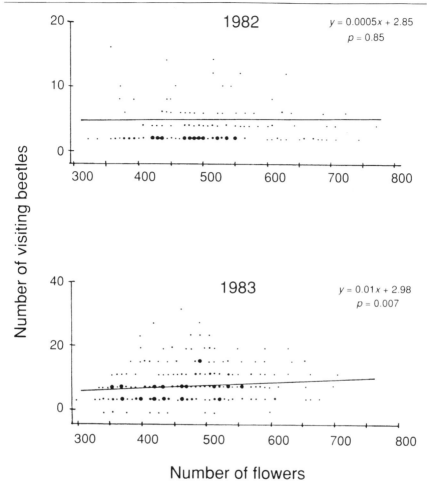

Figure 11.3 Number of beetles visiting inflorescences as a function of the total number of flowers. Only inflorescences visited by at least one beetle are included. In 1982, the relationship is not significant; in 1983, the two parameters are significantly positively related. N=168 for 1982; N=262 for 1983. The size of the symbol is representative of the number of points at each coordinate

neither year is male success for each inflorescence related to the number of male flowers on each inflorescence, nor is female success related to the number of female flowers (Table 11.4). Thus, despite the positive correlations between the number of male flowers and the number of visiting beetles, allocation to male function is a poor predictor of the number of seeds sired. This is because male success is limited by the number of female inflorescences in the population, as well as by the number of beetles carrying pollen away from a male-phase inflorescence. It should be

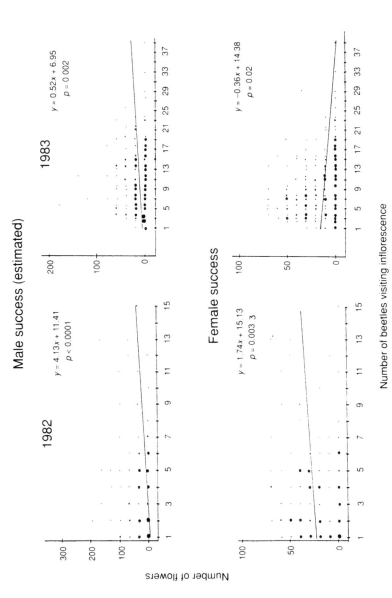

Figure 11.4 Male and female success as functions of the number of visiting beetles. The unit for female success is number of fruits (seeds) on an inflorescence. The unit for male success is the estimated number of seeds sired. Only visited inflorescences are included. The size of the symbol is representative of the number of points at each coordinate

Table 11.4   Results from regression analysis of success on flower number

	Inflorescences visited by beetles	
	1982	1983
Female success vs. no. female flowers	$F = 0.16$	$F = 1.53$
	$p = 0.69$	$p = 0.22$
Male success vs. no. male flowers	$F = 0.17$	$F = 0.19$
	$p = 0.68$	$p = 0.67$

Only inflorescences visited by beetles are included (N = 168 for 1982; N = 262 for 1983). F-values and the probability values associated with them are given

emphasized that I did not perform genetic analyses of the adults and seeds in the population, and, therefore, I do not know which male flowers sired which seeds. The number of seeds sired was estimated from parameters assumed to be correlated with male success (the number of pollinators, the number of female flowers available, the number of seeds produced).

DISCUSSION

The reproductive biology of *Dieffenbachia* is characteristic of cantharophilic species: inflorescences are protogynous, the inflorescence volatilizes odour by increasing in temperature, non-pollen food serves as the reward, and the changes in floral receptivity are nocturnal (Beach, 1982; Cramer *et al.*, 1975; Gottsberger and Amaral, 1984; Prance and Arias, 1975; Valerio, 1984). Although visited by numerous taxa of insects, *D. longispatha* is pollinated only by scarab beetles.

Because the density of flowering ramets is low and flowering occurs over an extensive period, inflorescences in any particular sexual phase are geographically quite distant. This feature results in extensive movements of beetles between inflorescences. Neighbourhood sizes are large as a consequence of long distances between available mates and long distance pollinator movements. These large areas of random mating preclude the possibility of adaptation to local habitats. It is also likely that random changes in gene frequency will not play an important role in the evolution of this species because mating occurs among so many individuals and over such a wide area.

Although it is known that pollen is required for fertilization of ovules, the factors affecting pollen donation and pollen success are little understood. There is good reason for the paucity of knowledge concerning male success. Microgametophytes are small; their journey from anther to stigma to ovule is difficult to follow. Genetic studies involving electrophoresis have

161

contributed substantially by revealing the extent of pollen flow (Handel, 1983; Schaal, 1980). But to determine the success of an individual plant as a paternal parent requires knowledge of the genetic identity of all seeds produced in a panmictic population.

Paternity exclusion studies (Ellstrand, 1984; Meagher, 1986; Schoen and Stewart, 1986) have indicated that successful determination of paternity of seeds can only be accomplished with very small populations of plants and populations that are electrophoretically polymorphic. For a large plant population, male success can only be estimated, using information about the proportion of all male flowers (or pollen grains) in the population that a given plant possesses (Devlin and Stephenson, 1987; Garnock-Jones, 1986; Lloyd, 1980; Primack and Lloyd, 1980; Primack and McCall, 1986), which assumes that all pollen grains produced in the population have an equal probability of siring the seeds produced. Modifications of this estimate include incorporating the temporal effects of changing sex ratio in the population (Thomson and Barrett, 1981). None of these studies consider the effect of pollinators on male success: the number of pollinators visiting a flower, the number of pollen grains removed by each pollinator visit, the floral constancy of that pollinator taxa to the plant species under study.

Each inflorescence of *Dieffenbachia* has the capability to act as both a maternal and a paternal parent (as the source of both ovules and pollen for the next generation). Male success in *D. longispatha* has been estimated in this study using information on the number of beetles visiting an inflorescence, the probability that each beetle will visit an inflorescence in female phase after leaving a male-phase inflorescence, the number of female-phase inflorescences in the population and the distances to those inflorescences. Surprisingly, neither male nor female success of inflorescences is related to the number of male and female flowers an inflorescence possesses. Producing more male flowers (and thus more pollen) does not increase the estimate of the number of seeds sired. This is because the main factors affecting male success are the number of beetle visitors and the number of female inflorescences available to receive pollen (Young, 1986*b*). It appears that, instead of gametes limiting the number of seeds produced, beetles are the limiting factors. Beetle number is positively related to the estimated number of seed sired in 1982 and 1983, and positively related to female success in 1982. However, the regression coefficient of beetle number and female success is negative in 1983.

There appears to be a conflict between selection on male and female flowers. Male success can be hypothesized to increase with increases in the size of the male part of the spadix because larger inflorescences attract more beetles, which will disperse more pollen. However, large numbers of beetles do not guarantee high female success (as the negative relationship between beetle number and number of fruits produced in 1983 shows). Selection is acting on total fecundity, without regard to relative numbers of seeds sired or mothered by a particular plant. If natural selection has influenced the size of inflorescences and the sex ratio of flowers on inflorescences, then the phenotypic gender of inflorescences of *D. longispatha* may reflect maximum fecundity. Yet, a major factor influencing male success is stochastic with respect to male flower allocation (i.e. the number of female flowers receptive at any given time), so selection may not be acting to 'fine-tune' allocation to male gametes. Inflorescences with no male flowers will not sire seeds, but, beyond that, because male success is not limited by the number of pollen grains produced, selection may not be a strong force in determining allocation to male flowers in *Dieffenbachia*.

What are the factors that influence the sex ratio of *Dieffenbachia* inflorescences? Both male and female functions play a role in pollinator attraction. Floral odours are produced primarily at the base of the male portion of the spadix (Nagy *et al.*, 1972) and are important in long-distance attraction of pollinators. Once the pollinators arrive at an inflorescence, they are rewarded with fleshy staminodia that are located in the female region of the spadix. A male sterile inflorescence will not attract pollinators; a female sterile inflorescence will not offer rewards that lure pollinators to stay. For inflorescences of *Dieffenbachia*, reproductive success in a particular sexual function is dependent on allocation to both male and female flowers. Heat production and staminodia may be considered "shared costs" to male and female function, without which no reproductive success in either sexual function would be achieved.

ACKNOWLEDGEMENTS

Many thanks are extended to J. Thomson, D. Futuyma, B. Bentley, R. Primack, M. Stanton, G. Schatz, J. Beach, L. Goldwasser, B. Ratcliffe, T. Young, and D. Stratton for help of all kinds. This study was supported by a fellowship from the Jessie Smith Noyes foundation (administered through the Organization for Tropical Studies) and a grant from Sigma Xi.

REFERENCES

Antlfinger, A.E. (1982). Genetic neighborhood structure of the salt marsh composit, *Borrichia frutescens*. *Journal of Heredity*, **73**, 128-32
Bannister, M.H. (1965). Variation in the breeding system of *Pinus radiata*. In Baker, H.G. and Stebbins, G.L. (eds) *The Genetics of Colonizing Species*, pp. 353-72 (New York: Academic Press)
Beach, J.H. (1982). Beetle pollination of *Cyclanthus bipartitus* (Cyclanthaceae). *American Journal of Botany*, **69**, 1074-81
Beattie, A.J. and Culver, D.C. (1979). Neighborhood size in *Viola*. *Evolution*, **33**, 1226-9
Bell, G. (1985). On the function of flowers. *Proceedings of the Royal Society of London*, Series B, **224**, 223-65
Cramer, J.M., Meeuse, A.D.J. and Teunissen, P.A. (1975). A note on the pollination of nocturnally flowering species of *Nymphaea*. *Acta Botanica Neerlandica*, **24**, 489-90

Devlin, B. and Stephenson, A.G. (1987). Sexual variation among plants of a perfect-flowered species. *American Naturalist*, **130**, 199–218

Ellstrand, N.C. (1984). Multiple paternity within the fruits of the wild radish, *Raphanus sativus*. *American Naturalist*, **123**, 819–28

Essig, B.F. (1971). Observations on pollination of *Bactris*. *Principes*, **15**, 20–4

Garnock-Jones, P.J. (1986). Floret specialization, seed production and gender in *Artemisia vulgaris* L. (Asteraceae, Anthemideae). *Biological Journal of the Linnean Society*, **92**, 285–302

Gibbs, P.E., Semir, J. and Cruz, N. (1977). Floral biology of *Talauma ovata* St. Hil. (Magnoliaceae). *Ciencia e Cultura*, **29**, 1436–41

Gottsberger, G. and Amaral, A. (1984). Pollination strategies in Brazilian *Philodendron* species. *Berichte der Deutschen Botanischen Gesellschaft*, Band **97**, 391–410

Handel, S.N. (1983). Contrasting gene flow patterns and genetic subdivision in adjacent populations of *Cucumis sativus* (Cucurbitaceae). *Evolution*, **37**, 760–71

Lloyd, D.G. (1980). Sexual strategies in plants. III. A quantitative method for describing the gender of plants. *New Zealand Journal of Botany*, **18**, 103–8

Meagher, T.R. (1986). Analysis of paternity within a natural population of *Chamaelirium luteum*. I. Identification of most likely male parents. *American Naturalist*, **128**, 199–215

Mora Urpi, J. and Solis, E. (1980). Polinizacion en *Bactris gasipaes* H.B.K. (Palmae). *Revista de Biologia Tropical*, **28**, 153–74

Nagy, K.A., Odell, D.K. and Seymour, R.S. (1972). Temperature regulation by the inflorescence of *Philodendron*. *Science*, **178**, 1195–7

Prance, G.T. and Arias, J.R. (1975). A study of the floral biology of *Victoria amazonica* (Poepp.) Sowerby (Nymphaeaceae). *Acta Amazonica*, **5**, 109–39

Primack, R.B. and Lloyd, D.G. (1980). Sexual strategies in plants. IV. The distributions of gender in two monomorphic shrub populations. *New Zealand Journal of Botany*, **18**, 109–14

Primack, R.B. and McCall, C. (1986). Gender variation in a red maple population (*Acer rubrum*; Aceraceae): a seven-year study of a "polygamodioecious" species. *American Journal of Botany*, **73**, 1239–48

Queller, D.C. (1983). Sexual selection in a hermaphroditic plant. *Nature*, **305**, 706–7

Schaal, B.A. (1980). Measurement of gene flow in *Lupinus texensis*. *Nature*, **284**, 450–1

Schatz, G.E. (1985). A new *Cymbopetalum* (Annonaceae) from Costa Rica and Panama with observations on natural hybridization. *Annals of the Missouri Botanical Garden*, **72**, 535–8

Schmitt, J. (1980). Pollinator foraging behavior and gene dispersal in *Senecio* (Compositae). *Evolution*, **34**, 934–42

Schoen, D.J. and Stewart, S.C. (1986). Variation in male reproductive investment and male reproductive success in White Spruce. *Evolution*, **40**, 1109–20

Smyth, C.A. and Hamrick, J.L. (1987). Realized gene flow via pollen in artificial populations of musk thistle, *Carduus nutans* L. *Evolution*, **41**, 613–19

Stanton, M.L., Snow, A.A. and Handel, S.N., (1986). Floral evolution: attractiveness to pollinators increases male fitness. *Science*, **232**, 1625–6

Thomson, J.D. and Barrett, S.C.H. (1981). Temporal variation in gender in *Aralia hispida* Vent. (Araliaceae). *Evolution*, **35**, 1094–107

Valerio, C.E. (1984). Insect visitors to the inflorescence of the aroid *Dieffenbachia oerstedii* (Araceae) in Costa Rica. *Brenesia*, **22**, 139–46

Wright, J.W. (1953). Pollen-dispersal studies: some practical applications. *Journal of Forestry*, **51**, 114–18

Wright, S. (1943). Isolation by distance. *Genetics*, **28**, 114–38

Wright, S. (1946). Isolation by distance under diverse systems of mating. *Genetics*, **31**, 39–59

Wright, S. (1969). *Evolution and the Genetics of Populations*. **Volume 2**. *The Theory of Gene Frequencies*. (Chicago: University of Chicago Press)

Young, H.J. (1986*a*). Beetle pollination of *Dieffenbachia longispatha* (Araceae). *American Journal of Botany*, **73**, 931–44

Young, H.J. (1986*b*). Pollination of *Dieffenbachia longispatha*: effects of beetles on reproductive success, gene flow, and gender. Ph.D. Dissertation. State University of New York, Stony Brook, New York.

Young, H.J. (1988). Neighborhood size in a beetle pollinated tropical aroid: effects of low density and asynchronous flowering. *Oecologia*, **58**, 373–7

CHAPTER 12

POLLEN GRAIN DEPOSITION PATTERNS AND STIGMA STRATEGIES IN REGULATING SEED NUMBER PER POD IN MULTI-OVULATED SPECIES

R. Uma Shaanker and K.N. Ganeshaiah

ABSTRACT

Prefertilization strategies may lead to negatively skewed distribution of seeds per pod in species with multiple ovules per ovary (multi-ovulated species). Comparing a set of multi-ovulated species that exhibited negative skew of seeds per pod with that of uni-ovulated species, we found that the pollen grain deposition patterns and the stigmatic inhibition of pollen grain germination regulate the seed number per pod in multi-ovulated species. Stigmas of multi-ovulated species generally receive more than enough pollen grains to fertilize all the ovules in an ovary, thereby preventing the formation of pods with few seeds. Number-dependent stigmatic inhibition of pollen grain germination prevented the development into fruits of flowers containing less than a critical load of pollen grains. Data on stigma pollen loads indicate that the pollen grain deposition patterns probably serve more as a means to ensure economic packing of seeds than as mechanisms for coping with sexual selection. Female choice of paternal genotype is shown to occur at the postfertilization stage in multi-ovulated species.

INTRODUCTION

Many multi-ovulated species exhibit negatively skewed distribution of seed number per fruit, with infrequent few seeded fruits (Bawa and Webb, 1984; Lee, 1984; Ganeshaiah *et al.*, 1986; Ganeshaiah and Uma Shaanker, 1988*a*). This distribution has been attributed to the economy of packing the seeds in fruits (Ganeshaiah *et al.*, 1986; Janzen, 1982; Lee and Bazzaz, 1982*a*, 1982*b*; Willson and Schemske, 1980). The negatively skewed distribution is accomplished by the selective abortion of few seeded fruits (Bookman, 1984; Lee and Bazzaz, 1982*a*; Stephenson, 1978). However, fruit abortion is energetically expensive as the maternal parent incurs a cost proportional to the extent of development of the aborted fruits. In contrast, pre-fertilization regulatory mechanisms are relatively less expensive as they can completely prevent the formation of few seeded fruits (Ganeshaiah *et al.*, 1986; Ganeshaiah and Uma Shaanker, 1988*a*).

The pre-fertilization regulatory mechanisms may involve the pollen grain deposition patterns and/or the stigmatic regulation of pollen grain germination. First, as regards pollen grain deposition patterns, the prevention of formation of few seeded fruits in multi-ovulated species might be facilitated by the deposition of large loads of pollen grains so that pollen grain number per stigma is equal to, or more than, the ovule number per ovary.

Second, in respect to stigmatic regulation of pollen grain germination, the seed number per pod can also be regulated by preventing pollen grain germination until a critical load (equal to or more than the ovule number) of pollen grains is deposited on the stigma. This state might be attained by delaying the stigma receptivity until it receives the critical load of pollen grains (Murdy and Carter, 1987) or by a number-dependent inhibition of pollen grain germination by the stigma (Ganeshaiah *et al.*, 1986; Ganeshaiah and Uma Shaanker, 1988*a*). Thus, flowers receiving more than this critical number of pollen grains develop into fruits while those receiving less do not.

We show that, in contrast to uni- or few-ovulated species, multi-ovulated species have evolved specific pre-fertilization regulatory (PFR) mechanisms involving distinct patterns of pollen grain deposition and stigmatic control over pollen grain germination to regulate the seed number in fruits. These PFR mechanisms can also be interpreted as female strategies for inciting male competition because they generate an intense gametophytic competition on the stigmatic surface. We discuss also the alternative selective forces affecting the evolution of PFR mechanisms.

MATERIALS AND METHODS

The study was conducted from May 1984 to April 1986 on species available in the open grasslands and shrub forests adjoining the University Campus, Bangalore (12°58'N, 77°35'E) and at Sangama Valley, Kanakapura, about 100 km east of Bangalore. The species studied and their sample sizes are given in Tables 12.1 and 12.2. Data on the pollen grain number per stigma of the uni-, oligo- (2–3 ovules per ovary) and multi-ovulated (> 4 ovules per ovary) species were collected from randomly selected flowers which were completely open and about to dry. Excised stigmas were stained in 1% acetocarmine and mounted, and pollen grains were counted by squeezing the stigmas under the microscope. The skewness value for the frequency distribution of log pollen grain number on the stigma was computed, following Sokal and Rohlf (1969). The ovule number per ovary of the same flowers was also counted by dissecting the flower under a Trycon 100 Z stereo-microscope. From this number, the pollen grain number per stigma, and per ovule, and the percentage of pollen grains germinated on the stigma, were computed for each species. As an estimate of the extent of male gametophytic competition for every ovule fertilized, the number of pollen grains on the stigmas of 15 to 20 old flowers (potential fruits) that persisted even after vigorously tapping the inflorescence were counted and expressed as pollen grain number per stigma per ovule. The seed to ovule ratio was estimated to indicate the extent of embryo abortion (Wiens, 1984).

RESULTS

Because the number of pollen grains per stigma was highly variable, the distribution patterns were plotted on a log scale (Figure 12.1 to 12.5). The pollen grain deposition patterns of the multi-ovulated species were distinctly different from that of the uni- or few-ovulated species. Whilst most of the uni- or few-ovulated species considered were wind pollinated and exhibited a log-normal distribution (Figure 12.1), the multi-ovulated species exhibited the following three distinct patterns.

Negatively skewed distribution

This pattern (Figure 12.2) was seen in *Cassia sericia, Peltophorum feruginium* and several other species (Table 12.1). It is characterized by the relative absence of flowers whose pollen loads per stigma were less than the ovule number per ovary, thus preventing the formation of pods with few seeds. The modal pollen load exceeded the number of ovules per ovary. Interestingly, this pattern is a general feature of buzz-pollinated species (Buchmann, 1983).

Discontinuous bimodal distribution

This pattern is essentially similar to the negatively skewed distribution except that a certain proportion of flowers do not receive pollen grains at all; pollinated flowers always received pollen grains in excess of the number of ovules per ovary, resulting in pods with many seeds. This distribution is found in buzz-pollinated species such as *Cassia fistula* (Figure 12.3) and others (see Table 12.1), and is characterized by a simultaneous deposition of pollen grains.

Table 12.1 Pollen grain deposition patterns, pollination vectors, pollen grain number per stigma (PGS) and PGS per ovule in the species studied

Sl.No.	Species	N	PGS X±SD	Skewness	Deposition pattern*	Mode of pollination	Ovule number/ovary	PGS/ovule in persistent flowers
1	Poaceae							
	Heteropogon contortus	42	87.02 ± 69.46	− 0.403	Normal	Wind	1	87.02
	Panicum maximum	61	19.52 ± 13.78	− 0.515	Normal	Wind	1	19.52
	Rhynchelytrum repens	40	39.52 ± 36.00	+ 0.103	Normal	Wind	1	39.52
2	Asteraceae							
	Hymenantherum tenuifolium	51	14.64 ± 13.21	− 0.259	Normal	Wind	1	14.64
	Tridax procumbens	41	30.34 ± 17.27	− 0.186	Normal	Wind	1	30.34
3	Euphorbiaceae							
	Croton bonpladianum	96	12.52 ± 12.12	− 0.002	Normal	Wind	3	4.16
	Euphorbia geniculata	13	32.00 ± 32.04	− 0.510	Normal/Negative	Wind/Insect	3	10.60
	Phyllanthus asperculatus		S	− 0.001	Normal	Wind	6	4.29
	P. gardenarianus		S	− 0.003	Normal	Wind	6	2.50
	P. madraspatensis		S	− 0.003	Normal	Wind	6	6.59
	P. niruri		S	− 0.002	Normal	Wind	6	3.32
	P. polyphyllus	31	10.38 ± 21.93	− 0.001	Normal	Wind	6	1.73
	P. rheedi		S	− 0.001	Normal	Wind	6	7.50
	P. urinaria		S	− 0.001	Normal	Wind	6	5.00
	P. virgatus		S	− 0.001	Normal	Wind	6	2.49
	P. sp.1.		S	− 0.002	Normal	Wind	6	3.0
	P. sp.2.		S	− 0.002	Normal	Wind	6	3.33
4	Verbenaceae							
	Lantana tiliaefolia	20	88.55 ± 73.04	− 0.829	Negative	Insect	2	14.27
5	Moringaceae							
	Moringa oleifera	100	17.84 ± 41.33	+ 0.839	CBM/positive	Insect	20	4.00
6	Faboideae							
	Clitoria ternatea	13	226.51 ± 53.61	− 0.533	Negative	Insect	35	6.47

(continued)

168

Table 12.1 *continued*

Sl.No.	Species	N	PGS X±SD	Skewness	Deposition pattern*	Mode of pollination	Ovule number/ovary	PGS/ovule in persistent flowers
	Dalbergia sissoo	20	116	−0.540	Normal	Wind/Insect	4.41	26.60
7	Caesalpinioideae							
	Caesalpinia coraria	22	61.82 ± 38.15	−0.148	Negative	Insect(!)	10	6.18
	C. pulcherrima	114	8.78 ± 36.7	+0.451	CBM	Insect	7	3.67
	Cassia fistula	23	28.76 ± 46.18	−1.85	DBM	Insect(!)	120	7.22
	C. sericia	8	210.62 ± 148.83	−0.493	Negative	Insect(!)	8.8	23.86
	C. siami	14	38.15 ± 85.69	−0.543	CBM	Insect(!)	25	4.08
	C. sp.1	35	11.57 ± 35.49		DBM	Insect(!)	80–100	1.03
	C. sp.2	62	39.30 ± 43.04	−1.709	DBM	Insect(!)	22	2.60
	Delonix regia	19	75.05 ± 85.60	−0.561	Negative	Bird/Insect	48.66	2.41
	Peltophorum ferruginium	8	481.25 ± 106.3	−0.983	Negative	Insect	3.41	96.25
8	Solanaceae							
	Solanum pubescens	25	203.88	−0.681	Negative	Buzz	26.0	7.84
	S. xanthophyllum	17	253.50 ± 76.9	−0.459	Negative	Insect	35	7.44
9	Mimosoideae							
	Albizzia odaratissima	21	0.52 ± 0.81	+0.719	Positive	Insect	13	1.79
	Lucaena leucocephala	111	21.21 ± 24.4	+1.103	CBM	Wind/insect	28	1.66
	Pithecellobium saman	12	0.83 ± 1.02	+0.851	Positive	Insect	19	2.93
10	Bignoniaceae							
	Jacaranda mimosifolia	15	96 ± 97.78	–	Negative	Insect		
11	Malvaceae							
	Hibiscus rosa-sinesis	31	54.67 ± 34.27	–	Negative	Insect		

CBM – continuous bimodal distribution; DBM – discontinuous bimodal distribution. * – The pollen grain deposition patterns were classified based on the frequency distribution of pollen grains on stigma on log scale and the skewness value. $ – Data from Uma Shaanker and Ganeshaiah, 1984b. ! – these species are buzz pollinated

Table 12.2   Ovule number per ovary and seed to ovule ratio in a few explosively dispersed species

Sl.No.	Family and species	Ovules per ovary		Seed to ovule ratio	
		n	x	n	x
	Faboideae				
1	*Atylosia lineata*	10	2.0	41	0.98
2	*A. scarabaeoides*	10	6	59	0.77
3	*Cajanus cajan*	10	43	55	0.89
4	*Crotalaria striata*	10	41.6	106	0.80
5	*Crotalaria* sp.1	12	18	50	0.71
6	*C.* sp.2	10	5	20	0.84
7	*Desmodium incanum**	–	6	–	0.81
8	*Gliricidia sepium**	–	10	–	0.45
9	*Lab-lab niger*	16	3.86	69	0.91
10	*Macrotyloma accilare*	21	7	25	0.84
11	*Phaseolus acontifolius*	14	13.7	43	0.81
12	*P.* sp.	20	9.6	25	0.96
13	*Rhincosia sp.*	10	2	60	0.97
14	*Tephrosia purpurea*	23	5	25	0.78
15	*T.* sp.				
		17	7.5	30	0.81
	Caesalpinioideae				
16	*Bauhinia purpurea*	30	11	50	0.90
17	*B. racemosa*	20	19	50	0.54
18	*B. ungulata**	–	18.67	–	0.59
19	*Cassia hirsuta*	5	82.5	30	0.77
20	*C. sericea*	11	8.8	50	0.89
21	*C.* sp.	53	30.4	40	0.71
22	*Caesalpinia decapetala*	14	6.64	25	0.96
23	*C. eriostachys**	–	6.17	–	0.46
	Mimosoideae				
24	*Acacia* sp.	12	8	30	0.93
	Euphorbiaceae				
25	*Kirganelia reticulata*	10	6	35	0.86
	Caparidaceae				
26	*Cleome monophylla*	6	49.66	55	0.73
	Cruciferae				
27	*Brassica nigra*	15	12.0	121	0.78
	Convolvulaceae				
28	*Ipomea muricata*	15	4	39	0.89
29	*I. pericarpa*	12	4	64	0.83

* Data from Bawa and Buckley (1989)

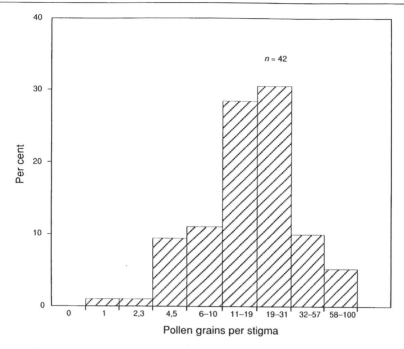

Figure 12.1 Normal distribution of pollen deposition on stigmas (on log scale) in *Panicum maximum*. Note that most flowers receive pollen grains greater than the ovule number per flower (=1). N = Number of flowers

Positively skewed distribution

This pattern was observed in species that exhibited a less efficient reception of pollen grains on the stigma, and showed up either as highly positively skewed (e.g. *Moringa*) or as a continuous bimodal distribution (e.g. *Caesalpinia pulcherrima, Cassia siami*). In *Moringa* (Figure 12.4), the deposition of pollen grains on the stigma ranged from 0 to 156, with as many as 68% of flowers receiving pollen loads that did not exceed the number of ovules per ovary. This load might lead to the formation of pods with few seeds. However, as shown in Figure 12.4, stigmas of this species, as in *Leucaena leucocephala* (Ganeshaiah *et al.*, 1986), exhibit a number-dependent inhibition of germination such that pollen grains are allowed to germinate only when there are more than a critical number on the stigma (Figure 12.4). This regulation prevents the formation of pods with few seeds, leading to a negatively skewed distribution of seeds per pod.

This pattern was also found in *Pithecellobium saman* and *Albizzia lebbek* where pollen grain dispersal occurs through polyads (Figure 12.5). Because, in these species, a single pollen dispersal unit is sufficient to effect complete

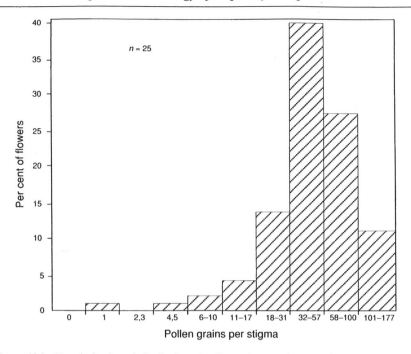

Figure 12.2 Negatively skewed distribution of pollen grains on stigma (on log scale) in *Cassia sericia*. The arrow indicates the number of pollen grains on stigma equal to ovule number per ovary (=8). Note that most flowers receive pollen grains in excess of the ovule number. N = Number of flowers

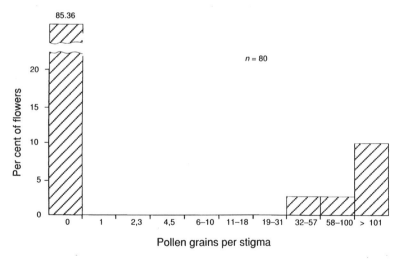

Figure 12.3 Discontinuous bimodal distribution of pollen grain deposition (on log scale) on stigma in *Cassia fistula*. Note that a significant percentage of flowers do not receive pollen grains at all. N = number of flowers

172

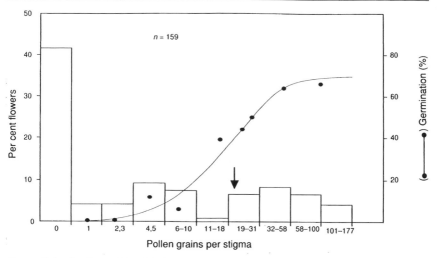

Figure 12.4 Continuous bimodal distribution of pollen grains on stigma on log scale in *Moringa oleifera*. The arrow indicates pollen grains equal to ovule number per ovary (=20). The figure also shows the sigmoidal relation between grain number per stigma and their germination percentage. Flowers receiving less than 20 pollen grains may not set into fruits. N = number of flowers

fertilization of the ovule (Kenrick and Knox, 1982), the positively skewed distribution of a pollen dispersal unit does not lead to the formation of pods with few seeds. For instance, in *P. saman*, a single polyad containing 32 pollen grains can bring about the fertilization of all the 20–25 ovules in the ovary, leading to a negatively skewed distribution of seeds per pod.

As shown in Figure 12.6, the pollen grains per stigma per ovule in the persistent flowers decreased with increase in ovule number per ovary ($r = -0.77$; $p < 0.01$). The seed ovule ratio decreased with increase in ovule number (Figure 12.7) for a set of species with ballistic or explosive seed dispersal mechanism ($r = -0.61$; $p < 0.01$). Only the species with ballistic or explosive dispersal mechanism were considered because the magnitude of abortion differs significantly between these species and those dispersed by either animals or wind (Ganeshaiah and Uma Shaanker, 1988*b*; Uma Shaanker *et al.*, 1988).

DISCUSSION

Regulation of seed number per pod in multi-ovulated species is generally explained by the selective abortion of the pods with few seeds (Bookman, 1984; Lee and Bazzaz, 1982*a*, 1982*b*; Stephenson, 1978). Our results suggest the existence of certain PFR mechanisms that are relatively less expensive in regulating the seed number per pod in the multi-ovulated

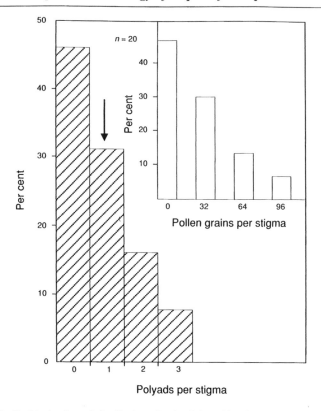

Figure 12.5 Positively skewed distribution of polyad deposition in *Pithcellobium saman*. The arrow indicates the pollen grains in the polyad equal to or more than the ovule number per ovary (=19). Inset shows the frequency of flowers receiving various pollen grain numbers consequent to the polyad deposition pattern. N = number of flowers

species. The negatively skewed or discontinuous bimodal deposition of pollen grains on the stigma prevents the formation of pods with few seeds. Further, deposition of pollen grains in groups (e.g. polyads) of discrete numbers equal to or more than the ovule number assures that each flower receives the minimum number of pollen grains needed to fertilize all its ovules in a single pollination event. In few species (e.g. *Moringa*), where pollen grain deposition is continuous, there exists a number-dependent stigmatic inhibition of pollen grain germination such that the flowers receiving less than a certain number of pollen grains (generally equal to ovule number) do not set fruits. Such number-dependent germination of pollen grains has also been reported in other species, such as *Leucaena leucocephala* (Ganeshaiah *et al.*, 1986; Ganeshaiah and Uma Shaanker, 1988*a*) and *Epilobium* (Snow, 1986).

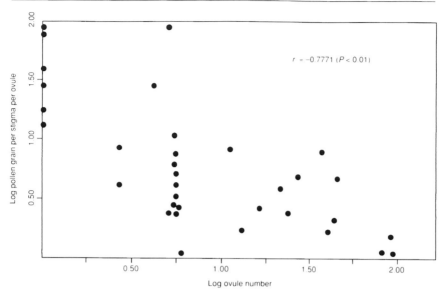

Figure 12.6 Relation between the log pollen grain number per stigma per ovule and log ovule number per ovary

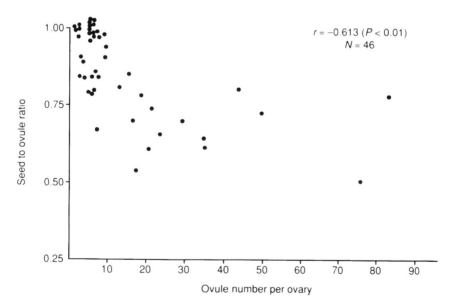

Figure 12.7 Relation between the seed to ovule ratio and the ovule number per ovary. N=number of species; data for seven species are from Bawa and Buckley 1989; for details see Table 12.2. To avoid overlap, a few of the points in the top left with the ratio of approximately 1.00 are plotted beyond the range of the y-coordinate

In contrast to the multi-ovulated species, the deposition of pollen grains in the uni- or few-ovulated species generally follows a log-normal distribution with a low variance to mean ratio. This difference in the pattern of pollen grain deposition appears to be due to differences in pollination strategies. For instance, all buzz-pollinated systems were found to exhibit a negatively skewed or discontinuous bimodal deposition pattern. A majority of the wind-pollinated species show log-normal distribution, compared to the negative skewness resulting from the clumped deposition of pollen grains in insect pollinated systems.

The distinct pollen deposition patterns and the number-dependent stigmatic inhibition of pollen grain germination might have evolved to regulate the seed number per pod, ensuring economic packing of seeds in the multi-ovulated species. However, since all these mechanisms result in deposition of pollen grains on stigma in excess of the ovule number per ovary, a severe gametophytic competition is generated among the germinating pollen grains to fertilize the ovules (Snow, 1986; Willson and Burley, 1983). Hence, it can also be argued that these mechanisms have evolved as female strategies to incite competition among the pollen grains (Ganeshaiah *et al.*, 1986; Ganeshaiah and Uma Shaanker, 1988*a*). By using the pollen grains per stigma per ovule in persistent flowers that set pods as an index of the extent of mate competition generated (Uma Shaanker and Ganeshaiah, 1984, 1986; Namai and Ohsawa, 1986; Vasudev *et al.*, 1987), we found that its magnitude decreases with increase in the number of ovules per ovary (Figure 12.6). In other words, the pollen grain deposition patterns and stigmatic inhibition of pollen grain germination observed in the multi-ovulated species might not reflect sexual selection as much as selection for the economic packing of seeds. This reflection is also clear from the observation of Kenrick and Knox (1982), who showed a close correspondence between the pollen grain number per polyad and the ovule number per ovary in several species of *Acacia* (Pollen:ovule ratio = 1.2). Such association could be better explained as a mechanism evolved to attain complete fertilization of ovules rather than as a result of sexual selection.

Though the foregoing indicates that gametophytic competition decreases in the multi-ovulated species, it might be manifested as zygotic competition in the developing fruits. This is clear from the negative association between the seed to ovule ratio and the ovule number per ovary indicating an increase in embryo abortion in the multi-ovulated species (Figure 12.7). Embryo abortion is advantageous in the multi-ovulated species because it is relatively inexpensive as compared to that in the uni-ovulated species. Abortion of every embryo in the latter is associated with the loss of maternal resources invested on the floral structures containing the embryo, while, in the former, the cost of abortion is just an ovule. Besides, the performance of the paternal genome in the sporophytic generation can be more effectively evaluated at the embryo stage because these embryos are subjected to a severe competition with other sibs in the uniform milieu of the developing fruit. Such embryo selection, as a manifestation of sexual selection at the post-fertilization stage (Mazer, 1987), could be a common feature of the multi-ovulated species.

ACKNOWLEDGEMENTS

The authors thank Dr Kamal Bawa for critically commenting on the manuscript and improving it, and to the Department of Science and Technology, New Delhi, for financial support.

REFERENCES

Bawa, K.S. and Webb, C.J. (1984). Flower, fruit and seed abortion in tropical forest trees: implications for the evolution of paternal and maternal reproductive patterns. *American Journal of Botany*, **71**, 736–51

Bawa, K.S. and Buckley, D.P. (1989). Seed-ovule ratios, selective abortion and mating systems in Leguminosae. In Stinton, C.H. and Zerucchi, K.L. (eds) *Advances in Legume Biology*. Missouri Botanical Garden Monographs in Systematic Botany, **29**, 243 62

Bookman, S. (1984). Evidence for selective fruit production in *Asclepia*. *Evolution*, **38**, 72–86

Buchmann, S.L. (1983). Buzz pollination in angiosperms. In Jones, C.E. and Little, R.K. (eds) *Handbook of Experimental Pollination Biology*, pp. 73–113. (New York: Van Nostrand Reinhold)

Ganeshaiah, K.N., Uma Shaanker, R. and Shivashanker, G. (1986). Stigmatic inhibition of pollen grain germination – its implications for frequency distribution of seed number in pods of *Leucaena leucocephala* (Lam) de Wit. *Oecologia*, **70**, 568–72

Ganeshaiah, K.N. and Uma Shaanker, R. (1988a). Regulation of seed number per pod and female incitation of mate competition by a pH-dependent proteinaceous inhibitor of pollen grain germination in stigmatic fluid in *Leucaena leucocephala*. *Oecologia*, **75**, 110–13

Ganeshaiah, K.N. and Uma Shaanker, R. (1988b). Embryo abortion in a wind dispersed pod of *Dalbergia sissoo*: maternal regulation or sibling rivalry? *Oecologia*, **77**, 135–9

Janzen, D.H. (1982). Variation in average seed size and fruit seediness in a fruit crop of a Gunacaste tree (*Enterolobium cyclocarpum*). *American Journal of Botany*, **69**, 1169–78

Kenrick, J. and Knox, R.B. (1982). Function of the polyad in reproduction of *Acacia*. *Annals of Botany*, **50**, 721–7

Lee, T.D. (1984). Patterns of fruit maturation: a gametophyte competition hypothesis. *American Naturalist*, **123**, 427–32

Lee, T.D. and Bazzaz, F.A. (1982a). Regulation of fruit and seed production in an annual legume, *Cassia fasciculata*. *Ecology*, **63**, 1363 73

Lee, T.D. and Bazzaz, F.A. (1982b). Regulation of fruit maturation pattern in an annual legume, *Cassia fasciculata*. *Ecology*, **63**, 1374–88

Mazer, S.J. (1987). Maternal investment and male reproductive success in angiosperms: parent-offspring conflict or sexual selection. *Biological Journal of the Linnean Society*, **30**, 115–33

Murdy, W.H. and Carter, M.E.B. (1987). Regulation of timing of pollen germination by the pistil in *Talinum mengesii* (Portulaceae). *Americal Journal of Botany*, **74**, 1888–92

Namai, H. and Oshawa, R. (1986). Variation of reproductive success rates of ovule and pollen deposited upon stigmas according to the different number of pollen on a stigma in angiosperms. In Mulcahy, D.L., Mulcahy, G.B. and Ottaviano, E. (eds) *Biotechnology and Ecology of Pollen*, pp. 423–8. (New York: Springer-Verlag)

Snow, A.A. (1986). Pollination dynamics of *Epilobium canum* (Onagraceae): consequences for gametophytic selection. *American Journal of Botany*, **73**, 139–51

Sokal, R.R. and Rohlf, F.J. (1969). *Biometry*. (San Francisco: W.H.Freeman)

Stephenson, A.G. (1978). The flowering and fruiting strategy of *Catalpa speciosa* (Bignoniaceae). Ph. D. Thesis, University of Michigan, Ann Arbor, Michigan.

Uma Shaanker, R. and Ganeshaiah, K.N. (1984). Age specific sex ratio in a monecious species, *Croton bonplandianum* Baill. *New Phytologist*, **93**, 523–31

Uma Shaanker, R. and Ganeshaiah, K.N. (1986). Does pollination efficiency shape the pollen grain to ovule ratio? *Current Science*, **53**, 751–3

Uma Shaanker, R., Ganeshaiah, K.N. and Bawa, K.S. (1988). Parent-offspring conflict, sibling rivalry and brood size reduction in plants. *Annual Review of Ecology and Systematics*, **19**, 177–205

Vasudev, R., Vinayak, K., Ganeshaiah, K.N. and Uma Shaanker, R. (1987). Sex ratio variations in *Acalypha fruiticosa* Frosk along plant height and altitude. *Proceedings of Indian Academy of Sciences* (*Plant Sciences*), **97**, 11–15

Wiens, D. (1984). Ovule survivorship, brood size, life history, breeding systems and reproductive success in plants. *Oecologia*, **64**, 47–53

Willson, M.F. and Schemske, D.W. (1980). Pollinator limitation, fruit production and floral display in Pawpaw (*Asimina triloba*). *Bulletin of the Torrey Botanical Club*, **107**, 401–8

Willson, M.F. and Burley, N. (1983). *Mate Choice in Plants*. (Princeton: Princeton University Press)

Section 4

Seed and fruit dispersal

CHAPTER 13

SEED AND FRUIT DISPERSAL – COMMENTARY

John Terborgh

INTRODUCTION

Seed dispersal and seedling establishment represent the most critical and sensitive stages in the life history of plants. Since tropical forests are prominently represented among the world's most diverse plant communities, it can be anticipated that the processes of seed dispersal and seedling establishment in them will be accordingly diverse. This diversity is suggested by the presence of many classes of disperser organisms in most tropical forests, and in the varied consequences and levels of pre- and post-dispersal seed predation. However, neither the simple identification of mechanisms, nor even the elucidation of their workings, will necessarily serve to answer more remote and fundamental questions about the density dependent interactions that control the compositional stability and predictability of particular forest types. Nevertheless, these distant goals are likely to remain elusive until we achieve a detailed understanding of the proximal mechanisms involved.

Contributions at the Bangi workshop on the topic of "Seed and Fruit Dispersal" drew on research results from the neotropics (Henry Howe), equatorial Africa (Annie Gautier-Hion) and South-east Asia (Mark Leighton*). Although each offered a different perspective on dispersal processes in their respective forests, the presentations contain sufficient common ground to allow some points of comparison. One is impressed that, in certain ways, the dispersal biology of these forests is similar, while, in others, it seems very different. This question of similarities and differences, and their possible underlying causes, provides a forum for discussion.

* Paper by Mark Leighton was not received in time to be included in this volume.

In offering some synthetic remarks, my commentary is contained in three sections. The first of these sections examines hypotheses about the mechanisms driving the pronounced phenological cycles of fruiting observed in virtually all tropical forests. Second, in a more speculative vein, I wonder out loud about some possible morphological relationships between fruits and the animals that feed on their seeds and pulp. Finally, I have a word of advice to managers concerned about maintaining the diversity of tropical forests.

PHENOLOGICAL PATTERNS IN FRUIT PRODUCTION

Over the past 20 years, numerous studies of fruiting phenology have been conducted in tropical forests around the world. With unfailing consistency, the results indicate that, wherever one takes the trouble to measure it, fruit production fluctuates widely, usually with an unambiguous seasonal rhythm. Strongly seasonal behaviour is found in forests growing in a wide range of climates showing markedly different types and degrees of seasonality of rainfall (Terborgh, 1986; Terborgh and van Schaik, 1987). This finding points to the suggestion that factors other than, or in addition to, climate may be driving these rhythms. Gautier-Hion (this volume) suggests three hypotheses.

'Competition avoidance hypothesis'

What can be called the 'competition avoidance hypothesis', originally proposed by Snow (1966), holds that sympatric species of plants that share a common pool of dispersers should stagger their fruiting seasons so as to minimize competition among themselves for dispersers. While this proposal may indeed account for the species of *Miconia* (Melastomataceae) that were the focus of Snow's attention, subsequent evidence has not greatly extended its generality. Moreover, stated in the above form, the hypothesis is nearly impossible to test.

First, it requires that one define a set of plant species that share a common pool of dispersers. Observers have maintained vigils at countless tropical fruiting trees and found enormous variation, both within and between tree species, in the number and species composition of potential dispersers. Indeed, one of the points Gautier-Hion makes most strongly about the M'Passa forest in Gabon, is that most species of fruit are taken by many species of consumers, and that it is very seldom that a particular fruit can be associated with a particular disperser or even group of dispersers. Thus, the occurrence of sets of plant species sharing common pools of dispersers is likely to be more the exception than the rule.

Another difficulty intrinsic in this hypothesis is that its prediction of staggered fruiting seasons is sensitive to one's ability to define the sets of plant species that share dispersers, and hence may possibly compete for them. If one fails to include some of the appropriate species, seasonal gaps will be evident in what may truly be a uniform temporal staggering of fruiting periods, and, conversely, if too many species are included, temporal staggering among some of them can be swamped by the more seasonal behaviour of other species extraneous to the interacting set.

Finally, to the extent that competition among plants for dispersers really does lead to staggered fruiting seasons, the trend will result, at the community level, in a pattern that will most likely be indistinguishable from a random one.

This brings us full circle to our opening observation that fruit production in tropical forests is seasonally concentrated, and hence decidedly non-random. We cannot reject the possibility that, in many of these forests, certain plant species may mutually avoid each other's fruiting periods, but, wherever one looks, the overall statistical pattern is non-uniform.

'Predation satiation hypothesis'

Complementary to the competition avoidance hypothesis is the 'predator satiation hypothesis'. This hypothesis states that trees should adjust their fruiting seasons to coincide in order to overwhelm the appetites of seed predators. The prediction is of a clumped, rather than a temporally uniform, distribution of fruiting periods.

Gautier-Hion offers a test of this hypothesis with data from the M'Passa forest. If the need to satiate seed predators were paramount in selecting for fruiting seasons, then one might expect to observe different patterns of seasonality in sets of species that are heavily versus lightly attacked by seed predators. Gautier-Hion identifies two 'large guilds' of fruits. One is made up of species that are brightly coloured, possessing pulp or arils rich in sugar or lipid, which are dispersed by large birds and monkeys 'without significant predation'. The other consists of fruits that are 'dull with a fibrous and nutritionally poor flesh and well-protected seeds' that suffer from pre-dispersal seed predation by squirrels and ruminants. In Gabon, both types of fruits show marked seasonal fruiting peaks, so the comparison fails to resolve the issue.

In a slightly different approach to the question, Gautier-Hion examined the phenological behaviour of zoochorous versus non-zoochorous species (anemochorous plus autochorous species), reasoning that zoochorous species should be under selection to avoid disperser competition (hypothesis 1, above), while non-zoochorous species should cluster their fruiting seasons to satiate seed predators (hypothesis 2). Again, both classes of fruits showed strongly aggregated fruiting seasons, so no conclusion could be drawn.

More convincing support for the predator satiation hypothesis is provided by Leighton, who offers the first measurements of seed predation rates in a masting versus non-masting year in a South-east Asian forest. In a non-masting year (1986), the depredations of arboreal seed predators (principally squirrels and primates) at the Gunung Palung site in West Kalimantan (Borneo) were so systematic that very few viable seeds reached the ground (<1 m^{-2} month^{-1}). Then, in the early months of 1987, there was a major masting event, the first in several years, and scores of viable seeds m^{-2} rained on to the forest floor, with the subsequent appearance of lawns of seedlings. Leighton's valuable observations raise some important questions to which I shall later return.

'Optimal time of ripening hypothesis'

The third suggestion offered by Gautier-Hion we may term the 'optimal time of ripening hypothesis'. It affirms the possibility that climatic conditions may determine the time of ripening of fruit crops. Gautier-Hion presents suggestive evidence in the finding that dehiscent fruits tend to mature in the late dry season when atmospheric conditions may favour desiccation of their outer walls, and that fleshy fruits more often mature in the main rainy season, which in Gabon is a time of high insolation that could, in the presence of ample moisture, promote the rapid accumulation of carbohydrates and lipids. In further support of this possibility, she points out that the flowering times of the species belonging to a given morphological type tend to extend over a longer season than the subsequent fruiting periods.

What are we to make of all this? The competition avoidance model clearly does not apply at the community level to any tropical forest yet studied. Panama and Gabon, with strong annual fruiting rhythms and relatively low seed predation rates, seem to be driven by climate. Borneo, in contrast, does not experience a major fruiting season every year, and the pronounced masting behaviour of its forests, along with intense seed predation outside of masting episodes, supports the predator satiation hypothesis.

Aggregated fruit production schedules thus seem to result from different forcing mechanisms in different portions of the tropics, in opposition to whatever tendencies towards uniformity might be furthered by competition among dispersers. One cannot conclude that competition avoidance is negligible or non-existent, but rather that it is a weaker force in selecting for the timing of fruiting than either of the other two.

As a footnote to the above discussion, it is important to stress that neither the predator satiation hypothesis nor the optimal timing hypothesis has yet been subject to rigorous testing. In fact, since both predict aggregated fruiting peaks, it is not clear how they may be conclusively discriminated. One improvement would be to compare the dispersion of fruiting periods among species known to suffer heavy seed predation with that shown by species that are largely free of seed predation. The comparison of fruits belonging to different morphological categories, as in the Gautier-Hion contribution, is a first step, but the results are evaluated qualitatively without the benefit of statistical criteria. In the end, it may prove difficult to distinguish between the two hypotheses because the evolution of tightly aggregated fruiting peaks for the avoidance of seed predators is compatible with an evolved timing that takes advantage of the most propitious climatic conditions. Obviously, we are far from having any final answers to these questions.

FRUIT MORPHOLOGY IN RELATION TO DISPERSERS

A point emphasized by both Howe and Gautier-Hion is that relationships between the taxonomic identity of consumers and fruit morphological traits are loose at best. Both discount the existence of strong co-adaptive links in their forests. In Howe's study, oily *Virola* arils were taken mainly by toucans and other birds, while sugary *Tetragastris* arils were favoured by primates. Nevertheless, primates harvested some *Virola* fruits, and birds some *Tetragastris fruits*. A far more extreme example is presented by Gautier-Hion in the case of *Trichilia gilgiania* (Meliaceae), the fruits of which are taken by ruminants, squirrels, monkeys, porcupines, hornbills and other birds. We may suspect that the frequency with which inappropriate species harvest the fruits of a given tree will vary greatly between species and from one occasion to another, in accordance with the availability of alternative resources. It is often presumed that uncommon visitors are seldom effective as dispersers, but this is generally an unproven contention.

A contrasting picture is painted by Leighton of the lowland dipterocarp forest he has studied in Borneo. A sizeable fraction of the fruits there are subject to heavy attack by pre-dispersal vertebrate seed predators which consume seeds in the milk just prior to the hardening that accompanies final maturation. A legion of avid seed predators, including numerous squirrels, rats and primates, seems to impose a strong selection on plants to evolve means of protection: masting, morphological resistance and chemical defences. All three types of protection seem to be developed in the Bornean flora to a degree that surpasses what has been reported for African and neotropical sites.

185

Many Bornean fruits are protected by heavy fibrous husks. Such formidable armatures will predictably reduce the number of potential dispersers, increasing the specificity of dispersal in parallel with the increased cost of ancillary structures. One large class of heavily protected dehiscent fruits, comprising some 75 species of predominantly Meliaceous and Burseraceous trees, appears to be dispersed exclusively by hornbills, as only their strong cuneate bills possess the capacity to open the thick husks and extract the large arilate seeds from within. More generally, fruits belonging to the bird and primate morphological syndromes seem to affect greater disperser specificity in the Bornean forest, though this could be a consequence of masting or more potent chemical deterrents. There is much to be learned from pursuing such inter-regional comparisons, though a lack of standard data gathering protocols so far precludes anything beyond impressionistic speculation.

Another impression gained from Leighton's presentation is that the fruits and seeds of bird-dispersed species in families common to my own study site in Amazonian Peru are consistently larger at his Bornean site. This appeared to be so in families producing dehiscent fruits – Burseraceae, Meliaceae, Myristicaceae, Sapindaceae – as well as in the genus *Ficus*. A striking, albeit anecdotal corollary of this observation can be found in comparing dispersers. There are eight species of toucans at the Amazonian site, and eight hornbills filling the equivalent ecological roles in Borneo. Yet the toucans are of modest size, ranging in weight from 200 to 700 g, while the Bornean hornbills, in contrast, are comparatively gigantic, the smallest of them weighing 1 kg, and the largest more than 4 kg. Is this merely a chance outcome of independent throws of the evolutionary dice? Perhaps, but I would rather think that Borneo's ponderous hornbills were adapted to opening its equally prodigious Burseraceous and Meliaceous fruits, which, in turn, may have evolved their present remarkable dimensions in response to the unceasing attentions of arboreal seed predators. The Bangi workshop presents us with many more intriguing questions of this type than we can presently answer.

Gautier-Hion stresses that the distinction between seed predators and seed dispersers may often be blurred, and cites compelling data to bring the point home. Seeds recovered from stomach contents of *Cercopithecus pogonius* (a guenon) were > 50% broken, while those eaten by a close relative, *C. cephus*, were < 20% broken, despite close similarities in body size and dentition. Although rodents and ruminants more commonly destroy seeds in the M'Passa forest, rodents frequently serve as critical dispersers through scatter-hoarding (Emmons, 1980), while ruminants (duikers) have been found to regurgitate seeds during rumination (Dubost, 1984). Scatter-hoarding by seed predators is also an important mechanism of dispersal in the neotropical forest (Smythe, 1970; Kiltie, 1981), but Leighton finds little evidence of it in

Borneo. If arboreal, pre-dispersal seed predators are as prevalent in Borneo as Leighton's results indicate, then the abundance of seeds on the ground may not be sufficient to support a scatter-hoarding guild.

Similarly, one can wonder whether the Bornean forest supports a guild of secondary dispersers. In neotropical forests, seeds are often redispersed from faeces by mice (Janzen, 1986) or dung beetles (Estrada and Estrada-Coates, 1986), while Africa enjoys a certain renown for the extraordinary diversity and size of its dung beetles. With so much yet to be learned about primary dispersal mechanisms, it is no surprise that secondary mechanisms have been looked at in only a few places.

A MESSAGE FOR MANAGERS

Seed dispersal biology is highly relevant to the future management of tropical forests. Emerging generalities about seed dispersal mechanisms can lead potentially to the conscious manipulation of species compositions, with the consequent enhancement of economic values.

The principal value of primary tropical forests to the logging industry has been in the exploitation of hardwoods, usually only a few species in any given region. Typically, these high value species belong to what are called mature phase species, and very often these are dispersed by large birds and mammals.

Under natural conditions, the balance of pioneer versus mature phase species in any given tract is believed to reflect the size and frequency of disturbance events. Natural disturbances include the mortality of adult trees due to natural causes, as well as induced mortality resulting from windstorms, fires, flooding, landslides, etc. (Clark, this volume). The overwhelming majority of such natural disturbances are small in scale, involving only one to a few individual trees, and the near absence of gap phase pioneer species in many primary forests reflects this scale. Indeed, the most heavily stocked primary forests are often ones having very low rates of natural disturbance.

Logging of any kind results in disturbance, and, the less selective and more extensive the logging operation, the greater the disturbance. Clear cutting is, of course, the extreme case. Inevitably, the regrowth that follows logging is more biased in its composition toward pioneer (gap phase) species than was the original primary stand. Large-scale clear cutting typically leads to the complete dominance of soft-wooded pioneer species (Jordan, 1986). Although these may grow rapidly, the wood they produce is of a lower average value than the wood that was removed from the primary forest. As more and more primary forest is exploited, the availability of prime hardwood species will inevitably decline, inducing a consequent rise in the price per unit volume. At some point, the rising value of hardwood should compensate for its slower rate of growth and create incentives for management directed toward increased production. It is expressly this kind of management to which I address the following remarks.

A growing body of research is pointing to the likelihood that the densities of many tree species in natural stands are limited by pre- and post-dispersal seed predation. This is particularly true of the large-seeded species that predominate in the mature phase. In the main, these are dispersed by large birds and mammals, with some assistance from bats. Seeds that are not dispersed, that is, those that fall under or near the parent tree, have a vanishingly small chance of escaping predation, as Howe (this volume) has so vividly demonstrated. Either they are bored by larvae, eaten by rodents or, upon germination, are damped by fungi. Study after study has now shown that the highest chances of survival are possessed by seeds that have been dispersed many metres away from the parent tree.

To obtain good regeneration of such species, it is thus essential to retain dispersers in the system. Large birds and/or mammals are often the only dispersers of mature phase species, but, even in species attracting a wide range of potential dispersers, the larger dispersers are generally found to be more effective because:

(1)   They tend to be more selective of large seeded mature phase species;

(2)   They consume more fruits per feeding bout; and,

(3)   They tend to carry the seeds farther before regurgitating or defaecating them.

Such observations firmly establish the indispensable role of large vertebrates in the perpetuation of mature phase species in forest stands.

Future management plans for tropical forests will therefore have to consist of two components: (1) strategies for maintaining large vertebrate dispersers in the ecosystem, and (2) strategies for increasing the representation of tree species of particular economic importance. Let us consider these two stipulations.

Systematic over-hunting frequently extirpates the large vertebrate fauna of tropical forests long before the first advent of loggers. Where some game remains, loggers are likely to eliminate it for their own needs while they remove the trees. Regeneration is, thus, likely to begin with a deficiency of dispersers, a situation that can only be remedied by controlling hunting. The presence of smaller scatter-hoarders, such as squirrels and other rodents, may, to a degree, be able to mitigate the absence of monkeys or hornbills, but this is as yet an unstudied possibility.

The maintenance of a fauna of large vertebrates will depend not only on controlling hunting, but also on retaining a high plant species diversity in managed forests. The large vertebrates that play major roles as seed dispersers require a plentiful year round supply of suitable fruits and/or seeds. A forest composed of only a few tree species will create an environment of boom and bust, brief surges of abundance offsetting long periods in which no species is in fruit. Animals cannot survive such conditions. Only through diversity can they obtain the continuous food supply needed to support growth and reproduction.

This conclusion cautions us that management should not be too intensively directed toward one or a few species of special interest. Instead, to maintain adequate plant diversity and the interdependent animal community, management should be based on the exploitation of many species. Heretofore, use of timber in the tropics has been largely focused on export markets for cabinet and veneer woods, but taking such a narrow view does a disservice to local markets. People in less developed countries use and need wood for many purposes: fuel, thatch, building materials, tool handles, fence posts, etc. In many areas, different species are exploited for each of these purposes. Production that is exclusively export-oriented overlooks much of the potential of the forest resource, and consequently leads to waste on a large scale. Residents of the exporting country are deprived of resources that could be theirs.

By including species that have value on local as well as international markets, the evaluation of management options could be altered radically. The perceived worth of a given forest will inevitably be enhanced by the inclusion of additional marketable species. Animals, as well as plants, should enter in the account in recognition of their value as game, pollinators and dispersers. Such economic aspects of tropical forest management have been even less explored than some of the arcane biological topics touched upon above.

The point was made above that the disturbances produced by logging in a primary stand will bias the regrowth toward pioneer species of low commercial value. Herein lies the greatest challenge to the manager who wishes to restock the stand with high value hardwoods. In effect, he has to swim against the tide. Is this going to be a practical proposition, even if we grant the presence of large vertebrate dispersers?

The question is an important one to consider seriously, because the success or failure of future management efforts will depend on it. Certainly a haphazard, laissez faire approach is doomed to failure from the start. My feeling is that successful management for hardwoods is a serious possibility, though it will require the application of restraint during extraction procedures, and a level of technical expertise that most tropical foresters do not now possess.

In principle, it should be possible to exploit the existing age structure of primary stands to promote the regeneration of mature phase species. Saplings and pole-sized, immature individuals of gap phase species tend to be rare in primary stands. The pole stage individuals that crowd the understorey of many primary forests belong mostly to mature phase species. They are following a 'sit-and-wait' strategy of persisting in the shade at negligible growth rates while waiting for openings in the canopy above. Through carelessness and indifference, such pole-size trees are commonly ravaged in logging operations. If spared judiciously, they could provide the key to successful management. Once released from the overtopping shade of a higher canopy, such trees, being half grown already, should reach maturity quickly. The period between successive harvests could be dramatically less than for trees grown from seed. Over the long run, there might be difficulties in perpetuating a favourable structure in twice or thrice harvested stands, especially where dispersers were scarce or absent, but, from the perspective of today, this long run is a far distant horizon. By then, there may be no tropical forests, or in a rosier projection, we may have learned enough to manage them effectively.

REFERENCES

Dubost, G. (1984). Comparison of the diets of frugivorous forest ruminants of Gabon. *Journal of Mammalogy*, **65**, 298–316

Emmons, L.H. (1980). Ecology and resource partitioning among nine species of African rain forest squirrels. *Ecological Monographs*, **50**, 31–54

Estrada, A. and Estrada-Coates, R. (1986). Frugivory in howling monkeys (*Alouatta palliata*) at Los Tuxtlas, Mexico: dispersal and the fate of seeds. In Estrada, A. and Fleming, T. (eds) *Frugivores and Seed Dispersal*, pp. 93–104 (The Hague: Dr. W. Junk Publishers)

Janzen, D.H. (1986). Mice, big mammals, and seeds: it matters who defecates what where. In Estrada, A. and Fleming, T. (eds) *Frugivores and Seed Dispersal*, pp. 251–72 (The Hague: Dr. W. Junk Publishers)

Jordan, C. (1986). Local effects of tropical deforestation. In Soulé, M.E. (ed.) *Conservation Biology: The Science of Scarcity and Diversity*, pp. 410–26 (Sunderland, Massachusetts: Sinauer Associates)

Kiltie, R.A. (1981). Distribution of palm fruits on a rain forest floor: why white-lipped peccaries forage near objects. *Biotropica*, **13**, 141–5

Snow, D.W. (1966). A possible selection factor in the evolution of fruiting species in a tropical forest. *Oikos*, **15**, 274–81

Smythe, N. (1970). Relationships between fruiting season and seed dispersal methods in a neotropical forest. *American Naturalist*, **104**, 25–36

Terborgh, J. (1986). Community aspects of frugivory in tropical forests. In Estrada, A. and Fleming, T. (eds) *Frugivores and Seed Dispersal*. pp. 371–84 (The Hague: Dr. W. Junk Publishers)

Terborgh, J. and van Schaik, C.P. (1987). Convergence vs. non-convergence in primate communities. In Gee, J.H.R. and Giller, P.S. (eds) *Organization of Communities Past and Present*, pp. 205–26 Twenty-Seventh Symposium of the British Ecological Society (Oxford: Blackwell Scientific Publications)

CHAPTER 14

SEED DISPERSAL BY BIRDS AND MAMMALS: IMPLICATIONS FOR SEEDLING DEMOGRAPHY

Henry F. Howe

ABSTRACT

Patterns of seed and seedling survival reflect conditions imposed by different forms of seed dissemination by animals. Most (54%) of the seeds of the toucan-dispersed Virola surinamensis *(Myristicaceae) are scattered singly in the forest, while 46% drop under the parent trees. Seed and seedling mortality under the parental crown exceeds 99.99%. Only seeds regurgitated by birds away from parents have any appreciable chance of survival for 15 months in the understorey (0.3–0.5%). Most establishment requires well-watered slopes under a broken canopy. Less than a fourth (23%) of the seeds of monkey dispersed* Tetragastris panamensis *(Burseraceae) are removed from the vicinity of the parent trees, and these are dropped in clumps in monkey scats. Dispersed and undispersed seeds occur in high densities. A high proportion of seeds survive for 15 months under the crown (44%), in clumps in former monkey scats (34%), or singly away from parent trees (72%). Different dispersal agents impose different selective conditions on seeds and seedlings. Seedling demography reflects adaptive responses to these conditions. The only* Virola *seedlings likely to survive near parents will be at the edge of the crown, where heavy seed fall and reduced seed predation by weevils leave a few survivors, and in light gaps far from adult* Virola *trees. Trees with seed that are scatter-dispersed by birds or bats are especially vulnerable to local extinction if dispersal agents are over-hunted, because such trees cannot recruit substantial numbers of seedlings near parent trees.* Tetragastris *normally recruit heavily under and near parent trees, as well as in clumps in monkey scats.*

INTRODUCTION

The behaviour of dispersal agents determines patterns of seed distribution and consequently defines the conditions under which seedlings live or die. Janzen (1970) stimulated interest in the issue by hypothesizing that tropical frugivores helped seeds escape from devastating seed or seedling predators under parent trees. More recently, ecologists have recognized that escape from seed predators or pathogens under the parent tree is only one of several possible advantages to seed dispersal (Howe and Smallwood, 1982). Colonization of light gaps and directed dispersal of seeds to sites that are particularly suitable for germination and establishment are other advantages that might be important for tropical plants.

Speculation about the possible effects of seed dispersal by different birds and mammals has far outstripped empirical corroboration (Clark and Clark, 1984; Howe and Estabrook, 1977; Wheelwright and Orians, 1982). Beyond such classic cases as the directed dispersal of mistletoe (Loranthaceae) seeds by Asian flowerpeckers (Dicaeidae) that scrape seeds off on to the bark of host trees (Docters van Leeuwen, 1954), the roles of different dispersal agents in plant recruitment are very poorly understood. Zoologists interested in frugivore diets rarely know whether seeds regurgitated or defaecated by their study animals actually germinate and successfully establish as seedlings. Botanists interested in plant demography rarely know which of dozens or hundreds of possible dispersal agents are responsible for successful seedling recruitment. No one has evaluated carefully the ecological pressures imposed upon different plant species by consistent dispersal by different kinds of animal seed vectors.

The objective of this chapter is to compare consequences of seed dispersal for seedling survival in the toucan-dispersed tree *Virola surinamensis* (Rol.) Warb. (Myristicaceae) with those of the monkey-dispersed tree *Tetragastris panamensis* O. Kuntze (Burseraceae) in central Panama. Patterns of fruit production and its depletion by birds and mammals have been reported earlier for *Virola* (Howe, 1986a) and for *Tetragastris* (Howe, 1980). Salient features of these background studies are reviewed. The original contribution of this paper is the synthesis of this earlier work with ongoing and largely unpublished studies of seedling survival. I ask:

(1)   Do these two tree species benefit from escape from seed and seedling mortality near parents?

and the novel corollary question

(2)   Do patterns of seed dissemination imposed by different dispersal agents strongly influence seed and seedling demography?

Tropical dispersal ecology is in its infancy, and generalizations made from the limited information at hand must be tentative. However, a sense of urgency requires that a synthesis be attempted for the relatively comprehensive studies of *Virola* and *Tetragastris* in order to evaluate two alternative possible patterns of mutualism that have profound – and entirely different – implications for tropical conservation.

Dispersal mutualisms may be especially important in tropical forests, where up to 90% of tree and shrub species bear fruits adapted for animal dispersal (Frankie *et al.*, 1974). Chance extinction of critical food resources, or dispersal agents, from either natural causes or human interference could remove "pivotal" or "keystone" mutualists from fragile tropical communities (Howe, 1977; Gilbert, 1980). Disruption of local dependencies between animals and plants could lead to widening circles of local extinctions, precipitated by the demise of trees that support a variety of animals and frugivores that effectively disseminate a variety of plant species (Howe, 1977, 1984*a*) On the other hand, dispersal mutualisms may be so general that many trees can persist for millennia without effective dispersal (Janzen and Martin, 1982). It would be good to know the relative importance of these views before logging, hunting, and natural catastrophies in isolated reserves simultaneously prove the first scenario correct and cause extinction of a substantial proportion of the tropical flora and fauna.

METHODS

An evaluation of the relative effectiveness of different dispersal agents requires evaluation of:

(1)   Fruit production, removal, and waste;

(2)   Patterns of seed dissemination attributable to different animals; and,

(3)   The relationship between patterns of seed distribution and seed and seedling survival.

More detail for items (1) and (2) may be found in Howe (1980, 1982, 1983) and Howe and Vande Kerckhove (1981).

Study site

My collaborators and I have studied dispersal ecology since September 1977 at the Smithsonian Tropical Research Institute field station on Barro Colorado

Island (9°09'N, 79°51'W), Panama. This island of 15 km² was separated from the mainland by flooding of Gatun Lake during the building of the Panama Canal in 1914. Approximately 2 km² on the top of the island are a flat basaltic cap; the remainder is heavily dissected with ravines. Approximately half of the island, including most of the cap, is "Old Forest" that has not been extensively disturbed by humans for at least 450 years (Foster and Brokaw, 1982). The remainder is "New Forest", now a tall canopy 70–85 years old, that was largely cleared during the building of the Panama Canal. The climate is that of a seasonal moist forest, with virtually all of the annual rainfall of 2500 mm falling between late April and December. Biotic and abiotic factors influencing the ecology of the Barro Colorado forest are discussed in Leigh *et al.* (1982).

Twenty-five *Virola surinamensis* trees used in various phases of this study are scattered throughout a 23 ha study area in remnants of Old Forest in ravines and surrounding ridges. Nineteen fruiting females of this dioecious species ranged from 19–80 cm diameter at breast height (dbh) (51 ± 13 cm [means and 95% confidence intervals used throughout]) and 14–34 m (25 ± 2 m) in height. The 19 *Tetragastris panamensis* trees sampled for fruit production and frugivore activity are located in what is now the 50 ha Hubbell/Foster plot (Hubbell and Foster, 1983, this volume) on the basaltic plateau of the island. These trees ranged from 11–60 cm dbh (33 ± 7 cm) and 10–38 m (21 ± 4 m) high. The nine *Tetragastris* trees selected for seedling studies are in remnants of Old Forest and in New Forest in the watersheds of Allee, Lutz, and Shannon Creeks. The original *Tetragastris* study area was avoided because of potential trampling during recensus of the Hubbell/Foster plot.

Fruit estimation

Fruit-fall was estimated using a system of 1 m² fruit traps that captured fruit husks, undispersed seeds, and other debris. *Virola* traps sampled $12 \pm 2\%$ (mean ± 95% confidence intervals, throughout) of the area under the crown, *Tetragastris* traps sampled approximately $9 \pm 2\%$ of the area under the crown. Traps were placed at random (co-ordinates chosen from a random number table) under each crown, and monitored 1–5 times per week, depending on fruit fall and the threat of depredations by terrestrial mammals such as coatimundis (*Nasua narica*, Procyonidae). Trap contents were discarded in place.

Frugivore activity

Animal activity was evaluated with extended observations at fruiting individuals of both tree species. Eight 5-hour watches (from 06:00–11:00 h) at each of eight *Virola surinamensis* individuals determined which animals visited, how many seeds they dropped, how many seeds were removed, and, when possible, how far seeds were taken before they were regurgitated. Two 10-hour watches at each of ten *Tetragastris panamensis* trees yielded similar information for this species. The schedules differed because activity at *Virola* was greatest in the morning when fruits dehisced, whereas *Tetragastris* trees were frequented throughout the day. Night censuses at both species allowed a relative estimate of nocturnal mammal activity.

Depletion and dissemination

Fruit traps permitted estimates of the number of fruits produced, and the number and proportions of seeds taken or dropped underneath fruiting trees. The number of husks for *Virola* or core locules for compound fruits of *Tetragastris* recovered in traps permitted an estimation of fruit fall by dividing the number of items caught under each tree by the proportion of area sampled under each tree. A similar calculation indicates the number of seeds dropped under the tree; the difference between husk (or locule) numbers and undispersed seed numbers estimates the proportion dispersed. Fruit traps estimate what falls under the crown. Seeds not captured estimate those dropped beyond the crown edge.

Fruit traps allow an estimate of absolute numbers of fruits produced and removed from each tree. The lifetime dispersal success of an individual tree is the sum of its annual dispersal successes. An instantaneous evaluation of success, the percentage of seeds taken in a given year, allows an additional measure of intraspecific variation in success (Howe, 1986*b*, page 160).

Mortality and survival

Methods for evaluating seedling survival differed substantially. *Virola* seed fall to 45 m (20 m in this chapter) was estimated from a 10° wedge randomly directed away from each of five trees. Seeds counted every 2–3 days in 3 m segments during the peak 2 weeks of fruit production (estimated to be half of the fruits produced during the season for these plants) indicated natural seed fall. Seedling survival was estimated by controlled placement of seeds at marked flags on circles at 5, 15, 25, 35, and 45 m for freshly fallen seeds (40 seeds/circle for 17 trees in 1982), for 6-week-old germinating seeds (25 seeds/circle for 13 trees in 1983), and for 12-week-old seedlings (25 seedlings/circle for seven trees in 1984). Greenhouse space for growing protected seedlings and logistics determined the sample sizes. In each experiment, the seeds or seedlings were checked biweekly for 12 weeks and were then checked every 6 weeks thereafter.

Tetragastris seedlings were found and marked with numbered rings in 10° wedges directed randomly for 20 m from each of nine fruiting trees in 1984. The carpet of seedlings was discovered in August and September, approximately 3 months after fruit fall. Consequently, initial seed survival is not known, although earlier observations suggest it to be very high (Howe, 1980). Marked seedlings were checked monthly. Sixteen clumps of seedlings in former mammal scats were similarly found and marked, as were 68 solitary seedlings of the year > 20 m from fruiting *Tetragastris* trees.

RESULTS

Fruit production

Virola surinamensis and *Tetragastris panamensis* both produce encapsulated fruits roughly the size of a lime, but they differ markedly in character (Figure 14.1). The *Virola* fruit has a thick yellow-orange husk that dehisces in the morning to expose a single seed covered by a brilliant red aril. The *Tetragastris* fruit has a thin, purple-green husk that dehisces throughout the day and night to expose 1–6 (5 is the mode) white arillate seeds that stand out against a bright purple core. The arillate seeds of both are roughly similar in bulk, but different in the relative proportions of aril and seed. The fatty *Virola* aril has approximately twice the energetic content per unit weight of the watery and sugar-rich *Tetragastris* aril (Table 14.1)

Virola surinamensis and *Tetragastris panamensis* also differ in fruiting phenologies. In central Panama, some *Virola surinamensis* trees are in fruit during every month of the year, but the vast majority bear fruit between March and August, with a strong peak in June or July (Howe, 1982). Most females produce fruit crops yearly, with a two- to threefold range in median annual fecundity (Table 14.2). *Tetragastris panamensis* bears fruit intermittently. Between 1969 and 1987, large crops have been recorded in 1975, 1978, and 1984 (R. Foster, pers. comm.). Fruit production during other years is moderate to negligible, although trees flower annually. The *Tetragastris* fruiting season extends from January to June, with a strong peak in April. Crop sizes in 1978 ranged from 165–99 221, with a median of 11 102 (mean of 22 951).

Table 14.1 Aril constituents of two neotropical trees

	Protein	Lipid	Sugars	Kcal/g
	(%)	(%)	(%)	
Virola surinamensis	3	53	6	7
Tetragastris panamensis	1	4	56	4

Source: Howe (1982)

Figure 14.1 Fruits of *Virola surinamensis* and *Tetragastris panamensis*, as presented to fruigivores. Left: *Virola surinamensis* capsules dehisce to expose a seed 21 x 16 mm. The seed averages 2.0 g dry weight, the aril 0.8 g. Right: *Tetragastris panamensis* capsules dehisce throughout the day and night to expose an aromatic white arillate seed against a bright purple/red core. The diaspore averages 18 x 14 mm, with the dry weight of the seed 0.2 g and that of the aril 0.4 g

Table 14.2   Annual variation in fruit production in *V. surinamensis* trees

Year	N	Range	Median (25–75% quartiles)
1979	17	214–10412	2082 (1326–3584)
1980	25	428–31006	8579 (4161–12493)
1981	25	638–26163	3990 (2687–6687)
1982	25	78–14075	5612 (1945–8008)
1983	25	92–14450	2420 (509–5333)

Source: Howe (1986*a*)

Frugivore activity

Extended watches at these two tree species showed that both attracted a small subset of 80 fruit-eating mammals and birds on Barro Colorado Island (Enders, 1935; Willis, 1980; see Table 14.3). Data from regular census schedules, not repeated here, confirmed the compositions of the two assemblages (Howe, 1980, 1986*a*).

Virola surinamensis fruit were taken by six fruit-eating birds ranging in size from the slaty-tailed trogon (Trogonidae; 145 g) to the black-crested guan (Cracidae; 2050 g). Most fruits were taken by these birds and by rufous motmots (Motmotidae; 185 g), and keel-billed (339 g) and chestnut-

Table 14.3   Relative contributions to dispersal and waste of arillate seeds handled by frugivores visiting *Virola surinamensis* and *Tetragastris panamensis* trees

Common name	Binomial	Virola system Wasted (%)	Virola system Removed (%)	Tetragastris system Wasted (%)	Tetragastris system Removed (%)
Howler monkey	*Alouatta palliata*	–	–	59	6
White-faced monkey	*Cebus capucinus*	–	–	19	1
Spider monkey	*Ateles geoffroyi*	9	3	–	–
Coatimundi	*Nasau narica*	–	–	10	<1
Black-crested guan	*Penelope purpurascens*	0	9	–	–
Slaty-tailed trogon	*Trogan massena*	<1	10	0	<<1
Black-throated trogon	*Trogon rufus*	–	–	0	<<1
Rufous motmot	*Baryphthengus martii*	2	14	–	–
Collared aracari	*Pteroglossus torquatus*	0	1	<1	<<1
Chestnut-mandibled toucan	*Ramphastos swainsonii*	2	35	0	<<1
Keel-billed toucan	*Ramphastos sulfuratus*	2	8	<1	<<1
Masked tityra	*Tityra semifasciata*	5	<1	0	<<1
Fruit crow	*Querula purpurata*	–	–	0	1

Data are derived from 200 hours of observation at *Tetragastris* and 320 hours at *Virola*.
Source: Howe (1982)

mandibled (640 g) toucans (Ramphastidae). A tityra (Tyrannidae), too small (85 g) to swallow the fruits, picked off the arils and dropped seeds. The most conspicuous visitors were large and vocal spider monkeys (*c*: 8000 g), which ate surprisingly few fruits and dropped or knocked down three times as many fruits as they consumed.

Tetragastris panamensis attracted many of the same frugivores as *Virola surinamensis*, but with substantially different effect. Arboreal mammals visited the trees in approximate rank of abundance, removing 97% of the seeds taken by animals. As argued elsewhere (Howe, 1980), the contribution of mammals to effective dispersal must be calculated by discounting the effects of excessive knockdown under the trees and by seed dissemination in faecal clumps from which no more than one seedling can ultimately survive. There is no evidence of effective secondary dispersal of seeds after they are deposited by mammals. By such calculations, howler monkeys, which remove 74% of the seeds taken by animals, effectively disperse only 8% of the seeds handled by animals. This is a much higher proportion than any other species, however. Birds take a very small proportion of seeds, which are scattered singly in the forest and consequently do not compete in dense clumps.

Night censuses showed that nocturnal mammals used both tree species regularly during at least some seasons. Nocturnal mammals (mostly the kinkajou *Potos flavus*, Procyonidae) regularly visited *Virola surinamensis* during seasons of heavy fruit fall in 1980 and 1982 when many fruits were uneaten by nightfall. These are not effective dispersal agents because seeds are defaecated in masses, often underneath trees with hollows in which the animals sleep during the day. No seedlings of > 500 per year survived weevil attacks under one such roost during 4 years of observation (Howe, 1986*a*). Kinkajous and other arboreal mammals regularly visited *Tetragastris panamensis* during 1978, and they probably had the same effects on dispersal as monkeys. No fruit-eating bats were seen at either of these tree species, although chewed *Tetragastris* cores occasionally appear under *Artibeus jamaicensis* (Phyllostomatidae) feeding stations when figs and other preferred fruits are in short supply (C. Handley, unpublished).

Fruit depletion and seed dissemination

Fruit traps demonstrated that fruit depletion is highly variable. As many as 91% or as few as 13% of *Virola surinamensis* fruits may be removed from fruiting trees, leaving as few as 9% or as many as 87% directly under the crown. The percentage taken from productive trees also varies considerably from year to year (Table 14.4). Several thousand fruits fall under some trees, dozens under others. Removal from *Tetragastris panamensis* trees was less

Table 14.4 Annual variation in the percentage of fruits taken from 15
V. surinamensis trees

Year	Range	Mean (±95% C.I.)
1979	13–91	60 ± 10
1980	24–73	46 ± 8
1981	40–77	59 ± 5
1982	13–64	41 ± 6
1983	27–90	65 ± 8

Source: Howe (1986*a*)

complete, with only 1–56% (23 ± 7%) of the fruits removed (Howe, 1982).
Hundreds to thousands of seeds fall directly underneath large fruiting trees
during productive years.

Direct measures show that seed fall is heavy under parental crowns, much
lighter a few metres away (Figure 14.2). Most notable is the extreme

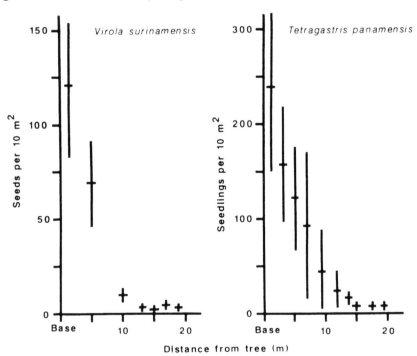

Figure 14.2 Natural seed and seedling densities near fruiting *Virola surinamensis* and *Tetragastris
panamensis* trees. *Virola* data were collected by walking a 10° wedge transect away from five trees
every 3 days during the peak 2 weeks of the fruiting season (Howe *et al.*, 1985). *Tretragastris* data
are from seedlings marked on nine similar transects approximately 3 months after fruit fall

200

variability in seed fall under both species. Both the mean and variance of seed density decline sharply at the crown edge, largely because density is so low. Area in which seeds may fall increases as πr^2, with radius (r) the distance from the tree base. Few of the 54% of *Virola surinamensis* seeds or 23% of *Tetragastris panamensis* seeds that are carried beyond the crown edges are likely to fall within 20 m of the trees.

Dispersed seeds are deposited singly or in clumps beyond the crown edge, which varies from 4 to 12 m in these two trees (Figure 14.3). *Virola surinamensis* seeds that are removed by birds are generally scattered through the forest, where they germinate and establish far from neighbours. A small minority of *Tetragastris panamensis* seeds are also scattered in the forest, but the majority of seeds that are eaten by animals are deposited, germinate, and establish in dense clumps of 5–60 derived from mammal scats. The 16 scats marked in this study contained 5–27 freshly-germinated seedlings (17 ± 4) in 1984.

Mortality and survival

Although data on early seedling survival are not strictly comparable, it is clear that *Virola surinamensis* and *Tetragastris panamensis* seedling demographies differ sharply. *Virola surinamensis* suffers devastating mortality through the first 15 months of life, whereas *Tetragastris panamensis* suffers by comparison only moderate attrition.

An earlier experimental study demonstrated that less than 2% of *Virola surinamensis* seeds survive as seedlings 3 months after fruit fall (Figure 14.4; see Howe *et al.*, 1985). The most important source of mortality is a weevil (*Conotrachelus* spp., Curculionidae), which oviposits on seeds during or shortly after germination (Figure 14.5). A larva hatches within days and burrows through the seed, eventually killing the embryo directly or admitting pathogens that have the same effect. In 1982, 99.5% of 1323 seeds infested with *Conotrachelus* larvae died. Weevil infestations are much more pronounced under than away from fruiting trees, leading to a 22-fold immediate advantage to dispersal 25 m as compared with 5 m from the tree base, and a 44-fold advantage to dispersal 45 m away. Mammals also eat seeds, but account for a smaller proportion of mortality and do not eat a disproportionate number under tree crowns. Overall, 99.96% of the seeds that fall under *Virola surinamensis* trees die within 3 months, while 1% 25 m away and nearly 2% 45 m away survive. Seedlings are rare under parent trees.

Comparable data are not available for *Tetragastris panamensis*. However, casual observation during an earlier study of frugivore activity and carpets of thousands of 1–3-month-old seedlings under *Tetragastris* adults in productive years indicate minimal mortality during germination

Figure 14.3 Seedlings. Above: *Virola surinamensis* rarely survive to establishment unless they are isolated from conspecifics. This individual is 3 months old and 14 cm high. Below: *Tetragastris panamensis* seedlings occur in dense aggregations under parent trees, or in clumps that represent former mammal droppings. Seedlings in this clump are one year old and 12–16 cm high

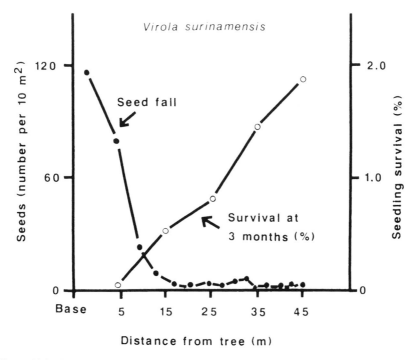

Figure 14.4 Summary of *Virola surinamensis* seed fall and probability of seedling survival during the first 3 months after fruit fall. Seed predators kill 99.96% of the seeds that fall under parent trees, approximately 98.3% 45 m away, resulting in a more than 40-fold advantage to local seed dispersal. Small birds such as trogons and motmots (< 200 g) leave most of these bulky seeds under or near the trees. Larger toucans and guans are more valuable because they carry most seeds that they eat at least 45 m from *Virola* trees. Data from Howe *et al.* (1985)

and establishment. No common insect or vertebrate seed predators seem to kill many seeds between fruit fall and germination 1–4 weeks later.

Most notable in the comparison of survival between 3 and 15 months are (1) the absence of any clear advantage to dispersal and (2) the remarkable variability in survival of both species (Figure 14.6). If anything, *Tetragastris* survival is lower at 20 m than at 2 m. The primary source of seedling mortality after independence from parental endosperm (3 months) is herbivory by mammals, which probably include rodents (*Agouti paca*, perhaps *Dasyprocta punctata*), deer (*Dama virginiana*), and tapirs (*Tapirus bairdii*). Drought further kills *Virola surinamensis* seedlings, limiting survivors to moist ravines or light gaps (Howe, 1986*a*). Attrition of *Tetragastris panamensis* seedlings in clumps appeared to be due to competition, perhaps complicated by pathogen attack.

Figure 14.5 Insect seed predation on *Virola surinamensis* seeds. Weevils (*Conotrachelus* spp., Curculionidae) oviposit on seeds during or shortly after germination. Larvae kill 99.5% of the seeds that they burrow through, leading to disproportionate seed and seedling mortality under fruiting trees (Figure 14.4). Mammals eat *Virola* seeds, but they do not contribute to the disproportionate mortality under the crowns

Other aspects of seedling survival between 3 and 15 months of age differ. Overall, only 3% of the *Virola* seedlings planted within 25 m of fruiting trees survived 1 year to 15 months of age. Survival of relatively isolated seedlings planted at 35 and 45 m (not shown), was 3 ± 2% and 4 ± 2%,

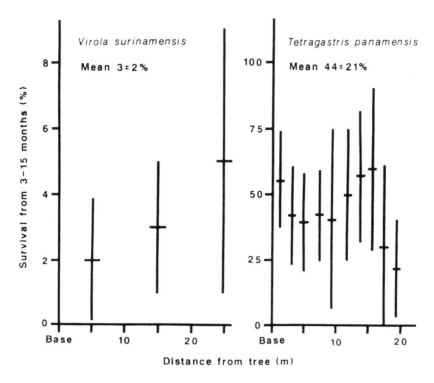

Figure 14.6 Survival of seedlings from 3 15 months of age as a function of distance from fruiting trees. Survival over this 12 month period is variable, and shows no signigicant effect of distance from fruiting trees. Most mortality was due to herbivory in both cases. Means are accompanied by 95% confidence intervals

respectively. These are comparable to isolated *Tetragastris* seedlings away from parent trees (see below). No *Virola* seedlings survived in clumps. In contrast, 44% of the *Tetragastris* seedlings within 20 m of fruiting adults lived 1 year (to 15 months of age), while 72% of the isolated seedlings away from conspecifics lived as long. Only 34% of the seedlings in former monkey scats survived this interval, and, of course, the maximum ultimate survival per clump is 1 (a mean of 6% in this sample).

Evidence of density-dependent seedling mortality between 3 and 15 months of age is limited for these tree species. *Virola* seedlings are too rare for such an effect, and *Tetragastris* mortality is too variable (Figure 14.6). The one suggestive datum is that 72% of the scattered solitary *Tetragastris* seedlings survived to 15 months, as compared with 44% within 20 m of adults, and 34% in mammal scats. These are in increasing order of density.

Too much variability exists for any trend to show with increasing distance from fruiting *Tetragastris*, which is associated with decreasing seedling density (Figure 14.6). Contrary trends may show within different *Tetragastris* transects. For instance, a negative correlation of the proportion surviving against original density is highly significant for one tree (no. 84–6; $r = -0.70$, $p < 0.01$), the correlation does not approach significance for another nearby (no. 84–8; $r = -0.03$), and the correlation is positive for a third (no. 84–10; $r = 0.56$, $p < 0.01$). Local heterogeneity in initial density, light, and slope probably obscures patterns in comparisons between individual transects.

Gross differences in survival between species are not due to procedural differences (planting *Virola* and using naturally dispersed *Tetragastris*). Herbivores, not planting stress, killed *Virola* seedlings over several months. Ongoing experiments with transplanted *Virola* seedlings in mammal exclosures show that planted seedlings survive well if they are protected from herbivores (Howe, unpublished).

DISCUSSION

Differences in frugivore activity have profound effects on the conditions under which seeds and seedlings must survive, and consequently should influence the evolution of tree demographies. Known differences in the ways in which frugivores handle fruit, the striking contrast in demographies demonstrated here, and the implications of these demographies for tree distributions indicate an important interplay of animal digestive physiology and behaviour with tree life histories (Table 14.5). A critical evaluation of these contrasts points to general implications for both the ecology and evolutionary history of animal-dispersed tree species.

Frugivore behaviour and seed deposition

Fruit-eating animals have potentially disparate effects on seed and seedling survival (Howe, 1986*b*; Levey, 1987). Seeds may be digested, scarified, or discarded unchanged, singly or in large, or small clumps, near or far from other concentrations of seeds. Digestive physiology and behaviour play critical roles as diverse as the animal taxa themselves. Of interest here is the fundamental distinction between frugivores that scatter seeds singly and those that leave seeds in dense clumps.

Birds and mammals that eat *Virola* and *Tetragastris* fruits are among the many highly frugivorous species with digestive physiologies geared to rapid seed passage (see Moermond and Denslow, 1985). Highly frugivorous birds

Table 14.5 Comparison of the fruiting ecology of two canopy tree species, bird-dispersed *Virola surinamensis* and monkey-dispersed *Tetragastris panamensis* on Barro Colorado Island, Panama

Virola surinamensis	*Tetragastris panamensis*
Diaspore 21x 16 mm; aril 0.8 g, seed 2.0 g dry weight	Diaspore 18x14 mm; aril 0.4 g, seed 0.2 g dry weight
Aril 3% protein, 53% lipid, 6% sugar by dry weight	Aril 1% protein, 4% lipid, 56% sugar by dry weight
Annual crop 78–31 000/tree; 5-year median 3990	Intermittent crops of 165–99 000/tree; 1-year median 11 102
13–91% of seeds removed by frugivores; 5-year grand mean 54% removed	1–59% of seeds removed by frugivores; 1-year mean 23% removed
Dispersed by birds (toucans and guans); monkeys conspicuous but remove few seeds	Dispersed by arboreal mammals (monkeys); birds conspicuous but remove few seeds
Seeds scattered by birds and germinate in isolation	Seeds deposited and germinate in faecal clumps of 2–60
Almost all seeds and seedlings under parents die	High seed and seedling survival under parents

Source: Howe (1980, 1982, 1986*a*)

and mammals must process enormous quantities of fruits to extract enough protein for life processes. Mechanically or chemically protected seeds are unwelcome ballast to be disposed of by defaecation or regurgitation as quickly as possible.

Virola frugivores handle seeds quickly and gently. Large guans defaecate *Virola* seeds within 15–45 min after consumption, while smaller motmots, toucans and trogons regurgitate seeds within 10–25 min after feeding. Dispersal distances range from a few metres for smaller (< 200 g) species, such as trogons and motmots, to 50 m or more for larger (> 350 g) guans and toucans (Howe *et al.*, 1985). There is no indication that any of these birds attempt to digest *Virola* seeds. Seeds are regurgitated or defaecated singly, in pairs, or occasionally in threes from perches in the forest and fall to the soil surface and germinate in 2–4 weeks. Away from parental crowns, *Virola* seeds rarely land within 2–5 m of each other. It is clear from patterns of seed deposition and seedling survival that toucans are 30–60 times as efficient, from the plant perspective, as smaller motmots and trogons that leave most seeds under *Virola* crowns.

Similarly, mammals responsible for *Tetragastris* dispersal treat seeds without enhancing or diminishing viability (Howe, 1980). Most seeds are knocked down or drop directly under parental crowns, and those that are carried away germinate dense clumps of five to 60, depending on the mammal species involved. Scats of mammals that feed on *Tetragastris* are white and sweet smelling, and show no sign of seed digestion. Almost all seeds recovered from such droppings germinate quickly. Dispersal distances of similar arboreal mammals elsewhere range from 100–300 m in the relatively sedentary howler monkey (*Allouatta palliata*; Estrada and Coates-Estrada, 1986) to a potential of several km in wide-ranging spider monkeys (*Ateles belzebuth*; see Milton and May, 1976). As with *Virola*, *Tetragastris* seeds taken by birds are regurgitated singly in the forest. However, the vast majority of *Tetragastris* seeds germinate in close proximity to others of their species, either under the parent tree or in faecal clumps.

In short, frugivores eating *Virola* and *Tetragastris* fruits impose starkly different challenges on seedlings. Subtle quantitative differences in the dispersal of each species are eclipsed by the fundamental distinction between seedlings that are usually isolated from conspecifics and those that virtually always germinate in dense aggregations.

Seed and seedling tactics

Plant adaptations for attracting different fruit-eating animals have recently been reviewed, and require no further elaboration here (Estrada and Fleming, 1986; Howe, 1986b; Wheelwright, 1985). In contrast, seed and seedling adaptations to different conditions that are imposed by different dispersal agents are virtually unknown. The *Virola* and *Tetragastris* studies discussed here offer insights to what may be a wide variety of unexplored but important demographic phenomena in tropical forests.

As is common in tropical trees, *Virola surinamensis* produces a substantial seed that is a valuable resource for insect seed predators (see Janzen, 1969). Weevil infestations are so extreme under and near fruiting trees that virtually no seeds or seedlings survive near their parents. Because secondary removal by rodents is negligible (Larson and Howe, 1987), undispersed seeds are as good as dead when they fall. Unlike *Tetragastris panamensis* seeds, *Virola surinamensis* seeds with arils intact do not germinate. Because *Virola surinamensis* recruitment depends on scattered seeds, in or near canopy gaps, there is no premium on adaptations for survival under intense seedling competition. Clumps of *Virola surinamensis* seedlings rarely if ever occur, or survive if they do occur. The demographic tactic of this animal dispersed plant, like some wind-dispersed species in the same forest (Augspurger, 1983*a*, 1983*b*), is one of escape from the parent and establishment elsewhere under a broken canopy.

Tetragastris panamensis must constantly contend with intense seedling competition. Birds scatter a few of the small seeds to good effect, but virtually all *Tetragastris panamensis* seeds drop or are knocked down under parent trees or are deposited in clumps in mammal droppings. In either situation, they persist for years as suppressed juveniles until minor breaks in the canopy allow them to grow. Whereas *Virola surinamensis* seedlings cannot survive in dense aggregations, *Tetragastris panamensis* seedlings rarely exist outside of them. More than 99% experience the 34–44% survival under parental crowns or in faecal clumps. Fewer than 1% of these seeds are scattered by birds, and consequently experience the unusually high 72% survival recorded here.

Density-dependent seedling mortality occurs under large *Tetragastris* trees (Howe, 1980), and rare isolated seedlings survive better than those under trees or in monkey scats. But density-dependent mortality is clearly not the overriding factor in contemporary *Tetragastris* demography (Figure 14.6), even though it must have been an overriding factor in the evolution of *Tetragastris* life history. Without adaptation for resistance to seed predation, pathogen infection, herbivory, and intraspecific competition, essentially no *Tetragastris* seeds and seedlings would survive a dispersal process that produces dense aggregations of both disseminated and undisseminated seeds.

In short, adaptation or lack of adaptation to high seedling densities places different ecological constraints on these species. *Virola surinamensis* probably could not survive in the presence of seed predators without guans, toucans, or other large avian dispersal agents (Howe, 1984*a*, 1986*a*). *Tetragastris panamensis* might well persist with inefficient dispersal; its capacity to survive in high seedling densities pre-adapts it to do so.

Ontogeny of spatial pattern

Patterns of seed fall and seedling and sapling mortality determine adult spatial pattern. Disproportionate seed or seedling mortality under and near parent trees could produce spaced distributions of adult trees (Connell, 1971; Janzen, 1970). If mortality under parents is less than complete, is random with respect to parents, or is limited to unfavourable environments, populations of surviving trees are clumped (Becker *et al.*, 1985; Hubbell, 1980). What insights might these seedling studies offer for explaining characteristics of the adult populations?

Rare *Virola* seedlings that survive anywhere near the parent will most likely be near the crown edge, 10–15 m from the tree base (Figure 14.7). Complete mortality under the crown, and low seed density combined with

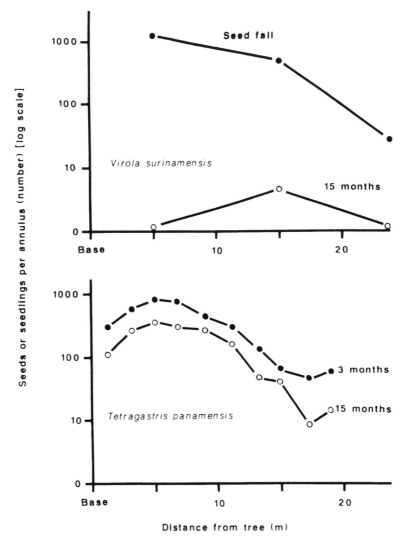

Figure 14.7 Expected seedling recruitment 15 months after fruit fall on 360° annuli around fruiting trees. *Virola surinamensis* seedlings that survive weevil depredations and later herbivory by mammals are rare, and are concentrated at the crown edge 10–15 m from the tree base. Moderate seed fall and moderating seed predation permit a few seedlings to survive. Calculations for an entire season assume twice the density of seeds found during the 2 peak weeks of fruit fall that were sampled. Much more abundant *Tetragastris panamensis* seedlings easily survive directly under the crown, as well as beyond the crown edge. However, most seedlings should be found at the crown edge, 5–10 m from the tree base. Animals carry 54% of the *Virola* seeds and 23% of *Tetragstris* seeds > 25 m from fruiting trees

substantial mortality beyond the crown edge, make survivors unlikely anywhere other than the immediate vicinity of the crown edge. Note that the average number of seedlings expected in a 360° annulus from 5–15 m (an area of 638 m²) is low (5), and that variance is high (Figure 14.6; also Howe *et al.*, 1985). In nature, none or several 15-month-old or older seedlings may be found. Even if they are found, 15-month-old seedlings near the crown edge might die before growing to sapling size. Calculations from seedling data can only give the qualitative prediction that *Virola surinamensis* seedlings near adult fruiting trees are likely to be at the crown edge. Despite the fact that about half of the seeds of this species are carried well beyond the 25 m radius (Howe, 1986*a*), the mean nearest neighbour distance of adult trees of this species on Barro Colorado Island is 19 m (Table 14.6). Continued juvenile and early adult mortality, which is probably density-independent, extends the nearest-neighbour distance to 34 m for mature trees. The species is infrequent, but clumped in low densities as these seedling data might predict.

On the other hand, *Tetragastris panamensis* might be expected to be both highly aggregated and more common (Figure 14.7). This is the case. Adults (≥ 11 cm dbh) are at least twice as abundant as *Virola surinamensis* adults (≥ 19 cm dbh) (see Hubbell and Foster, 1983). The average nearest neighbour distance is 5 m; mature trees average only 9 m apart (Table 14.6). Perhaps more to the point in the contemporary forest, juvenile *Tetragastris panamensis* are far more abundant near adults than juvenile *Virola surinamensis* are near conspecific adults.

Extensions and predictions

How general are these patterns? The question must be rhetorical because no answer is yet possible. The demographic dichotomy between widely scattered *Virola* seeds and seedlings and their consistently clumped *Tetragastris* counterparts is probably as distinct as can be found, and suggests general traits that might be expected of similar plants (Table 14.7). Definition of seedling syndromes then permit predictions about unknown aspects of *Virola surinamensis* and *Tetragastris panamensis* ecology, as well as predictions for other species. The fundamental prediction is that plants with the "scatter syndrome" should produce relatively unprotected seeds and seedlings that normally recruit as isolated individuals, whereas plants with the "clump syndrome" should produce seeds and seedlings well defended by allelochemicals, lignification, or mechanical protection.

The first benefit of a general syndrome is that it generates predictions about known dispersal systems in unknown places. For instance, *Virola surinamensis* occurs in the Amazon basin, where it is thought to be primarily monkey-

Table 14.6 Nearest neighbour distances of *Virola surinamensis* and *Tetragastris panamensis* trees to conspecifics in the completely censused 50 ha Hubbell-Foster plot on Barro Colorado Island, Panama

	Nearest neighbour distance (m)	
Species	To any conspecific (N)	From trees ≥ 15 cm dbh to conspecifics < 15 cm dbh (N)
V. surinamensis	19.3 ± 1.8 (239)	34.5 ± 4.2 (131)
T. panamensis	4.8 ± 0.1 (3477)	9.3 ± 0.6 (229)

Shown are means ± 9.5% confidence intervals
Source: Courtesy of S.P. Hubbell and R. Foster, unpublished data.

Table 14.7 Predicted seedling syndromes of tropical tree species that produce (≥ 1 g wet weight) frugivore-dispersed seeds with little or no dormancy

Scatter syndrome	Clump syndrome
Usually dispersed by airborne bats or birds (< 3 kg)	Usually dispersed by terrestrial birds or mammals or arboreal mammals (often >> 3 kg)
Seeds scattered singly	Seeds deposited in clumps
Seeds and seedlings vulnerable to insects, herbivores, pathogens, and seedling competition	Seeds and seedlings resistant to insects, herbivores, pathogens, and seedling competition
Seedling recruitment near parent trees rare	Seedling recruitment near parent trees common
Often establish in gaps	Often establish in shaded understorey

dispersed (e.g. Foster *et al.*, 1986). If the "scatter syndrome" is general for this species, seedling recruitment in monkey droppings and under parental crowns should be negligible in Peru and Brazil, as on Barro Colorado Island, whether or not *Conotrachelus* weevils are present. Less conspicuous toucans, guans, or other large avian frugivores should be the principal dispersal agents. Likewise, *Tetragastris panamensis* seedling biology in South America should resemble that on Barro Colorado Island. If howler monkeys are not the primary dispersal agents, other arboreal mammals should deposit seeds in clumps that produce bouquets of highly resistant seedlings. As in

Panama, substantial *Tetragastris* seedling recruitment should be expected near fruiting adult trees. In short, seedling strategy should be independent of intraspecific variation in dispersal agents and seed predators.

Other species lend some credence to the notion of general seedling syndromes that are linked to dispersal syndromes. For instance, bird-dispersed *Casearia corymbosa* (Flacourtiaceae) seedlings rarely survive under or near fruiting trees, but are more common 15–30 m away (Howe, 1977; Howe and Primack, 1975). In contrast, up to 80% of the seedlings from large-seeded and normally mammal-dispersed *Gustavia superba* (Lecythidaceae) survive in stands of this tree in Panama (Sork, 1985). Distributions of sapling and adult tropical trees are also consistent with an interplay between dispersal and seedling biology. Hubbell (1979) found that trees of Costa Rican dry forest that were dispersed by arboreal or terrestrial mammals were much more aggregated than those dispersed by birds or bats.

Can these syndromes accommodate other reproductive adaptations, such as extended seed dormancy, or deposition in immense clumps containing hundreds of seeds of dozens of species? Evidence is sketchy, but suggests that the fundamental distinction between scattering and clumping holds.

Many seeds, both large and small, are dormant when they are defaecated by vertebrates (e.g. Gautier-Hion *et al.*, 1985; Janzen, 1982, 1986). Even if deposited in clumps, there is likely to be secondary scattering of dormant seeds by insects, rodents, or water in the intervening months between fruit fall and germination. The demographic consequences are virtually never known, but may not have much influence on critical seed and seedling attributes. For instance, seed and seedling survival of *Faramea occidentalis* (Rubiaceae), an abundant understorey tree in Panama, generally fits the mammal-dispersed mode even though the seeds are dormant for several months (Schupp, 1987, 1988). Thirty-week survival (through establishment) averages approximately 7% under *Faramea* crowns, 24% 5 m away. Like *Tetragastris*, *Faramea* produces persistent carpets of seedlings in densities far higher than any observed in *Virola surinamensis* or in rain forest populations of *Casearia corymbosa*.

Large animals often retain seeds for weeks or even months, digest the majority, and pass the remainder in viable condition. Janzen (1982, 1986) has documented both digestive seed predation and seed dispersal by cattle and horses for a variety of Central American seeds, and similar processes are likely to occur in many species of wild antelope, buffalo, cattle, elephants, horses, and tapirs, as well as in large flightless birds such as rheas and cassowaries (see Howe, 1986*b* and references therein). Does seed processing by large animals impose different conditions on survivors than seed processing by monkeys or kinkajous?

Digestion and partial digestion of seeds, and deposition in immense clumps, distinguish dispersal by seed digesters from that by animals that digest pulp but not seeds. Digestive seed predation may select for extreme mechanical protection, and partial digestion in the guts of large herbivores often breaks dormancy (e.g. Lieberman *et al.*, 1979). This effect could be advantageous in seeds adapted for mammal consumption, or be catastrophic in seasonal climates in which seeds are forced to germinate during a severe dry season (see Garwood, 1983). Premature or not, germination in the dung of large birds or mammals sets up much more severe conditions of intra- and interspecific competition than those faced by *Tetragastris* seedlings in monospecific clumps. For instance, Alexandre (1978) records hundreds of seeds and ultimately seedlings of dozens of species of shrubs and trees in dung heaps of west African elephants (*Loxodonta africana*). Seeds of megafaunal fruits should be even more resistant to seed predators, pathogens and competition than those dropped in much smaller aggregations by arboreal mammals.

The foregoing discussion suggests genetic as well as ecological predictions for animal-dispersed plants well beyond those suggested by population geneticists (Loveless and Hamrick, 1984). The absence of large numbers of nearby offspring should decrease inbreeding and family structure, and increase overall outcrossing in scatter-dispersed species, resulting in relatively high levels of heterozygosity. Clump-dispersed species with many offspring neighbours should be expected to show higher levels of inbreeding, and consequently reduced heterozygosity, although consistent seed dispersal from several to several hundred metres will maintain a certain degree of outcrossing and consequently heterozygosity. Finally, gravity-dispersed trees that bear fruits not eaten by animals should show both the greatest degree of clumping of offspring around parents, the greatest degree of seed and seedling persistence, and the most pronounced inbreeding. Such gravity-dispersed plants should show lower heterozygosity than scatter-dispersed or clumped-dispersed species, all things (e.g. seed size, dormancy, pollination systems) being equal.

Implications for conservation

The relationships between seed dispersal and seedling demography have profound implications for tropical conservation policy (Howe, 1984*a*). Seasonal lows in fruit production often concentrate frugivore species on a very few tree species. Variously called "pivotal" or "keystone" species (Howe, 1977, and Gilbert, 1980, respectively), trees that provide critical resources during annual periods of fruit scarcity may support a variety of frugivores that disperse seeds of many other plant species at other times of the year. One could imagine that a chance disappearance of such a pivotal resource could force the extinction of other plants that depend on these animals for seed dissemination (Howe, 1977, 1984*a*). Are such scenarios likely? The extent to which plants depend on dispersal agents is inversely related to the ability of seeds and seedlings to survive in dense aggregations (Howe, 1985). Without proper dispersal, seeds fall in dense aggregations underneath parent trees. By this logic, species with seeds and seedlings that fit the scatter syndrome (Table 14.5) are highly vulnerable, while those with seeds that fit the clump syndrome or have extended seed dormancy are less vulnerable. Chance extinction of either avian frugivores or their preferred food plants might, or might not, quickly precipitate a series of interconnected extinctions, depending on the seedling ecology of the plants involved.

How vulnerable are species with seedlings that survive in dense aggregations? This is a question of demographic statistics, for which relevant data do not exist. Plants adhering to the seedling clump syndrome might persist if offspring replace parents. Persistence should be easiest for plants adapted for dispersal by large mammals that regularly deposit seeds in large clumps (perhaps a "megaclump syndrome"). However, there is no demographic evidence to support a claim by Janzen and Martin (1982) that many common tropical trees can persist for thousands of years without consistent dispersal (Howe, 1985). In fact, Alexandre (1978) found that west African trees deprived of elephant dispersal agents disappeared in decades, not millennia.

An emerging consensus holds that contemporary dispersal mutualisms rarely reflect long-term co-evolutionary adjustment of particular species of plants and animals (Fleming, 1989; Herrera, 1985; Howe, 1984*b*). It does not follow that ecological relationships between plants and frugivores are inconsequential. Tree species thrive only if seeds and seedlings survive and ultimately reproduce. In a developing world in which fruit-eating animals are frequently hunted to local extinction, and fruit-bearing trees are cut without regard to frugivore needs, many tree species will not persist, much less thrive.

SUMMARY

(1) In Panama, *Virola surinamensis* seeds are normally scattered by birds, while *Tetragastris panamensis* seeds are normally dropped under parental crowns or deposited in faecal clumps;

(2) Seed fall and seedling survival are highly variable under and near both tree species;

(3) Overwhelming seed and seedling mortality from weevils limit rare *Virola surinamensis* recruitment near parent trees to the edge of the tree crown;

(4) *Tetragastris panamensis* seedlings, adapted to intense competition in

faecal clumps, recruit profusely under and near parental crowns;

(5) Without toucans and other large avian frugivores that disperse its seeds, *Virola surinamensis* would be unable to recruit seedlings, and the species would face local extinction; and

(6) Seeds and seedlings adapted to survive in dense faecal clumps pre-adapts *Tetragastris panamensis* for at least temporary survival without normal dispersal agents.

ACKNOWLEDGEMENTS

This study would not have been possible without the sharp eys of Eugene W. Schupp, the diligent field assistance of Carlos Brandaris, and the co-operation of the Smithsonian Tropical Research Institute. The National Science Foundation (USA) provided financial support.

REFERENCES

Alexandre, D.Y. (1978). Le role disseminateur des elephants en foret de Taï, Côte d'Ivoire. *La Terre et La Vie*, **32**, 47–71

Augspurger, C.K. (1983a). Offspring recruitment around tropical trees: changes in cohort distance with time. *Oikos*, **40**, 189–96

Augspurger, C.K. (1983b). Seed dispersal of the tropical tree, *Platypodium elegans*, and the escape of its seedlings from fungal pathogens. *Journal of Ecology*, **71**, 759–72

Becker, P., Lee, L.W., Rothman, E.D. and Hamilton, W.D. (1985). Seed predation and the coexistence of tree species: Hubbell's models revisited. *Oikos*, **44**, 382–90

Clark, D.A. and Clark, D.B. (1984). Spacing dynamics of a tropical rain forest tree: evaluation of the Janzen-Connell model. *American Naturalist*, **124**, 769–88

Connell, J.H. (1971). On the role of natural enemies in preventing competitive exclusion in some marine animals and in rain forest trees. In Den Boer, P.J. and Gradwell, P.R. (eds) *Dynamics of Populations*, pp. 298–312. (Wageningen: Pudoc)

Docters van Leeuwen, W.M. (1954). On the biology of some Javanese Loranthaceae and the role birds play in their life-histories. *Beaufortia*, **41**, 105–206

Enders, R.K. (1935). Mammalian life histories from Barro Colorado, Panama. *Bulletin of the Museum of Comparative Zoology*, **78**, 85–502

Estrada, A. and Coates-Estrada, R. (1986). Frugivory in howling monkeys (*Alouatta palliata*) at Los Tuxtlas, Mexico: dispersal and the fate of seeds. In Estrada, A. and Fleming, T.H. (eds) *Frugivores and Seed Dispersal*, pp. 93–105. (The Hague: Dr. W. Junk Publishers)

Estrada, A. and Fleming, T.H. (eds). (1986). *Frugivores and Seed Dispersal*. (The Hague: Dr. W. Junk Publishers)

Fleming, T.H. (1989). The fruit-frugivore mutualism: the evolutionary theater and the ecological play. (in press)

Foster, R.B. and Brokaw, N. (1982). General character of the vegetation. In Leigh, E.G., Jr., Rand, A.S. and Windsor, D.S. (eds) *The Ecology of a Tropical Forest: Seasonal Rhythms and Long-term Changes*, pp. 67–82 (Washington: Smithsonian Press)

Foster, R.B., Arc, B., Javier and Wachter, T.S. (1986). Dispersal and sequential plant communities in Amazonian Peru floodplain. In Estrada, A. and Fleming, T.H. (eds) *Frugivores and Seed Dispersal*, pp. 357–70 (The Hague: Dr. W. Junk Publishers)

Frankie, G.W., Baker, H.G. and Opler, P.A. (1974). Comparative phenological studies of trees in tropical wet and dry forests in the lowlands of Costa Rica. *Journal of Ecology*, **62**, 881–919

Garwood, N.C. (1983). Seed germination in a seasonal tropical forest in Panama: a community approach. *Ecological Monographs*, **53**, 159–81

Gautier-Hion, A., Duplantier, J.M., Quris, R., Feer, F., Sourd, C., Decoux, J.P., Dubost, G., Emmons, L., Erard, C., Hecketsweiler, P., Moungazi, A., Roussilhon, C. and Thiollay, J.M. (1985). Fruit character as a basis of fruit choice and seed dispersal in a tropical forest vertebrate community. *Oecologia*, **65**, 324–37

Gilbert, L.E. (1980). Food web organization and the conservation of neotropical diversity. In Soulé, M.E. and Wilcox, B.A. (eds) *Conservation Biology: An Evolutionary-Ecological Perspective*, pp. 11–33 (Sunderland, Massachusetts: Sinauer Associates)

Herrera, C.M. (1985). Determinants of plant-animal coevolution: the case of mutualistic vertebrate seed dispersal systems. *Oikos*, **44**, 132–41

Howe, H.F. (1977). Bird activity and seed dispersal of a tropical wet forest tree. *Ecology*, **58**, 539–50

Howe, H.F. (1980). Monkey dispersal and waste of a neotropical tree. *Ecology*, **61**, 944–59

Howe, H.F. (1982). Fruit production and animal activity at two tropical trees. In Leigh, E.G., Jr., Rand, A.S. and Windsor, D.S. (eds) *The Ecology of a Tropical Forest: Seasonal Rhythms and Long-term Changes*, pp. 189–200 (Washington: Smithsonian Press)

Howe, H.F. (1983). Annual variation in a neotropical seed-dispersal system. In Sutton, S.L., Whitmore, T.C. and Chadwick, A.C. (eds) *Tropical Rain Forest: Ecology and Management*, pp. 211–27 (Oxford: Blackwell Scientific Publications)

Howe, H.F. (1984a). Implications of seed dispersal by animals for the management of tropical reserves. *Biological Conservation*, **30**, 261–81

Howe, H.F. (1984b). Constraints on the evolution of mutualisms. *American Naturalist*, **123**, 764–77

Howe, H.F. (1985). Gomphothere fruits: a critique. *American Naturalist*, **125**, 853–65

Howe, H.F. (1986a). Consequences of seed dispersal by birds: a case study from Central America. *Journal of the Bombay Natural History Society (Supplement)*, **83**, 19–42

Howe, H.F. (1986b). Seed dispersal by fruit-eating birds and mammals. In Murray, D.R. (ed.) *Seed Dispersal*, pp. 123–90 (Sydney: Academic Press)

Howe, H.F. and Estabrook, G.F. (1977). On intraspecific competition for avian dispersers in tropical trees. *American Naturalist*, **111**, 817–32

Howe, H.F. and Primack, R. (1975). Differential seed dispersal by birds of the tree *Casearia nitida* (Flacourtiaceae). *Biotropica*, **7**, 278–83

Howe, H.F. and Smallwood, J. (1982). Ecology of seed dispersal. *Annual Review of Ecology and Systematics*, **13**, 201–28

Howe, H.F. and Vande Kerckhove, G.A. (1981). Removal of wild nutmeg (*Virola surinamensis*) crops by birds. *Ecology*, 62, 1093–106

Howe, H.F. and Westley, L.C. (1988). *Ecological Relationships of Plants and Animals*. (New York: Oxford University Press

Howe, H.F., Schupp, E.W. and Westley, L.C. (1985). Early consequences of seed dispersal for a neotropical tree (*Virola surinamensis*). *Ecology*, **66**, 781–91

Hubbell, S.P. (1979). Tree dispersion, abundance, and diversity in a tropical dry forest. *Science*, **203**, 1299–309

Hubbell, S.P. (1980). Seed predation and the coexistence of tree species in tropical forests. *Oikos*, **35**, 214–29

Hubbell, S.P. and Foster, R. (1983). Diversity of canopy trees in a neotropical forest and implications for conservation. In Sutton, S.L., Whitmore, T.C. and Chadwick, A.C. (eds) *The Tropical Rain Forest: Ecology and Management*, pp. 25–41 (Oxford: Blackwell Scientific Publications)

Janzen, D.H. (1969). Seed eaters versus seed size, number, toxicity, and dispersal. *Evolution*, **23**, 1–27

Janzen, D.H. (1970). Herbivores and the number of tree species in tropical forests. *American Naturalist*, **104**, 501–28

Janzen, D.H. (1982). Removal of seeds from horse dung by tropical rodents: influence of habitat and amount of dung. *Ecology*, **63**, 1887–900

Janzen, D.H. (1986). Mice, big mammals, and seeds: it matters who defecates what where. In Estrada, A. and Fleming, T.H. (eds) *Frugivores and Seed Dispersal*, pp. 251–72 (The Hague: Dr. W. Junk Publishers)

Janzen, D.H. and Martin, P. (1982). Neotropical anachronisms: what the gomphotheres ate. *Science*, **215**, 19–27

Larson, D. and Howe, H.F. (1987). Agouti seed dispersal predation on *Virola surinamensis*:

appearances and reality. *Journal of Mammalogy*, **68**, 859–60

Leigh, E.G. Jr., Rand, A.S. and Windsor, D.S. (eds) (1982). *The Ecology of a Tropical Forest: Seasonal Rhythms and Long-term Changes*. (Washington, D.C.: Smithsonian Press)

Levey, D.J. (1987). Seed size and fruit-handling techniques of avian frugivores. *American Naturalist*, **129**, 471–85

Lieberman, D., Hall, J.B., Swaine, M.D. and Lieberman, M. (1979). Seed dispersal by baboons in the Shai Hills, Ghana. *Ecology*, **60**, 65–75

Loveless, M.D. and Hamrick, J.L. (1984). Ecological determinants of genetic structure in plant populations. *Annual Review of Ecology and Systematics*, **15**, 65–96

Milton, K. and May, M.L. (1976). Body weight, diet and home range area in primates. *Nature*, **259**, 459–62

Moermond, T.C. and Denslow, J. (1985). Neotropical avian frugivores: patterns of behavior, morphology, and nutrition with consequences for fruit selection. In Buckley, P.A., Foster, M.S., Morton, E.S., Ridgely, R.S. and Smith, N.G. (eds) *Neotropical Ornithology*. Ornithological Monographs **Volume 36**, pp. 865–97 (Lawrence: American Ornithologists Union)

Schupp, E.W. (1987). Studies on seed predation of *Faramea occidentialis*, an abundant tropical tree. Ph.D. Dissertation. University of Iowa, Iowa City, Iowa.

Schupp, E.W. (1988). Predation on seeds and early seedlings in the forest understory and in treefall gaps. *Oecologia*, **76**, 525–30

Sork, V.L. (1985). Germination response in a large-seeded neotropical tree species, *Gustavia superba* (Lecythidaceae). *Biotropica*, **17**, 130–6

Wheelwright, N.T. (1985). Fruit size, gape width, and the diets of fruit-eating birds. *Ecology*, **66**, 808–18

Wheelwright, N.T. and Orians, G. (1982). Seed dispersal by animals: contrasts with pollen dispersal, problems of terminology, and constraints on coevolution. *American Naturalist*, **119**, 402–13

Willis, E.O. (1980). Ecological roles of migratory and resident birds on Barro Colorado Island, Panama. In Keast, A. and Morton, E.S. (eds) *Migrant Birds in the Neotropics*, pp. 205–26, (Washington, D.C.: Smithsonian Institution Press)

CHAPTER 15

INTERACTIONS AMONG FRUIT AND VERTEBRATE FRUIT-EATERS IN AN AFRICAN TROPICAL RAIN FOREST

Annie Gautier-Hion

ABSTRACT

More than half of the mammal species of the tropical rain forest of north-eastern Gabon are primary consumers, and fruit and seeds make up the staple diet of 85% of these species. There exists a considerable overlap in fruit species taken by distant mammal groups (rodents, ruminants and monkeys) as well as between mammals and the large canopy birds. Interactions between 39 bird and mammal fruit-eaters and the fruit species they consumed were studied for 1 year. On the basis of simple morphological traits of fruits, it was possible to identify two large guilds. The first included large birds and monkeys and plant species whose seeds were dispersed without predation; the second guild included rodents and ruminants and fruit species which underwent predispersal seed predation. These results suggested some diffuse coadaptation between plants and fruit-eaters, which could result from the pressure of consumers acting as dispersal agents or seed predators. However, the comparative analysis of fruiting patterns according to the plant-life form, the morphological fruit type and the seed dispersal mode, did not suggest an evolutionary influence of dispersal agents on fruiting phenology and we found no evidence of close evolution between plant and consumer species.

INTRODUCTION

Among the 126 species of mammals recorded within 80 km of Makokou, a tropical rain forest in north-eastern Gabon, about 53% are primary consumers. Fruit and seeds make up the staple diet of 85% of these primary consumers; the remaining 15% are predominantly leaf-eaters. Fruits constitute the major food of seven out of 11 ruminant species (Dubost, 1984), eight out of nine

219

squirrels (Emmons, 1980), as well as the 13 diurnal primate species (Hladik, 1973; Gautier-Hion, 1978; Tutin and Fernandez, 1985). Frugivory is less developed among birds; however about 15%–20% of the species are estimated to be primary fruit-eaters. They are mainly medium-sized and large birds such as bulbuls, starlings, turacos and hornbills (Brosset and Erard, 1986).

Among all these frugivorous species, the overlap in fruit diets is quite high, even between distant taxa (Gautier-Hion *et al.*, 1980; Emmons *et al.*, 1983). Out of a sample of 112 fruit species found in the diet of ruminants, monkeys, and rodents, 35% were used by the three taxa, and at least 70% by two of them. Out of 44 fruit species eaten by seven species of large canopy birds, only 7% were not shared by some mammal consumers, and 50% were eaten by all consumer groups (Gautier-Hion *et al.*, 1985a).

This lack of specificity is exemplified by the guild of consumers observed around large fruiting trees. For example, in 61.5 hours spent observing day- and night-consumers at a single fruiting *Trichilia gilgiana* tree (Meliaceae), 22 species were seen eating fruit in more than 350 visits. This guild included two ruminants, eight rodents, 10 birds, and two monkey species (Gautier-Hion, unpublished data).

Thus, in the tropical rain forest of Makokou, there is a great extent of frugivory and considerable overlap in fruit utilization among consumers. This suggests that, in order to avoid simplistic generalizations on the existence of co-adaptions between plant and consumer species, studies on plant–animal interactions have to be based on trophically related guilds, rather than on taxonomic groups.

In seeking insights on possible co-adaptions between frugivorous vertebrates and their food plants, we addressed three questions. Are there broad morphological fruit characters which would account for the choice and partitioning of the available fruit spectrum among the vertebrate community? Do such characters evolve under the pressure of consumers considered as seed-dispersers or predators? Is there any evolutionary influence of dispersal agents on temporal patterns of fruiting?

The present contribution mostly summarizes the results obtained during a 1-year study on the vertebrate community of Makokou, conducted by a research team including zoologists and botanists. Details on methods and results can be found elsewhere (Gautier-Hion *et al.*, 1985a, 1985b).

RELATIONS BETWEEN FRUIT MORPHOLOGY AND FRUIT CHOICE BY CONSUMERS

The consumer community studied included seven large canopy birds, eight species of small rodents, nine squirrels, two large rodents, seven ruminants and six monkey species (a total of 39 species). The fruit morphology of 122 species of plants whose fruits were eaten by at least one consumer group was described in terms of simple characters that accounted for the energy needs of animals as well as their capacities of perception, manipulation and mastication. Such characters included, fruit and seed weight, fruit colour, the texture of the protective coat preventing access to the flesh and seeds, the type of edible tissue and the number of seeds. The seven categories of parameters defined included 25 variables. The overall relations between these variables and the six consumer groups were tested in a contingency table which was analysed by a multifactorial analysis (see details in Gautier-Hion *et al.*, 1985*a*).

It was shown (Figure 15.1) that the consumer groups were arranged, first, around the parameter of fruit weight which separated birds from large rodents and ruminants (axis 1): then, around the parameters of fruit colour, where monkeys diverged from squirrels (axis 2). Both birds and monkeys were found to be selective feeders. "Bird fruits" could be defined as small, red or purple, without protection, and more often as dehiscent fruit with arillate seeds. Monkeys mainly took red, orange and yellow fruit either with a succulent pulp or arillate seeds.

In contrast, small rodents appeared as opportunistic feeders and squirrels were not very selective. Large rodents preferentially took large-sized indehiscent fruit with fibrous flesh and seeds protected by hard kernels. Ruminants took a large variety of fruits but avoided the smallest. The overlap in fruit choice was not clearly based on taxonomic relatedness but more obviously on foraging levels and energy needs. In fact, we identified a "bird-monkey syndrome" characterized by brightly coloured fruit with succulent pulp or arillate seeds; and a "ruminant-large rodent syndrome" with large indehiscent fruit and fibrous pulp. Clearly, the diets of large rodents were more similar to those of ruminants than to those of related groups.

FRUIT CHARACTERS AND CONSUMER ACTION ON SEEDS

Depending on whether they eat pulp or seeds, vertebrates are commonly classified as seed-dispersers or seed-predators. Large rodents, squirrels and small rats are mostly seed-eaters that normally discard fruit flesh (Emmons, 1980; Duplantier, 1982). Ruminants eat entire fruits or only seeds (Gautier-Hion *et al.*, 1980). Finally, large birds and monkeys are mostly pulp-eaters (Gautier-Hion, 1980, 1984, Erard, pers.comm.). We found that large birds and monkeys respectively dispersed seeds of 90% and 80% of the species while ruminants and squirrels destroyed seeds of about 70% of the species (N = 82; Gautier-Hion *et al.*, 1985*a*).

Among ruminants, there are significant interspecific differences in the fruit parts eaten; seeds are the only fruit part taken in 30 to 55% of cases

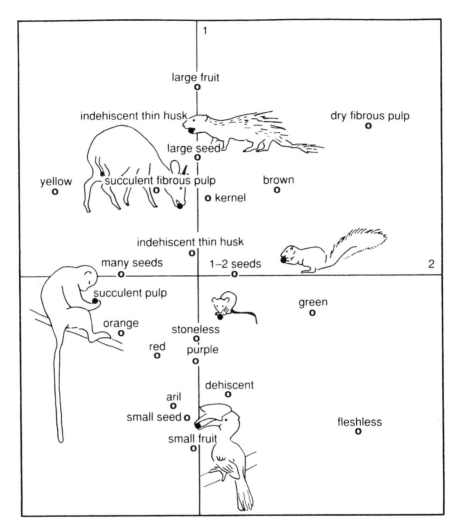

Figure 15.1   The interrelationships among the six groups of consumers and the fruit characters of their food. The factorial plane 1–2 accounts for 83% of the total inertia. Black circles: active variables for consumers; white circles: active variables for fruit (from *Oecologia*, see details in Gautier-Hion *et al.*, 1985*a*)

depending upon the ruminant species involved (Dubost, 1984). Furthermore, when the whole fruit is taken, seeds are not always destroyed, but can be spat out during rumination, away from the fruit source (Gautier-Hion *et al.*, 1985*a*; Feer, pers.comm.). As it is not easy to carry out such observations, the dispersal role of ruminants is probably underestimated.

Similarly, the estimate of the seeds dispersed by rodents through food hoarding cannot be easily determined, while the fate of stored seeds has been seldom investigated (Emmons, 1980; Duplantier, 1982).

Seed dispersal is also difficult to document in the so-called dispersers. In four monkey species, it was found that seeds were not swallowed in 61% of cases. In such situations, monkeys either dropped seeds under the parent tree, or filled their cheek pouches with whole fruits, before moving further to eat and dropping the seeds at a distance from the parent tree. When seeds were swallowed, they were found undamaged in 68% of cases. However, interspecific differences in handling seeds exist. For example, seeds were totally or partially broken in 54% of cases in *Cercopithecus pogonias* as opposed to 17% in *C. cephus* – despite the fact that these two monkeys are similar in body weight and tooth morphology. The fate of seeds also varies for the same plant species. Thus, the fruits of *Cissus dinklagei* (Vitaceae) were ingested *in toto* in 79 and 92% of cases respectively by the two species. However, in *C. cephus* stomachs, seeds were found crunched in 19% of cases, as compared to 69% in *C. pogonias* (Gautier-Hion, 1984).

Such results underline the difficulty of evaluating accurately the dispersal role of consumers without analysis of dispersal processes. Expected dispersers may indeed cause a significant amount of destruction upon seeds. Similarly, it is clear that seed-eaters are the exclusive dispersers of some of the fruit they eat, although the cost may be high.

Despite these limitations, we divided fruit species into two categories according to whether or not seeds are subject to predation before dispersal (Janzen, 1969). Species which did suffer seed predation before dispersal mainly bore heavy and dull fruit, which were either dehiscent without arillate seeds, or indehiscent with a dry fibrous pulp and seeds protected by a hard kernel (Gautier-Hion *et al.*, 1985a, Table 15.1). Conversely, fruit with no pre-dispersal seed predation were mostly small- or medium-weighted, and their seeds were not protected; they were brightly coloured and possessed either a succulent pulp or arillate seeds. These resources were found to be respectively richer in carbohydrates, and richer in proteins and fatty acids, than dry fibrous pulp (Table 15.2, Sourd and Gautier-Hion, 1986, unpublished data).

The simultaneous study of fruit characters and consumer actions on the seeds therefore suggests some diffuse coadaptation between plants and their guilds of consumers acting as seed-dispersers or predators. The selective pressure which could have been exerted by the predator-dispersers would have mainly led to a thickening of the protective coat, and to the development of a fibrous mesocarp (a useful nutrient for large ruminants and the African elephant). Conversely, plants whose seeds are dispersed without significant cost, have developed attractive displays and sugar- or lipid-rich resources for consumers (Thompson, 1982).

Table 15.1 Fruit characters which differ according to the mode of seed dispersal (+ + +: $p < 0.001$; +: $p < 0.05$)

		% of plant species with:	
Fruit characters		No seed predation before dispersal	Seed predation before dispersal
Colour	bright	87.5	20.5
+++	dull	12.75	79.5
Flesh	aril	29.0	0
+++	succulent pulp	71.0	20.5
	dry fibrous pulp	0	41.5
	no flesh	0	38.0
Weight	< 10g	58.5	32.0
+	> 10g	41.5	68.0
Kernel	absent	94.0	72.5
+	present	6.0	27.5

Table 15.2 Nutrient content of food items according to the type of fruit flesh: mean, standard deviation and range values; ( ): number of fruit species analysed. + : $p < 0.05$; + + : $p < 0.01$; + + + : $p < 0.001$; –, – –, – – –, indicate the figures which are significantly smaller; +, + +, + + + those which are greater (Student t test)

		Nutrient content, % of dry weight		
Type of flesh	Water content, %	Sugars	Proteins	Fatty acids
Succulent pulp,	81.4	22.3	3.1	7.8
coloured fruit (15)	+7.16	+24.2	+1.55	+4.4
	(69–93)	(4–79)	(1.3–5.9)	(2–16.9)
	+	+	– – –	– –
Arils,	62.7	12.2	5.8	23.2
coloured fruit (9)	+2.02	+13.1	+1.45	+9.9
	(35–87)	(1–38.7)	(4.3–8.2)	(10.3–36.7)
			+ + +	+ + +
Dry pulp,	70.3	4.85	3.3	5.25
dull fruit (6)	+12.3	+4.65	+2.75	+3.64
	(54–89)	(0.1–14)	(1.2–8.1)	(1–10.4)
	–	–		– – –

DETERMINANTS OF THE TIMING OF FRUIT PRODUCTION: SOME COMPETING HYPOTHESES

Fruiting in the Makokou forest is seasonal, with yearly maxima occurring during the two rainy seasons, while the main dry season (June to August) represents the period of fruit scarcity. However, large interannual variations do exist (see review in Gautier-Hion *et al.*, 1985*b*). To test the possible evolutionary influence of fruit consumers on the timing of fruit production, we compared the fruiting patterns according to the dispersal modes of the species and looked at other alternative constraints such as plant life forms and fruit types (Gautier-Hion *et al.*, 1985*b*).

Fruiting patterns and seed-dispersal modes

The amplitude of fruiting seasonality of zoochorous species was similar to that of anemochorous and 'autochorous' species and for the three dispersal modes, the maxima and minima of production were synchronous. Thus, no obvious desynchronization in fruiting patterns according to seed dispersal mode was observed (Figure 15.2). We then compared the fruiting patterns of plants which did not suffer from predation before seed dispersal (brightly coloured fruit with or without arils) and of those which suffered from seed predation (dull-coloured indehiscent fruit and dehiscent fruit without arils, see above). In both cases, fruiting was seasonal; its amplitude was comparable and peaks were approximately synchronous (Figure 15.2).

Therefore, species dispersed by members of the bird-monkey guild did not show fruiting patterns staggered in time (as observed by Snow, 1966; Smythe, 1970; Charles-Dominique *et al.*, 1981). Moreover, there was no evidence that plants whose seeds underwent pre-dispersal predation were more strongly seasonal than others, as expected from the satiation hypothesis (Janzen, 1969; Sabatier, 1985).

Fruiting patterns and plant life forms

Great differences were found in the amplitude and timing of fruit production in terms of plant life form (Figure 15.3). In respect to the number of fruiting species, middle-sized trees were found to be the most seasonal, and large trees and lianas the least. The greatest numbers of fruiting individuals occurred during the main rainy season and the short dry season, except for small-sized trees which gave fruit earlier in the main dry season. Seasonality was especially marked for middle-sized trees and lianas, and least so for tall trees.

Fruiting patterns and fruit types: the case of Myristicaceae

Plant species bearing dehiscent fruit with or without arils showed a similar pattern of fruiting, with maxima in dry seasons, although the former plants are zoochorous species which do not suffer from seed predation while the latter's seeds are destroyed by animals (Gautier-Hion *et al.*, 1985*b*).

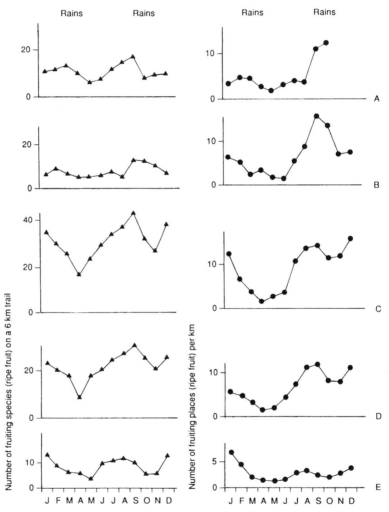

Figure 15.2 Fruiting patterns according to the mode of seed dispersal in terms of the number of species and of individuals giving ripe fruit (on a 6 km trail visited bi-monthly). A: 'autochorous species' (N = 29; seed dispersal mode frequently unknown); B: anemochorous species (N = 18); C: all zoochorous species (N = 99); D: zoochorous species with no seed predation before dispersal (N = 79); E: zoochorous species with seed predation before dispersal (N = 20)

Coelocaryon preussi, *Pycnanthus angolensis* and *Staudtia gabonensis* are three species of Myristicaceae which produce very similar dehiscent fruit, including one arillated seed, which begin to ripen during the main dry season. They are eaten by quite a large animal guild including birds, rodents, ruminants and monkeys, and might represent keystone plants for most frugivorous vertebrates (Gautier-Hion and Michaloud, 1989).

Arboreal birds and monkeys are efficient dispersers of Myristicaceae seeds and competition for seed-dispersers is likely. Indeed, we observed that, while the first trees seen to bear ripe fruit attracted a great number of consumers (up to 19 species in the first few days, pers. obs.), as the season

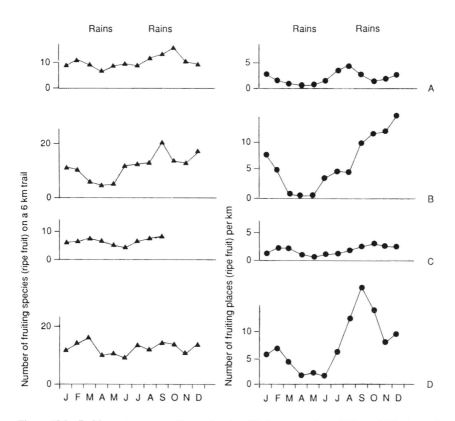

Figure 15.3 Fruiting patterns according to the plant life forms. (see legend Figure 15.2). A: small trees (<15 m; N = 36); B: middle-sized trees (15–30 m; N = 33); C: tall trees (>30 m; N = 21); D: lianas (N = 32)

went on, trees received fewer and fewer visits. Such competition for dispersers could be expected to have caused a character displacement of the timing of fruiting (e.g. Snow, 1966; Smythe, 1970). In fact, the phenological monitoring of 26 trees has shown that there was a good intra- and interspecific synchrony of fruiting. In contrast, a fourth Myristicaceae species which bears indehiscent drupes with fibrous pulp (*Scyphocephalium ochocoa*), mainly fruits 6 months later, when morphologically similar drupes belonging to distant taxa such as *Panda oleosa* (Pandaceae) or *Coula edulis* (Olacaceae) mature (Gautier-Hion*et al.*, 1985*b*).

Fruiting patterns: selective forces

We found no evidence that the availability of seed dispersers or seed predators shapes the fruiting patterns of their plant foods. In fact, the most obvious differences in fruiting patterns were observed when comparing different plant forms or morphologically different fruit types. Such differences could be related to abiotic constraints. For example, the ripening of dehiscent fruits, in the dry seasons, could result from the decrease of humidity, which should favour fruit dehiscence. Similarly, in the rainy season, the increase in moisture and sunshine should favour the ripening of succulent fleshy fruit.

The existence of such a favourable season for fruit ripening is supported by the allometry observed within a given plant category, between the seasonal changes in the number of fruiting species and in the number of fruiting individuals, the latter generally being more seasonal (see for example the case of lianas in Figure 15.3). Indeed, such allometry could partly result from the shortening of the maturation delay of fruit in a number of individuals, which should allow them to meet the most favourable period for ripening. This should induce an aggregated fruiting pattern even if the flowering is generally observed as being more scattered (Hecketsweiler pers.comm. for Makokou; Sabatier, 1985 for French Guyana; also see review by Wheelwright, 1985).

Similarly, microclimatic conditions (temperature, moisture, light access), which obviously differ for plant forms depending on their height and overall architecture, could account for the differences in fruiting phenology observed between small-, medium-sized and tall trees and between trees and lianas. On the other hand, we must recognize that the main constraints upon plant phenology most likely operate at other periods such as the flowering or the germination stages.

SUMMARY AND CONCLUSION

In summary, the following points can be made. First, the organization of the trophic community of vertebrates, studied at Makokou, shows a great overlap in fruit diets between distant taxa, and a lack of specificity between fruit and consumer species.

Second, it is nevertheless possible to identify two large guilds. The first guild includes large birds and monkeys, and plant species whose seeds are dispersed without significant predation: the fruits of these species are brightly coloured and possess a flesh rich in sugars or in fatty acids. The second guild includes rodents and ruminants (together with the elephant) and fruit species which suffer from pre-dispersal seed predation. Their fruits are dull with fibrous and nutritionally poor flesh and well-protected seeds. The occurrence of such guilds may account for diffuse coadaptation between plants and fruit eaters.

Third, the results of the comparative analysis of fruiting patterns in relation to dispersal mode do not substantiate the coadaptation concept as accounting for the timing of fruit production of zoochorous species.

Corner (1953) commented that 'Facts can be picked from among flowering plants to fit any theory'. Indeed, by selecting appropriate plant species subsets and considering a limited range of animal consumers, we could have succeeded in identifying coadaptive trends between plants and vertebrates. But, in the Makokou forest, the zoochorous plant species studied have a large set of alternative consumers which can take over from one another in seed dispersal; this outstanding lack of specificity precludes close coevolution between plant and vertebrate species. These results support the views of Herrera (1985) who stressed that a number of factors and especially 'the weak selective pressures of dispersers on plants, render close evolution between particular plant and disperser species unlikely.'

ACKNOWLEDGEMENTS

I would like to thank all colleagues who contributed to the field work of the "FVF-ECOTROP" programme in 1981. This research was supported by the Centre National de la Recherche Scientifique. We are grateful to the Ministère de la Recherche Scientifique of Gabon and to Dr. P. Posso, Director of the Institut de Recherche en Ecologie Tropicale at Makokou, for facilitating our study in the field.

REFERENCES

Brosset, A. and Erard, C. (1986). Les oiseaux des régions forestières du Nord-Est du Gabon. 1. Ecologie et comportement des espèces. *Revue d'Ecologie (Terre Vie)*, Suppl., **3**, 1–289

Charles-Dominique, P., Atramentowicz, M., Charles-Dominique, M., Gerard, H., Hladik, A., Hladik, C.M. and Prevost, M.F. (1981). Les mammifères frugivores arboricoles nocturnes d'une forêt guyanaise: interrelations plantes-animaux. *Revue d'Ecologie (Terre Vie)*, **35**, 341–436

Corner, E.J.H. (1953). The Durian theory extended. I. *Phytomorphology*, **3**, 465–76

Dubost, F. (1984). Comparison of the diets of frugivorous forest ruminants of Gabon. *Journal of Mammalogy*, **65**, 298–316

Duplantier, J.-M. (1982). Les rongeurs myomorphes forestiers du N.-E. Gabon. Thèse de Troisième Cycle. Université de Montpellier, Montpellier.

Emmons, L.H. (1980). Ecology and resource partitioning among nine species of African forest squirrels. *Ecological Monographs*, **50**, 31–54

Emmons, L.H., Gautier-Hion, A. and Dubost, G. (1983). Community structure of the frugivorous-folivorous forest mammals of Gabon. *Journal of Zoology, London*, **199**, 209–22

Gautier-Hion, A. (1978). Food niche and coexistence in sympatric primates in Gabon. In Chivers, D.J. and Herbert, J. (eds), *Recent Advances in Primatology*, pp. 269–86. (New York: Academic Press)

Gautier-Hion, A. (1980). Seasonal variations of diet related to species and sex in a community of *Cercopithecus* monkeys. *Journal of Animal Ecology*, **49**, 237–69

Gautier-Hion, A. (1984). La dissémination des graines par les cercopithèques forestiers africains. *Revue d'Ecologie (Terre Vie)*, **39**, 159–65

Gautier-Hion, A., Emmons, L.H. and Dubost, G. (1980). A comparison of the diets of three major groups of primary consumers of Gabon. *Oecologia*, **45**, 182–9

Gautier-Hion, A., Duplantier, J.M., Quris, R., Feer, F., Sourd, C., Decoux, J.P., Emmons, L., Dubost, G., Erard, C., Hechestweiler, P., Moungazi, A., Roussilhon, C. and Thiollay, J.-M (1985a). Fruit characters as a basis of fruit choice and seed dispersal in a tropical forest vertebrate community. *Oecologia*, **65**, 324–37

Gautier-Hion, A., Duplantier, J.-M., Emmons, L., Feer, F., Heckestweiler, P., Moungazi, A., Quris, R. and Sourd, C. (1985b). Coadaptation entre rythmes de fructification et frugivorie en forêt tropicale humide du Gabon: mythe ou réalité? *Revue d'Ecologie (Terre Vie)*, **40**, 405–34

Gautier-Hion, A. and Michaloud, G. (1989). Are figs always keystone resources for tropical frugivorous vertebrates? A test in Gabon. *Ecology*, **70**, 1826–33

Herrera, C.M. (1985). Determinants of plant-animal coevolution: the case of mutualistic dispersal of seeds by vertebrates. *Oikos*, **44**, 132–41

Hladik, C.M. (1973). Alimentation et activité d'un groupe de chimpanzés réintroduits en forêt du Gabon. *La Terre et la Vie*, **27**, 343–413

Janzen, D.H. (1969). Seed eaters versus seed size, number, toxicity and dispersal. *Evolution*, **23**, 201–28

Sabatier, D. (1985). Saisonnalité et déterminisme du pic de fructification en forêt guyanaise. *Revue d'Ecologie (Terre Vie)*, **40**, 289–320

Smythe, N. (1970). Relationships between fruiting season and seed dispersal methods in a neotropical forest. *American Naturalist*, **104**, 25–36

Snow, D.W. (1966). A possible selection factor in the evolution of fruiting species in a tropical forest. *Oikos*, **15**, 274–81

Sourd, C. and Gautier-Hion, A. (1986). Fruit selection by a forest guenon. *Journal of Animal Ecology*, **55**, 235–44

Thompson, J.N. (1982). *Interaction and Coevolution*. (New York: John Wiley)

Tutin, C.E.G. and Fernandez, M. (1985). Foods consumed by sympatric populations of gorillas and chimpanzees in Gabon. *International Journal of Primatology*, **6**, 27–43

Wheelwright, N.T. (1985). Competition for dispersers, and the timing of flowering and fruiting in a guild of tropical trees. *Oikos*, **44**, 465–77

Section 5

Seed physiology, seed germination and seedling ecology

CHAPTER 16

SEED PHYSIOLOGY, SEED GERMINATION AND SEEDLING ECOLOGY – COMMENTARY

Richard B. Primack

Seedling establishment is necessary for tree populations to complete their life cycle. Seedling establishment can be divided into the phase of seed germination and the subsequent growth of the seedling. The three papers in this section focus on this sequence in seedling development. All three papers take a broad view of their subject, and consider a wide range of examples. This survey perspective is appropriate because our general knowledge of seed biology in tropical trees is quite limited. Vázquez-Yanes and Orozco-Segovia report on recent research findings concerning seed dormancy and the requirements for seed germination, drawing mainly on results from the Los Tuxtlas station in Mexico. Hladik and Miquel develop a framework for categorizing seedling types and the processes of establishment in the African rain forest. This framework is then used in comparison with other tropical forests. The contribution by Augspurger is an original attempt to consider the role of fungal pathogens in tropical tree populations. While the three papers treat separate topics in seed and seedling biology, the basic ecological and morphological characters result in a certain continuity of ideas that unify the material.

Vázquez-Yanes and Orozco-Segovia point out that most trees of the mature tropical rain forest produce fruits with large, heavy seeds. These seeds have a high water content, and germinate quickly. Because of the large stored food reserves, the resulting seedlings have an extensive and often deep root development and a comparatively large leaf area. The seedlings are often well-adapted to shady or partially shaded conditions. If these seeds are kept in dry conditions or in some other way prevented from germinating, they will die. Since the seeds of these species lack dormancy, they are not found in the soil seed bank. A second group of trees in the mature-phase forest produce smaller seeds with a lower water content and often a hard seed coat. Certain species are adapted to dispersal by wind, while others have fruits that are ingested whole by vertebrate fruit dispersers. These species have a range of germination types but many may remain dormant in the soil if conditions are not suitable for germination. The third group is the pioneer species that invade gaps in the forest canopy and other disturbed areas. These species often produce small seeds of low water content. These seeds are adapted to remain dormant in the soil until conditions become favourable for germination. Many of these species have a light requirement for seed germination and are sensitive to red/far-red light ratios. These categories are, of course, generalizations, and many species do not fit into this simple framework. However, at this stage in our knowledge, such a framework is a starting point for further research.

Research on seed dormancy has considerable relevance to the management of tropical rain forest. Following large-scale disturbance of the forest by clear-cut logging, agriculture and forest fire, the primary source of seeds for recolonization of the area will come from dormant seeds in the seed bank. The pioneer species, and other species with dormant seeds, are pre-adapted to take advantage of such wide-scale disturbance by the high densities of seeds in the soil. When an area is disturbed, with the mineral soil exposed and sunlight falling on the ground, these seeds will germinate and quickly form a carpet of seedlings. A knowledge of seed dormancy and the seed bank will allow predictions to be made of the types of species likely to persist following forest disturbance.

A knowledge of seed dormancy is also relevant to forest managers and geneticists interested in storing seeds. Forest managers often wish to store large numbers of seeds for planting one or many years later during a reforestation programme. Forest geneticists may wish to store seeds of an economically valuable species from a number of localities in order to preserve the genetic diversity of the species for later use in a genetic improvement programme. For many tree species of the moist tropics, such seed storage is presently impossible or very difficult, seriously impeding efforts at reforestation and tree breeding. For example, species of dipterocarps are the most valuable timber trees in South-east Asia. The seeds are large, have a high water content, and lack dormancy. Because the trees only fruit every several years, the seeds of these important timber trees cannot be incorporated into the regular forestry planting programmes.

One potential reason why so many tree species can coexist in the tropical rain forest is the variety of specializations for establishment niches. Each species may have a particular type of seedling which is best adapted to a particular set of conditions, such as light, moisture, temperature, and herbivores. This issue is addressed by the innovative paper of Hladik and Miquel in which seedlings are classified into five morphological types. These classifications are based on whether the cotyledons are held in the air or at (or under) ground level, whether the cotyledons are thin and leafy or fleshy, and whether the cotyledons remain inside the seed coat or come out following germination.

In a comparison of four tropical forests, the overwhelming majority of seedlings belong to two common types: species with small seeds that develop into seedlings with upright, leafy cotyledons, and species with large seeds in which the cotyledons remain at ground-level. The small-seeded species include most of the pioneer species and vines and even some of the shade-tolerant tree species. These species also tend to be dispersed by a wide range of generalized animal dispersers or by wind. In contrast, species with large seeds are mainly trees of the mature forest, and dispersal is more by specialized vertebrates. Seedling surveys in the mature forest also show a surprising abundance of seedlings of species possessing fleshy cotyledons produced above ground. In many species, in families such as the Leguminosae, Meliaceae and Burseraceae, the seed coat does not cover the fleshy cotyledons; in other species, the upright, fleshy cotyledons remain inside the seed coat. Comparatively few species with leafy cotyledons are found within the forest, while such species predominate in disturbed areas.

These results are important in providing an ecological and evolutionary framework in which to classify seedling behaviour. By simple observations of seed size and seedling morphology, it is possible to predict the preferred habitat and light requirements of the seedlings of particular species. This information can be used by forestry nurseries in developing general methods for propagating different seedling types.

The chapter by Augspurger considers the importance of fungal pathogens in influencing tropical tree populations. Pathogens appear to be able to attack all phases in the lifecycle. The destructive effects of pathogens on tree seedlings is particularly well documented. In the wild, and in nurseries, the majority of seedlings can be weakened by fungal growth. The circumstances which favour the growth of fungi are unusually moist conditions, particularly when these follow some disturbance to the environment. Fungal pathogens are often quite specific in their host requirements, and a high density of a particular plant species facilitates the spread of the pathogen. So, in general, fungal pathogens are particularly damaging in forest nurseries where thousands of seedlings are grown together at high density. The poor growth of seedlings, "damping off", is the tangible evidence of fungal growth. The economic and ecological importance of fungal pathogens on tropical woody plants will certainly increase in the future, as the remaining forests are more intensively managed and lands are converted to fast-growing exotic timber species and tropical tree crops.

Fungal pathogens may play a role in restricting the population size of common tree species in the primary forest and in holding the density below a certain point. However, once the environment is disturbed by selective logging, with its exposure of the soil, piling up of organic matter, and injuring of the bark of the remaining trees, the opportunities for an outbreak of fungal pathogens may greatly increase. If one understands that the ecological niche of fungal pathogens in the tropical rain forest involves attacking populations that are at high densities, it makes sense that fungi would be particularly destructive in situations where the density of a species has been artificially increased, such as in a forest treated by a management programme to encourage certain timber species. Enrichment plantings and plantation forests are particularly susceptible to fungal attack because of the high density of particular species. Many of the extensive plantations of fast-growing exotic timber species in South-east Asia have had disappointing results due to pathogen attack. Enormous plantations of bananas, cacao, coffee, and other tropical crops have been destroyed by fungal pathogens that have been simply performing their ecological role. Understanding the ecology of fungal pathogens in primary forest can give insight into the control of fungal pathogens in environments affected by human activity.

These three papers describe specific aspects of the establishment of new trees in the forest. However, it is interesting to consider the continuity of reproductive characters, and to see how dormancy, seedling morphology, habitat requirements and resistance to fungal pathogens are interrelated. At one extreme are tree species which occur in the mature forest and produce large seeds. Only one or a few such seeds are contained in large fruits that are dispersed by specialized vertebrate dispersal agents. These seeds have a high water-content and germinate without any dormancy period. The resulting seedlings will typically have fleshy cotyledons that remain at ground level or below the ground. The mature seedlings will have extensive leaf tissue and a deep radicle that will allow them to persist for a long period even in deep shade and during periods with relatively little rainfall. The seedlings will quickly develop tough lignified tissue in the stem and root which will make them less susceptible to fungal pathogens. At the other extreme are species of early successional stages, particularly pioneer species with small seeds. The fruits of these species will tend to be small, dry, wind-dispersed fruits or fleshy fruits with small hard seeds that are swallowed whole by generalized fruit-eaters. These seeds when they land on the ground can remain dormant for long periods of time until they are exposed to the high-light, moist conditions necessary for germination. Their seedlings are typically small, with leafy cotyledons above ground. The photosynthetic capabilities of these leafy cotyledons are necessary for the rapid growth of the seedling because it only has a limited energetic reserve. These seedlings must grow rapidly or die. If they are exposed to shady, moist conditions, which might occur during a period of rainy weather, or if the canopy closes above them, then these seedlings may be susceptible to attack by fungal pathogens. These two common seed and seedling types represent the ends of a range, with many species possessing intermediate characteristics as well. It is a useful exercise to consider such inter-relationships among seedling characters and to realize that the plant is an integrated organism equipped to face a succession of challenges at each phase in its lifecycle.

CHAPTER 17

THE POTENTIAL IMPACT OF FUNGAL PATHOGENS ON TROPICAL PLANT REPRODUCTIVE BIOLOGY

Carol K. Augspurger

ABSTRACT

Pathogens inflict damage directly on flowers, fruits and seeds. By damaging leaves and roots of adult plants, pathogens restrict the resources available for plant reproduction. Pathogen-caused mortality potentially has consequences for a wide range of ecological and evolutionary phenomena, including the host's reproductive output, population dynamics, genetic variation, spatial distribution and interspecific interactions. Based on temperate zone observations, it is predicted that the incidence of pathogens will increase as tropical forests become managed more intensively to meet economic goals. Pathogens can have a major influence on breeding programmes and methodologies used in reforestation. Continued success in forest plantations depends on the availability of genotypes with disease resistance that are conserved in wild populations of undisturbed tropical forests.

INTRODUCTION

Plant pathology has been historically an applied discipline devoted to the study of disease in domesticated plants. Pathogens and their disease symptoms on tropical crop and plantation species are well documented (Weber, 1973; Wellman, 1972). In contrast, plant pathology as a basic science is in its infancy. The first book devoted to disease and plant population biology appeared only recently (Burdon, 1987a). Because of the overall dearth of studies, Burdon had to restrict his examples to temperate species and often used agricultural studies to make inferences about natural populations.

Pathogens in natural tropical communities have scarcely been studied. Interest in pathogens of tropical native plants has been restricted primarily to tracing the origins in natural vegetation of pathogens attacking plants of economic importance. For example, the virus causing swollen shoot disease of cacao (*Theobroma cacao*) arose from species in the Sterculiaceae and Bombacaceae in Africa (Harper, 1977). Some diseases in neotropical bananas (*Musa* sp.) originated in native *Heliconia* species (Buddenhagen, 1977).

The objectives of this contribution are to:

(1) Identify the type of damage pathogens inflict on those life stages that affect the reproductive biology of tropical plants;

(2) Consider potential consequences of pathogen-induced mortality for a variety of ecological and evolutionary phenomena; and,

(3) Comment on some of the implications of pathogen activity for the management and regeneration of tropical forests.

The emphasis throughout is on fungal pathogens with which the author is most familiar.

STAGES AFFECTED BY PATHOGENS

Adult plants

Airborne and insect-borne pathogens affect adult plants and either directly or indirectly restrict their production of flowers, fruits, and seeds (Burdon and Shattock, 1980). Direct effects include pathogen damage to flowers, resulting in a decrease in the number of functional flowers, a transformation of floral organs to non-reproductive masses, or a prevention of flowering altogether (Burdon, 1987*a*). Pathogens also directly attack developing fruits and cause a decrease in the number of mature fruits. Indirect effects of pathogens on plant reproduction include leaf damage that decreases the energy and carbon available for reproduction (Walters, 1985). Similarly, pathogen damage to roots decreases nutrients and water available for reproduction (Ayres, 1981). The net effect of such damage is to decrease the number of seeds produced by an adult plant and sometimes to decrease seed size and quality.

Seeds and seedlings

Seeds: pre-dispersal

Although insect seed predators are often the major mortality factor at this stage, pathogens can also be significant (Hong, 1981). In *Tachigalia versicolor*, a neotropical leguminous tree, pathogens ranked second in importance in seed destruction and caused nearly 25% of full-sized seeds to become non-viable (Kitajima and Augspurger, 1989).

Seeds: post-dispersal, pre-germination

Tropical seeds in the soil are exposed to the ever-present soil pathogens that may destroy them prior to germination. The length of seed dormancy varies widely among tropical species (Garwood, 1983; Ng, 1978; Vázquez-Yanes and Orozco-Segovia, this volume). Almost no data document the extent of pathogen damage to seeds prior to germination. It is unknown whether species with dormant seeds and delayed germination have greater protection against pathogens than species with non-dormant seeds and rapid germination. In one experiment, seeds of *Hybanthus prunifolius*, a neotropical shrub, were exposed to the soil for varying numbers of weeks in the dry season prior to the arrival of consistent rains. With a delay in germination, 23% to 65% of seeds were infected by fungi (Augspurger, 1979).

Germination of seeds and young seedlings

Pathogens at this stage include seed-borne fungi carried on the seed during dispersal (Neergaard, 1977) and soil-borne fungi such as *Pythium*, *Phytophthora*, and *Rhizoctonia* that cause damping-off (Bruehl, 1987). This disease can result in "pre-emergent" mortality of germinating seeds or "post-emergent" mortality of young seedlings (Halloin, 1986). Pre-emergent damping-off is rarely quantified but may account for a low percentage of germination of apparently viable seeds.

Damping-off was observed first in forest nurseries, where it can cause a major problem in regeneration programmes requiring healthy seedlings. The level of damping-off is affected by characteristics of the plant species as well as environmental factors. Ng (1978) suggested from his germination studies that small-seeded species were more vulnerable to damping-off than were large-seeded species. In a study of 18 neotropical tree species, Augspurger (1984a) concluded that successional status, and not seed size *per se*, was the critical variable; early-successional species, not necessarily small-seeded, were most prone to damping-off. Rapid maturation of lignified tissues deters attack by these fungi. Damping-off fungi flourish when environmental factors favour the growth of the pathogen over that of the seedling. The presence of other soil microflora can inhibit the growth of damping-off fungi. Damping-off is more prevalent in shady than sunny environments (Augspurger, 1984*b*), and is encouraged by water saturation of poorly drained soils and by inconsistent rains during seedling establishment (Augspurger, 1979).

ECOLOGICAL AND EVOLUTIONARY CONSEQUENCES

Reproductive output and seedling recruitment

Pathogens decrease the number of flowers, fruits, and seeds an individual produces. In *Hybanthus prunifolius*, a delay in fruit development due to drought was followed by fungal attack of immature fruits. As a result, the median seed production per individual declined to 25% of the value in a year with ample rain during fruit development (Augspurger, 1978). At the population level, the lower reproductive output reduces the birth rate, thus potentially affecting the population size in the future.

In terms of seedling recruitment, in the neotropical tree *Platypodium elegans*, damping-off was the major cause of seedling mortality for each of four parent trees (Augspurger, 1983*a*). In nine tree species in Panama, damping-off caused 0.2% to 74% of seedlings in the shaded understorey to die in their first 2 months (Augspurger, 1984*b*). It is unknown what percentage of seedlings would die from other causes in the absence of disease. Controlled studies are needed to identify compensatory mortality factors that act when pathogens are inactive. At the population level, seed and seedling mortality by pathogens can immediately reduce population size and alter the age structure.

Differential fitness and genetic variation

Pathogens may cause differential fitness among individuals in a population (Burdon, 1985; Antonovics and Alexander, 1989). In the temperate annual species *Amphicarpaea bracteata*, Parker (1986) found that individual fitness was inversely correlated with severity of infection by the rust *Synchytrium decipiens*. Furthermore, he demonstrated that both the probability of infection and the amount of damage resulting from infection had a genetic basis. This example illustrates the potential influence of pathogens on the evolution of plant populations. No comparable study is available for any tropical species.

Assuming that genetic variation for pathogen resistance exists in a plant population, it is possible that, over time, the genotype(s) with greater resistance would dominate the population. The alternative is that frequency-dependent selection favours rare, novel, resistant genotypes and therefore maintains a polymorphic population (Clarke, 1976, 1979). Evidence from studies of temperate plants favours the latter hypothesis, i.e. that genetic variation for pathogen resistance is common in populations. For example, in 1 m^2 of *Senecio vulgaris*, among 75 individuals exposed to five isolates of powdery mildew (*Erisyphe fischeri*), nine contrasting resistance phenotypes were found (Harry and Clark, 1986). The genetic structure for resistance is also quite complex. In breeding experiments on *Glycine canescens* exposed to the rust *Phakopsora pachyrhizi*, Burdon (1987b) estimated that 1–3 dominant genes for resistance occurred in each host line and that 10–12 resistant genes (or alleles) occurred within the local population.

Spatial distribution

When conditions favour the growth of the pathogen over that of the host, a pathogen might restrict the range of the host plant species. For example, the temperate conifer tree *Larix*, escapes large amounts of pathogen damage in its native high altitude zone of Europe. Attempts to grow it at lower altitudes were unsuccessful because it was overridden by its pathogen *Trichoscyphella willkommii* (Weir, 1918). However, there are no convincing examples of a native pathogen preventing a native plant from completing its lifecycle and thus controlling its range. In naturally occurring plant populations, pathogens appear to control population size and density, but not actual range.

On a more local scale, damping-off pathogens have been demonstrated to alter the spatial relation of seedlings relative to their parent tree (Augspurger, 1983b). Greater mortality by damping-off arose near four parent trees of *Platypodium elegans* (Augspurger, 1983a). Subsequent experiments determined that such mortality had both a density-dependent and a distance-dependent component. Disease levels were greater at higher densities, independent of distance from the parent tree; disease levels were greater closer to the parent tree, independent of density (Augspurger and Kelly, 1984). Additional studies of damping-off in nine tropical tree species showed that such mortality was greater in shaded areas than in light-gaps (Augspurger, 1984b). In some species, all seedlings were eliminated rapidly from shaded areas. Although it is customarily argued that these species require sun for physiological reasons, they may require sun because they lack defence against pathogens during early seedling establishment in the shade. It is unclear whether this microhabitat effect is transitory or long-term. Also, without pathogens present, these species may, for other reasons, persist eventually only in light-gaps.

The genetic specificity of host susceptibility can occur on quite a local scale. In one experiment, seeds of one parent tree of *Platypodium elegans* were distributed at various distances around the original parent and around two conspecific trees at least 1 km away. Mortality by damping-off pathogens was greater around the original parent than around the two non-parent, conspecific trees (Augspurger and Kelly, 1984). This may have resulted from a very local specificity between pathogen and host; alternatively, inoculum levels from previous years of infection may have been highest under the original parent. In a temperate study, Parker (1985) found that the specificity of a pathogen for its host changed over a very local scale, i.e. within metres.

The short-term effect of pathogen-induced mortality is that seedling distributions become discontinuous and recruitment occurs in a mosaic pattern. In some species, no seedlings survive very near the parent tree, and, in other species, seedlings are more common in or are totally restricted to light-gaps (Augspurger, 1984*b*). The long-term effect of this pattern of mortality is on the distribution of adult trees. Pathogen activity can contribute ultimately to a significantly less clumped distribution of adult trees.

Community composition

Pathogens are expected to influence the dominance hierarchy among species and enhance the community's species diversity in two ways. First, by attacking selectively the dominant species, pathogens may shift the competitive balance among species and thereby increase the number of individuals of rare species (Chilvers and Brittain, 1972). This outcome arose in chestnut-dominated forests attacked by blight in eastern US deciduous forests (Day and Monk, 1974; Stephenson, 1986). Second, pathogen attack of mature trees may act as a repeating disturbance to the community, thereby keeping it in a non-equilibrium state. Death of adult trees enhances light-gap formation, thereby providing appropriate habitat for light-demanding species and enhancing species diversity. This outcome has been documented in hemlock-dominated forests attacked by a root-rot fungus in western US coniferous forests (Copsey, 1985; McCauley and Cook, 1980). No comparable examples are known yet for tropical forests. Earlier investigations of die-back of *Metrosideros*-dominated forests in Hawaii were attributed to attack by the pathogen *Phythophthora cinnamomii* (Papp *et al.*, 1979). Longer-term studies concluded that pathogens played only a secondary role in the forests that were undergoing stand-level senescence and turnover (Mueller-Dombois, 1986).

IMPLICATIONS FOR FOREST MANAGEMENT AND CONSERVATION

Observations of temperate forests have led to the generalization that forest communities remain stable and their biological diversity is conserved because pathogens and their host populations are in equilibrium (Browning, 1974; Schmidt, 1978; Segal *et al.*, 1980, 1982). Pathogen damage is kept controlled to non-epidemic levels, and epidemics arise only when the community is altered. The most extreme alteration occurs when monospecific plantations are planted in high density and with low genetic diversity. The classical tropical example occurred in rubber plantations (*Hevea brasiliensis*) devastated by leaf blight (*Microcydus ulei*) in their native South American regions, thereby driving the entire rubber industry to other continents (Harper, 1977). Less extreme disturbances also appear to enhance disease epidemics. The increasing incidence of rust (*Cronartium fusiforme*) in southern US slash pine forests (*Pinus elliotii*) is attributed to increasing changes in forest structure brought about by logging (Schmidt, 1978). Finally, major epidemics arise when a community is disturbed by the introduction of a foreign pathogen for which the host plant has no defence. The accidental introduction of chestnut blight (*Endothia parasitica*) virtually eliminated the dominant chestnut trees (*Castanea dentata*) from their forests (Roame *et al.*, 1986). An increased invasion of natural tropical forests for logging purposes enhances the likelihood of the introduction of a foreign pathogen.

Pathogen activity has practical implications for the management of tropical forest regeneration (Evans, 1982). It affects tree reproductive output and thus seed availability for seeding programmes. Environmental conditions must be controlled to minimize pathogen activity during seed storage (Christensen, 1973; Harman, 1983). Some control of damping-off in seedling beds is obtained by using well-drained, non-sterile soils, planting seeds in relatively low density, and growing the seedlings in well-lit conditions.

The ever-present threat of pathogens will affect breeding programmes of tropical forest trees. A diversity of seed sources is needed in regeneration. Breeders must continually search plantations for individuals resistant to pathogens. Finally, intact tropical forests represent the ultimate source of genetic diversity to use for breeding resistance to pathogens into domesticated species. Only conservation of large reserves of tropical forests will provide such a reservoir of potential protection against pathogens.

CONCLUSIONS

Presently we are largely ignorant of the incidence and role of pathogens in undisturbed tropical forests. Evidence from natural communities in temperate regions and from domesticated tropical plants indicates that pathogens are likely to play a significant role in a wide variety of ecological and evolutionary phenomena. The incidence of pathogen damage may increase as our tropical forests become increasingly disturbed for economic management. Pathogens have a major influence over forest reforestation methods and breeding programmes. The continuing success of forest plantations may depend on the occasional introduction of resistance genes from wild populations in those limited conserved natural tropical forests that remain.

REFERENCES

Antonovics, J. and Alexander, H.M. (1989). The concept of fitness in plant-fungal pathogen systems. In Leonard, K.J. and Fry, W.E. (eds), *Plant Disease Epidemiology*. Volume II. *Genetics, Resistance and Management*, pp. 185–214. (New York: McGraw-Hill Publishing Company)

Augspurger, C.K. (1978). Reproductive consequences of flowering synchrony in *Hybanthus prunifolius* (Violaceae). Ph. D. Dissertation. University of Michigan.

Augspurger, C.K. (1979). Irregular rain cues and the germination and seedling survival of a Panamanian shrub (*Hybanthus prunifolius*). *Oecologia*, **44**, 53–9

Augspurger, C.K. (1983*a*). Seed dispersal by the tropical tree, *Platypodium elegans*, and the escape of its seedlings from fungal pathogens. *Journal of Ecology*, **71**, 759–71

Augspurger, C.K. (1983*b*). Offspring recruitment around tropical trees: changes in cohort distance with time. *Oikos*, **40**, 189–96

Augspurger, C.K. (1984*a*). Light requirements of neotropical tree seedlings: a comparative study of growth and survival. *Journal of Ecology*, **72**, 777–95

Augspurger, C.K. (1984*b*). Seedling survival of tropical tree species: interactions of dispersal distance, light-gaps, and pathogens. *Ecology*, **65**, 1705–12

Augspurger, C.K. and Kelly, C.A. (1984). Pathogen mortality of tropical tree seedlings: experimental studies of the effects of dispersal distance, seedling density, and light conditions. *Oecologia*, **61**, 211–17

Ayres, P.G. (1981). Effects of disease on plant water relations. In Ayres, P.G. (ed.) *Effects of Disease on the Physiology of the Growing Plant*, pp. 131–48 (London: Cambridge University Press)

Browning, J.A. (1974). Relevance of knowledge about natural ecosystems to development of pest management programs for agroecosystems. *Proceedings of American Phytopathology Society*, **1**, 191–9

Bruehl, G.W. (1987). *Soilborne Plant Pathogens*. (New York: MacMillan Publishing Company)

Buddenhagen, I.W. (1977). Resistance and vulnerability of tropical crops in relation to their evolution and breeding. In Day, P.R. (ed.), *The Genetic Basis of Epidemics in Agriculture*, pp. 309–26 (New York: Academic Press)

Burdon, J.J. (1985). Pathogens and the genetic structure of plant populations. In White, J. (ed.), *Studies on Plant Demography*, pp. 313–25. (New York: Academic Press)

Burdon, J.J. (1987*a*). *Diseases and Plant Population Biology*. (London: Cambridge University Press)

Burdon, J.J. (1987*b*). Phenotypic and genetic patterns of resistance to the pathogen *Phakopsora pachyrhizi* in populations of *Glycine canescens*. *Oecologia*, **73**, 257–67

Burdon, J.J. and Shattock, R.C. (1980). Disease in plant communities. *Applied Biology*, **5**, 154–219

Chilvers, G.A. and Brittain, E.G. (1972). Plant competition mediated by host-specific parasites – a simple model. *Australian Journal of Biological Science*, **25**, 749–56

Christensen, C.M. (1973). Loss of viability in storage: microflora. *Seed Science Technology*, **1**, 547–62

Clarke, B. (1976). The ecological genetics of host-parasite relationships. In Taylor, A.E.R. and Muller, R. (eds), *Genetic Aspects of Host-Parasite Relationships*, pp. 87–103. (Oxford: Blackwell Scientific Publications)

Clarke, B. (1979). The evolution of genetic diversity. *Proceedings of the Royal Society of London B.*, **205**, 453–74

Copsey, A.D. (1985). Long-term effects of a native forest pathogen, *Phellinus weirii*: changes in species diversity, stand structure, and reproductive success in a *Tsuga merensiana* forest in the central Oregon High Cascades. Ph.D. Dissertation. University of Oregon.

Day, F.P. and Monk, C.D. (1974). Vegetation patterns on a southern Appalachian watershed. *Ecology*, **55**, 1064–74

Evans, J. (1982). *Plantation Forestry in the Tropics*. (Oxford: Clarendon Press)

Garwood, N.C. (1983). Seed germination in a seasonal tropical forest in Panama: a community study. *Ecological Monographs*, **53**, 159–81

Halloin, J.M. (1986). Microorganisms and seed deterioration. In McDonald, M.B. and Nelson, C.J. (eds), *Physiology of Seed Deterioration*, pp. 89–98. CSSA Special Publication Number 11. Crop Science Society of America, Madison, Wisconsin.

Harman, G.E. (1983). Mechanisms of seed infection and pathogenesis. *Phytopathology*, **73**, 326–9

Harper, J.L. (1977). *Population Biology of Plants*. (London: Academic Press)

Harry, I.B. and Clarke, D.D. (1986). Race-specific resistance in groundsel (*Senecio vulgaris*) to the powdery mildew *Erysiphe fischeri*. *New Phytologist*, **103**, 167–75

Hong, L.T. (1981). A note on some seed fungi of dipterocarps. *Malaysian Forester*, **44**, 163–6

Kitajima, K. and Augspurger, C.K. (1989). Seed and seedling ecology of a monocarpic tropical tree, *Tachigalia versicolor*. *Ecology*, **70**, 1102–14

McCauley, K.J. and Cook, S.A. (1980). *Phellinus weirii* infestation of two mountain hemlock forests in the Oregon Cascades. *Forest Science*, **26**, 23–9

Mueller-Dombois, D. (1986). Perspectives for an etiology of stand-level dieback. *Annual Review of Ecology and Systematics*, **17**, 221–43

Neergaard, P. (1977). *Seed pathology*. Volumes 1 and 2. (New York: John Wiley and Sons)

Ng, F.S.P. (1978). Strategies of establishment in Malayan forest trees. In Tomlinson, P.B. and Zimmermann, M.H. (eds), *Tropical Trees as Living Systems*, pp. 129–62. (Cambridge: Cambridge University Press)

Papp, R.P., Kliejunas, J.T., Smith, Jr., R.S. and Scharpf, R.F. (1979). Association of *Plagithmysus bilineatus* and *Phytophthora cinnamomi* with the decline of 'ohi a-lehua' forests on the island of Hawaii. *Forest Science*, **25**, 187–96

Parker, M.A. (1985). Local population differentation for compatability in an annual legume and its host-specific fungal pathogen. *Evolution*, **39**, 713–23

Parker, M.A. (1986). Individual variation in pathogen attack and differential reproductive success in the annual legume, *Amphicarpaea bracteata*. *Oecologia*, **69**, 253–9

Roame, M.K., Griffin, G.J. and Rush, J.R. (1986). *Chestnut Blight, Other Endothia Diseases, and the Genus Endothia*. (St. Paul, Minnesota: American Phytopathology Society Press)

Schmidt, R.A. (1978). Diseases in forest ecosystems: the importance of functional diversity. In Horsfall, J.G. and Cowling, E.B. (eds), *Plant Disease*. Vol. II. *How Disease Develops in Populations*, pp. 287–315. (New York: Academic Press)

Segal, A., Manisterski, J., Fischbeck, G. and Wahl, I. (1980). How plants defend themselves in natural ecosystems. In Horsfall, J.G. and Cowling, E.B. (eds), *Plant Disease*. Vol. V. *How Plants Defend Themselves*, pp. 75–102 (New York: Academic Press)

Segal, A., Manisterski, J., Browning, J.A., Fischbeck, G. and Wahl, I. (1982). Balance in indigenous plant populations. In Heybroek, H.M., Stephan, B.R. and von Weissenberg, K. (eds), *Resistance to Disease and Pests in Forest Trees*, pp 361–70. (Wageningen: Centre for Agricultural Publishing and Documentation)

Stephenson, S.L. (1986). Changes in a former chestnut-dominated forest after a half century of succession. *American Midland Naturalist*, **116**, 173–9

Walters, D.R. (1985). Shoot:root interrelationships: the effects of obligately biotropic fungal pathogens. *Biological Review*, **60**, 47–79

Weber, G.F. (1973). *Bacterial and Fungal Diseases of Plants in the Tropics*. (Gainesville: University of Florida Press)

Weir, J.R. (1918). Notes on the altitudinal range of forest fungi. *Mycologia*, **10**, 4–14

Wellman, F.J. (1972). *Tropical American Plant Disease*. (Metuchen, New Jersey: Scarecrow Press)

CHAPTER 18

SEED DORMANCY IN THE TROPICAL RAIN FOREST

C. Vázquez-Yanes and A. Orozco-Segovia

ABSTRACT

Recent studies at Los Tuxtlas field research station in southern Veracruz (Mexico) have shed light on the ecophysiology of dormancy mechanisms in seeds of trees of the mature forest as well of pioneer plants. Moisture contents of seeds are variable among species, within the same species and the crop of an individual tree. Seeds with higher moisture content at dispersal tend to germinate more quickly. Soil moisture and air humidity play important roles in the regulation of germination among the seeds of many rain forest species. Seeds of pioneer trees and shrubs are smaller and drier, and are capable of germination immediately following dispersal. Light-regulated germination involves reactions to the red/far red ratio of light, and experiments reveal differential light sensitivity among Piper *species. Leaf litter may play an important inhibitory role in seed germination, while the physiology of dormancy imposed by light quality changes as seeds get older. Interactions of light conditions and other microclimatic factors such as temperature may also be an important germination trigger. Seed viability varies considerably among different species, seed crops, forest sites, burial times and seed densities.*

INTRODUCTION

Los Tuxtlas, the field station of the National Autonomous University of Mexico in southern Veracruz, is located in an area of high rainfall and covered by a rich tropical rain forest (Estrada and Coates-Estrada, 1983; Estrada *et al.*, 1985; Vázquez-Yanes and Orozco-Segovia, 1987*a*). The field station has provided us with the opportunity to perform many long-term experiments on seeds of pioneer forest trees and shrubs, and, to some extent, on seeds of tree species in the mature forest. These experiments have focused on understanding the physiological ecology and the nature of seed dormancy of rain forest plants.

It is well known that, in the almost continuously favourable conditions for growth, and the intense competition for light that exists at ground level beneath the rain forest canopy, seeds do not usually persist in the soil, as many of them do not have a period of dormancy (Willan, 1985). Most trees of the mature forest produce big fleshy seeds that germinate quickly and give rise to seedlings with extensive root and foliar surfaces (Foster, 1986). Most of these seeds cannot be stored or dehydrated. They can be classified as recalcitrant, according to the well-known definitions given by Roberts (1973). These seeds frequently show wide variability in size, weight and moisture content, not only within the same species, but also among seeds of single individuals (Puchet and Vázquez-Yanes, 1987).

Some tree species of the mature forest have dormant seeds, especially those species that show delayed seed germination due to the lack of water during the dry season (Garwood, 1983), or those with hard seeds coats. Knowledge about the germination physiology of these and other delay mechanisms has changed very little since the publication of a previous review paper (Vázquez-Yanes and Orozco-Segovia, 1984), mainly because of the lack of research on the subject.

In contrast to the trees of the mature forest, seeds produced by the so-called pioneer trees and shrubs have specialized dormancy mechanisms. These seeds are small and can tolerate drying and long-term storage, especially at low temperature, and they can be classified as orthodox (Roberts, 1973). They are the main components of the soil seed banks in many forests. The subject of seed dormancy in pioneer species has also been extensively reviewed recently (Whitmore, 1983; Garwood, 1989). In the present contribution, we focus on some aspects of the ecophysiology of dormancy in the tropical rain forest based on our own results from present research in Los Tuxtlas. We deal first with dormancy in trees of the mature forest, and second with dormancy in pioneer plants.

DORMANCY OF TREES IN THE MATURE FOREST

We know very little about dormancy mechanisms in mature forest tree seeds from the physiological point of view, although it is possible to delineate some aspects of this problem. Some seeds have hard coats that prevent germination until scarification occurs. Others, typical of more seasonal forests, have delayed germination due to the lack of water during the dry season with seedling development taking place as soon as water becomes available (Garwood, 1983). The most intriguing cases of dormancy are shown by seeds that have delayed germination even when they are fully

Table 18.1 Water content (wet basis) at the moment of dissemination in seeds of 17 plant species from the rain forest of Los Tuxtlas

Species	Number of seeds	% Water content
Astrocaryum mexicanum Liebm.	11	72.7
Pouteria durlandii (Standl.) Baehni	11	72.7
Calatola laevigata Standl.	10	69.2
Heliconia sp.	20	65.2
Siparuna andina (Tul.) DC	40	53.9
Nectandra ambigens (Blake) C.K. Allen	200	51.7
Pithecellobium arboreum (L.) Urban.	10	50.8
Pouteria sapota (Jacq.) Moore et Stearn	8	44.3
Couepia poliandra (H.B.K.) Ros	10	41.2
Chamaedorea tepejilote Liebm.	20	39.2
Calophyllum brasiliense Camb.	20	37.8
Cymbopetallum baillonii R.E. Fries	200	32.8
Desmocus ferox Bartlett	20	31.9
Bactris tricophylla Burret Repert	20	26.2
Cecropia obtusifolia Bertol.	200	18.7
Solanum diphyllum L.	200	16.3
Erytrina folkersii Krukoff and Moldenke	20	9.2

hydrated. This might indicate the existence of more complex mechanisms of endogenous dormancy, such as the presence of an immature embryo or an inhibitory substance in seeds. These seeds must at the same time be well protected against predators.

The moisture content of the seeds at the moment of dispersal is quite variable, as is shown in Table 18.1, where the gradient of moisture content decreases from very moist seeds of mature forest plants to relatively dry seeds of pioneer plants and also hard coated seeds. However, the moisture content also varies among the seeds from the same species, and even within the seed crop of the same tree, as for example in *Nectandra ambigens*, which is one of the dominant trees in Los Tuxtlas rain forest (Figure 18.1; data from Puchet, 1986).

Seeds from the same species, or even individual trees, often show differential dormancy, when part of the seed crop germinates several days, weeks, or even months before the rest of the seeds. Some preliminary experiments indicate that the seeds with the higher moisture content at the moment of dispersal germinate faster than the ones with the lowest moisture content. In *N. ambigens* the delay spans several weeks for the driest seeds (Figure 18.2), at least in trays placed in growth chamber conditions (Puchet, 1986).

Seeds of *Pithecellobium arboreum* have high initial moisture content and germinate rapidly inside the forest but cannot imbibe water when

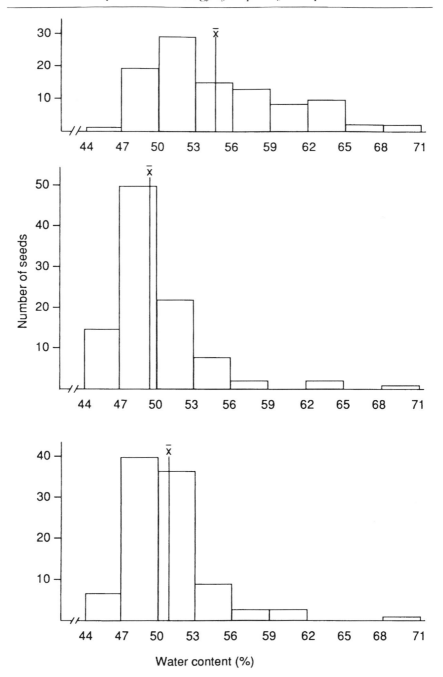

Figure 18.1 Variability of water content (wet basis) in samples of 200 seeds from three individual trees of *Nectandra ambigens*

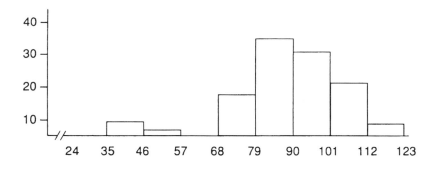

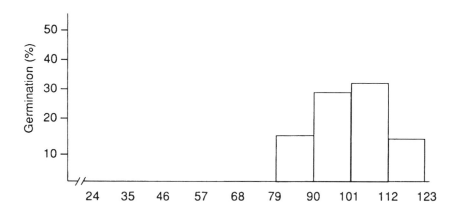

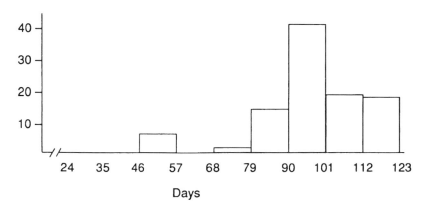

Figure 18.2 Distribution of germination flush through time in *Nectandra ambigens*

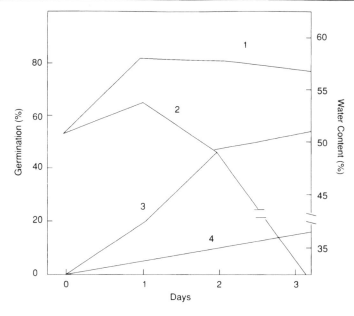

Figure 18.3 Water content and germination response in two groups of 100 seeds of *Pitecellobium arboreum* placed on the ground inside the forest (1, 3) and in an open sunny place (2, 4). Night rains kept the soil moist

temperatures rise and the relative humidity of the atmosphere drops, even if they lie on a wet surface. Thus, germination is more efficient in the rain forest microclimate than in open sites where seeds are exposed to full sunlight (Figure 18.3). This example illustrates the importance of soil moisture and air humidity for the regulation of germination among seeds of many rain forest species. Because most seeds in the rain forest germinate on the surface of the ground, it is probable that the balance between water gain from the soil and evaporation to the surrounding air regulates the speed of germination (Foster, 1986).

DORMANCY AMONG PIONEER PLANTS

In general, the pioneer trees and shrubs that invade light gaps produce small, dry seeds in large numbers (Vázquez-Yanes, 1976). These seeds have interesting and complex patterns of dormancy mechanisms. Almost all the species we have studied have seeds that are capable of germinating immediately after dispersal, and we have found only one case of endogenous dormancy that lasts several months, and one case of dormancy imposed by a hard coat (Vázquez-Yanes, 1974, 1981). All the rest of the seeds studied exhibited environmentally imposed dormancy determined by a light requirement, at least for fresh seeds taken from the mother plant. The mechanism of light regulated germination involves reactions to the red/far red ratio of the light that take place with the participation of the pigment phytochrome in the seeds (Smith, 1982).

The seeds are released with a low proportion of active phytochrome (Pfr) that does not permit the germination response threshold in darkness although seeds of some species may germinate under the low red/far red ratio of the forest, or in controlled experiments even in almost pure far red conditions. Others remain dormant for much longer times in such environments, as for example the seeds of *Piper* species, which are unable to germinate in the dark although some of them germinate quite fast beneath green canopies.

The effectiveness of light quality as an environmental cue for germination among heliophile species from the rain forest can be better understood through the study of the rain forest light regime (Vázquez-Yanes *et al.*, 1990). Spectro-radiometric determinations made with a portable spectro-radiometer (Licor, Lincoln, Nebraska) indicate a strong reduction of the red/far red ratio of the light beneath the green canopy in comparison with the open areas or gap areas (Figure 18.4).

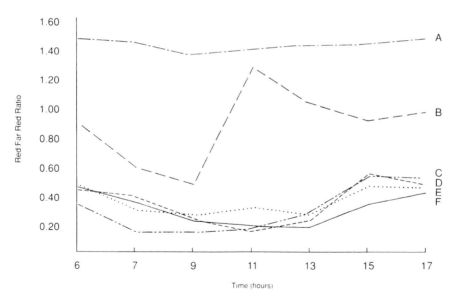

Figure 18.4 Time course of the red/far red ratio of the light (mean of three measurements) in the open (A), a medium size light gap (B), secondary vegetation (C, D), and two sites in the understorey of primary forest (E, F)

Table 18.2   Numbers of seedlings and species appearing under white light (WL), red light (R), far red light (FR) and in total darkness (D) in samples of superficial soil from five sites in the forest (250 ml of soil per treatment). Difference between WL, R and FR, D are statistically significant ($X < 2$)

			Site					
Light	1	2	3	4	5	Total	Mean	(SD)
WL								
No. of seedlings	31	20	8	27	25	111	22	(9)
No. of species	4	3	2	6	4	9	4	(1)
R								
No. of seedlings	16	9	16	12	18	71	14	(4)
No. of species	2	3	4	5	3	4	3	(1)
FR								
No. of seedlings	2	2	0	0	7	11	2	(3)
No. of species	1	2	0	0	1	2	1	(1)
D								
No. of seedlings	1	3	6	0	0	10	2	(2)
No. of species	1	1	1	0	0	1	1	

Total number of different species = 12

For most of the species present in the soil seed bank, light quality is the main environmental factor regulating dormancy. In an experiment with soil samples taken from different places in the forest, spread on trays kept properly watered in growth chambers and placed in several light conditions, the greatest number of seedlings and species emerged from the soil in white and red light. Few seeds germinated in darkness or far red light (Table 18.2). After germination, the seeds were placed in pots with soil and they were kept under white light to promote growth of the seedlings for further identification.

In four *Piper* species studied (Orozco-Segovia and Vázquez-Yanes, 1989), experiments performed with freshly collected seeds indicate that light sensitivity does not have the same ecological meaning among the four species. Two are strongly inhibited by the light inside the forest, while the other two can germinate inside the forest. Seeded Petri dishes taken in and out of the forest for different periods of time show a degree of phytochrome photo-reversibility in two of the species (*P. auritum* and *P. umbellatum*) and no indications of this effect in the other two (*P. aequale* and *P. hispidum*) (Figure 18.5). Experiments with the very common pioneer tree *Cecropia obtusifolia* also indicate strong phytochrome photo-reversibility (Vázquez-Yanes and Orozco-Segovia, 1987*b*).

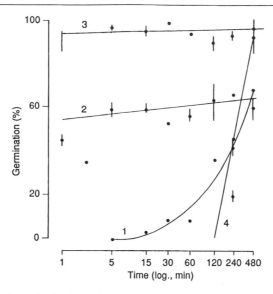

Figure 18.5 Final germination in Petri dishes seeded with: *Piper auritum* (1), *P. aequale* (2), *P. hispidum* (3) and *P. umbellatum* (4). The dishes were taken in and out of the forest for the periods stated in the x axis in order to induce phytochrome photo-reversion. After 13 days the dishes were placed in darkness for 15 days more before germination was registered

As the amount of litter accumulation in this rain forest is over 9 t ha^{-1} yr^{-1} (Alvarez and Guevara Sada, 1985), any experiment related to seed predation or seed dormancy in small seeds must take into consideration the fact that the seeds lying on the ground are soon covered by the fallen leaves. Measurements of the light transmittance of leaves taken from the forest litter show a peculiar light spectrum. The red/far red ratio is always lower than 1, and, when more than one leaf is used, the value is often extremely low (Figure 18.6). This finding indicates that, beneath the layer of litter inside the forest, the scarce available light must be even more inhibiting to germination than the diffuse light of the forest, and the effect of the sunflecks increasing the red/far red ratio of the light is eliminated.

One possible ecological significance of photoblastism in these species is to extend dormancy until the seeds become buried by the falling litter. Beneath the litter, the red/far red ratio is lower and more stable. This effect might have consequences on the duration of dormancy in the forest. However, we have also found that the physiology of the dormancy imposed by light quality changes as the seeds get older. In seeds recovered from nylon net bags buried in the forest, we have found changes in the response to light quality after several months of burial. Seeds become less sensitive, and some germinated in pure far red light or even in darkness when moved

255

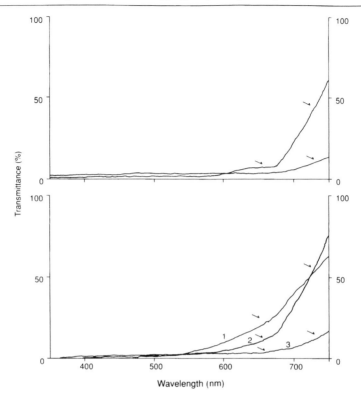

Figure 18.6 Spectral composition of the light transmitted through 1 and 2 dry litter leaves of *Nectandra ambigens* (upper graph) and 1, 2 and 3 dry litter leaflets of *Dussia mexicana* (lower graph). Leaves of the first species are thicker than the leaflets. Arrows indicate the critical 660 and 730 wavelengths

to Petri dishes (Orozco-Segovia and Vázquez-Yanes, 1989). In these experiments, the seeds were dormant when they were extracted from the buried bags; so, even though the response to light had changed, there must be other factors preventing germination in the buried seeds.

Interactions of light conditions and other microclimatic factors like temperature have also been found. In some species, temperature fluctuations reduce either the red/far red ratio or the exposure time required for germination (Orozco-Segovia *et al.*, 1987). In one case, the wide temperature fluctuations that take place in the soil of open areas, as compared to the stable temperature of the forest floor, act as a germination trigger without the effect of light (Vázquez-Yanes and Orozco-Segovia, 1982*a*).

The longevity of seeds buried in the forest soil is a direct consequence of their dormancy, so it is interesting to know for how long the seeds can remain alive and dormant when buried. Several longevity experiments have

Table 18.3 Mean germination in three samples of seeds buried in the forest.
A = samples of 200 seeds, B = samples of 1 g

		% of germination	
Species	Initial		After a year in the soil
Myriocarpa longpipes Liebm.	A 57 (6)		19 (4)
	B 11 (5)		0
Piper auritum H.B.K.	A 89 (2)		16 (6)
	B 97 (1)		4 (3)
P. hispidum Swartz	A 76 (13)		24 (20)
	B 97 (3)		0

been performed at Los Tuxtlas (Vázquez-Yanes and Smith, 1982; Vázquez-Yanes and Orozco-Segovia, 1982*a*, 1982*b*; Orozco-Segovia *et al.*, 1987; Orozco-Segovia and Vázquez-Yanes, 1989; Perez-Nasser and Vázquez-Yanes, 1986). The experiments with buried seeds entailed placing seeds mixed with sterilized soil or vermiculite inside nylon net bags that were buried in different places in the forest. The bags allow free circulation of water and soil micro-organisms between the content of the bag and the soil, but probably stop seed predators. The results indicate that many seeds can keep germinability for 1 year or more, although there are big differences in the proportion of seeds that survive among different species, seed crops, forest sites, times of burial, and seed densities inside the bags. As an example of this last factor, we compared the viability after 1 year between seeds of three species recovered from bags with only two hundred seeds and bags where the seeds were weighed instead of counted: 1 gram of seeds was placed inside each bag, so the amount of seeds was much greater than in the previous case (all of the seeds are very small). The seeds at high densities have much lower survival through time (Table 18.3)

DISCUSSION

The conclusions that we derive from our experiments are that dormancy and its effects on longevity in the soil remain a very complex subject that still requires much more experimentation. In pioneer rain forest plants that become established in the discontinuous and ephemeral habitats that appear randomly in the forest, dormancy is frequently environmentally imposed. Longevity of dormant seeds is affected by factors such as the characteristics of the reproductive event, the time of the year, the site in the forest, the period of time, the characteristics of the species itself, or even the type of dispersal agent (Vázquez-Yanes and Orozco-Segovia, 1986). Other factors that have an important effect on seed survival are the density and population dynamics of seed predators, which in the case of pioneer plants, depend upon the abundance of the prey seeds, the degree of disturbance, and probably the distance from the forest edge where the density of pioneer plants is frequently far greater than inside mature forest far from the edge. However, the available information on predation of small rain forest seeds is almost nil.

The only generalization about rain forest trees that can be made at present is that most produce recalcitrant seeds (in the sense of Roberts, 1973) although a small group of plants produce orthodox seeds with environmentally imposed dormancy. The orthodox seeds can undergo profound macro-molecular rearrangements in the cells that allow deep dehydration and profound long lasting dormancy in storage conditions. Part of these adaptations might be in some way related to dispersal. Most pioneer plants are either anemochorous or endozoochorous. For efficient wind dispersal, deep dehydration of the seeds is required, and deep dormancy might also be required for survival in animals' guts. These changes do not take place in the big fleshy seeds of most trees. The period between dispersal and germination presents a certain degree of metabolic activity and a more superficial dormancy very dependent on water availability (Chin and Roberts, 1980).

We have little information on the ecology of seed germination in rain forest plants. In the light of the importance of the regeneration and conservation of this threatened ecosystem, we hope that much more intensive research will occur in the future in this important area of plant ecology (Janzen and Vázquez-Yanes, 1988).

ACKNOWLEDGEMENTS

The research described in this paper has been supported in part by the Mexican Council of Science and Technology (CONACYT) through two grants: PCAFBNA-022636 and P220CCOR880134.

REFERENCES

Alvarez, J. and Guevara Sada, S. (1985). Caida de hojarasca en la selva. In Gómez-Pompa, A. and Del Amo, S. (eds), *Investigaciones Sobre la Regeneracion de Selvas Atlas en Veracruz, Mexico*, pp. 171–89 (Mexico: Editorial Alhambra)

Chin, H.F. and Roberts, E.H. (1980). *Recalcitrant Crop Seeds.*, (Malaysia: Tropical Press SDN. BHD)

Estrada, A. and Coates-Estrada, R. (1983). Rain forest in Mexico: research and conservation at Los Tuxtlas. *Oryx*, **17**, 201–4

Estrada, A., Coates-Estrada, R. and Martinez-Ramos, M. (1985). La Estaciön de Biologia Tropical Los Tuxtlas: un recurso para el estudio y conservacion de las selvas del tropico humedo. In Gómez-Pompa, A. and Del Amo, S. (eds), *Investigaciones Sobre la Regeneracion de Selvas Atlas en Veracruz, Mexico*, pp. 379–93. (Mexico: Editorial Alhambra)

Foster, S.A. (1986). On the adaptive value of large seeds for tropical moist forest trees: A review and synthesis. *Botanical Review*, **52**, 260–99

Garwood, N. (1983). Seed germination in a seasonal tropical forest in Panama: a community study. *Ecological Monographs*, **53**, 159–81

Garwood, N. (1989). Tropical soil seed banks: a review. In Leck, L.A., Simson, R.L. and Parker V.T. (eds), *Ecology of Seed Banks*, pp. 149–90. (London: Academic Press)

Janzen, D.H. and Vázquez-Yanes, C. (1988). Tropical forest seed ecology. In Hadley, M. (ed.) *Rain Forest Regeneration and Management*. Report of a Workshop. Biology Internationl Special Issue **18**: pp. 28–33. (Paris: IUBS)

Orozco-Segovia, A. and Vázquez-Yanes, C. (1989). Light effect on seed germination in *Piper* L. *Acta Oecologica Plantarum*, **10** (2), 123–46

Orozco-Segovia, A., Vázquez-Yanes, C., Coates-Estrada, R. and Perez-Nasser, N. (1987). Ecophysiological characteristics of the seed of the tropical forest pioneer *Urera caracasana* (Urticaceae). *Tree Physiology*, **3**, 375–86

Perez-Nasser, N. and Vázquez-Yanes, C. (1986). Longevity of buried seeds from some tropical rain forest trees and shrubs of Veracruz, Mexico. *Malaysian Forester*, **94**, 352–6

Puchet, C.E. (1986). Ecofisiologia de la Germinacion de Semillas de Algunos Arboles de la Vegetacion Madura de la Selva de "Los Tuxtlas", Veracruz, Mexico. Tesis. Facultad de Ciencias, UNAM, Mexico, D.F.

Puchet, C.E. and Vázquez-Yanes, C. (1987). Heteromorfismo criptico en las semillas recalcitrantes de tres especies arboreas de la selva tropical humeda de Veracruz, Mexico. *Phytologia*, **62**, 100–6

Roberts, E.H. (1973). Predicting the storage life of seeds. *Seed Science and Technology*, **1**, 499–514

Smith, H. (1982). Light quality, photoperception, and plant strategy. *Annual Review of Plant Physiology*, **33**, 481–518

Vázquez-Yanes, C. (1974). Studies on the germination of seeds of *Ochroma lagopus* Swartz. *Turrialba*, **24**, 176–9

Vázquez-Yanes, C. (1976). Estudios sobre la ecofisiologia de la germinacion en una zona calido humeda de Mexico. In Gomez-Pompa, A., Vásquez-Yanes, C., Del Amo, S. and Butanda, A. (eds), *Regeneracion de Selvas*. pp. 279–87. (Mexico, D.F.: Compania Editorial Continental)

Vázquez-Yanes, C. (1981). Germinacion de dos especies de Tiliaceas arboreas de la vegetacion secudaria tropical: *Belotia campbelli* y *Heliocarpus donnell-smithii*. *Turrialba*, **31**, 81–3

Vázquez-Yanes, C. and Orozco-Segovia, A. (1982a). Seed germination of a tropical rain forest pioneer tree (*Heliocarpus donnell-smithii*) in response to diurnal fluctuation of temperature. *Physiologia Plantarum*, **56**, 295–8

Vázquez-Yanes, C. and Orozco-Segovia, A. (1982b). Germination of the seeds of a tropical rain forest shrub, *Piper hispidum* Sw. (Piperaceae) under different light qualities. *Phyton*, **42**, 143–9

Vázquez-Yanes, C. and Orozco-Segovia, A. (1984). Ecophysiology of seed germination in the tropical humid forest of the World: A review. In Medina, E., Mooney, H.A. and Vázquez-Yanes, C. (eds) *Physiological Ecology of Plants of the Wet Tropics*, pp. 38–49. (The Hague: Dr. W. Junk Publishers)

Vázquez-Yanes, C. and Orozco-Segovia, A. (1986). Dispersal of seed animals: Effect on light controlled dormancy in *Cecropia obtusifolia*. In Estrada, A. and Fleming, T.H. (eds), *Frugivores and Seed Dispersal*, pp. 71–7. (Dordretcht: Dr. W. Junk Publishers)

Vázquez-Yanes, C. and Orozco-Segovia, A. (1987a). Fisiologia ecologica de semillas de la Estacion de Biologia Tropical de "Los Tuxtlas", Veracruz, Mexico. In Clark, D.A. and Fetcher, N. (eds) *Ecologia y Ecofisiologia de Plantas en los Bosques Mesoamericanos. Revista de Biologia Tropical* **35**, Suplemento, **1**, 85–96

Vázquez-Yanes, C. and Orozco-Segovia, A. (1987b). Light gap detection by the photoblastic seeds of *Cecropia obtusifolia* and *Piper auritum*, two tropical rain forest trees. *Biologia Plantarum*, **29**, 234–6

Vázquez-Yanes, C. and Smith, H. (1982). Phytochrome control of seed germination in the tropical rain forest pioneer trees *Cecropia obtusifolia* and *Piper auritum* and its ecological significance. *New Phytology*, **92**, 477–85

Vázquez-Yanes, C., Orozco-Segovia, A., Rincòn, E., Sànchez-Coronado, M.E., Huante, P., Barradas, V. and Toledo, J.R. (1990). Light beneath the litter in a tropical forest: effect on seed germination. *Ecology*, in press

Whitmore, T.C. (1983). Secondary succession from seed in tropical rain forest. *Forestry Abstracts*, **44**, 767–79

Willan, R.L. (1985). *A Guide to Forest Seed Handling with Special Reference to the Tropics*. Danida-FAO Forestry Paper **20** (2). (Rome: FAO)

CHAPTER 19

SEEDLING TYPES AND PLANT ESTABLISHMENT IN AN AFRICAN RAIN FOREST

Annette Hladik and Sophie Miquel

ABSTRACT

The morphological analysis of seedlings of 210 species from the Gabon forest reveals five seedling types based on the nature of cotyledons and length of hypocotyl. Among rain forests of the different continents, there is no significant difference in the proportion of these five seedling types (1 = 39%; 2 = 25%; 3 = 9%; 4 = 22%; 5 = 5%), and the occurrence of each type can be explained in part by specific conditions for germination and seedling growth in a heterogeneous environment. Seedling type is related to seed dispersal and seed size. Seedling morphology appears to be a critical parameter in determining whether a particular species will establish at a given site.

INTRODUCTION

Tropical rain forests are well known for their high plant species richness. In this environment, we encounter a great variety of morphological features in seedling development and a wide range of physiological adaptations to the heterogeneous structure of the tropical rain forest. Seed dispersal plays a critical role in seedling establishment. As a large proportion of seeds of the rain forest plant species are dispersed by animals and not by wind, relationships between seed size and efficient dispersal by a large variety of animals have been established in various groups. These relationships have been generally examined in relation to the specific animal syndrome (body weight, oral cavity and digestive tract, home range activity in time and space), and to the specific fruit and seed syndrome, but not to the seedling syndrome, which is so important for the success of the specific seedling establishment.

Morphological data on seedlings were obtained during a long-term study in plant ecology conducted at Makokou Field Research Station in Gabon, where the equatorial seasonal pattern of rainfall results in constant humidity on the forest floor (for details on climate, see Hladik and Blanc, 1987). For many years, seedlings of woody plant species of the Gabon rain forest have been raised in the nursery at Makokou with a view to building-up a key for seedling identification for use in inventory and survey (Miquel, 1985). In more recent programmes, seedlings have been introduced in mixed species plots in order to test various systems for future agroforestry management (Miquel and Hladik, 1984).

Here, we recognize five distinctive seedling types on the basis of the nature of cotyledons and the length of the hypocotyl. We also examine the relationship between seedling types and seed size, as well as the mode of seed dispersal. Finally, we discuss the relationship between seedling morphology and establishment.

SEEDLING TERMINOLOGY BASED ON SEEDLING MORPHOLOGY

Two contrasting terms have been used classically for seedling description: *epigeal*, concerning seedlings with cotyledons above ground level (Figure 19.1-1), and *hypogeal* for seedlings with cotyledons at (or under) ground level (Figure 19.1-4). In fact, these two terms refer to the length of the hypocotyl (reduced or absent hypocotyl in hypogeal). The terms are still used for plants of temperate zones because 95% of these species have epigeal seedlings (Vogel, 1980). In tropical zones, the terminology is also quite relevant for herbaceous plants, as illustrated by a study of African weeds (Merlier and Montégut, 1982).

Working on neotropical woody plant species, Duke (1965) introduced two further contrasting terms: *phanerocotylar*, applied to seedlings with exposed cotyledons, and *cryptocotylar* for those with cotyledons hidden in the seed coat. Duke discussed the occurrence of these two categories according to environmental conditions, especially light intensity and humidity.

Subsequently, Ng (1978), compiling the results of his thorough survey of Malayan tree species, proposed to combine these two sets of terms, the rationale being that the behaviour of the hypocotyl is independent of that of the cotyledons. He obtained four classes: to the first set of two terms (*epigeal* and *hypogeal*), he added the type *semi-hypogeal* (Figure 19.1-3) for seedlings with cotyledons exposed at ground level, and a "*Durian*" type (Figure 19.1-5) for seedlings with cotyledons hidden, but with a long hypocotyl. This last type had already been described in 1901 by Heckel as exceptional in *Ongokea gore*, an African tree species, and, in 1904, for the genus *Diospyros* in Sri Lanka, by Wright (in Ng, 1983).

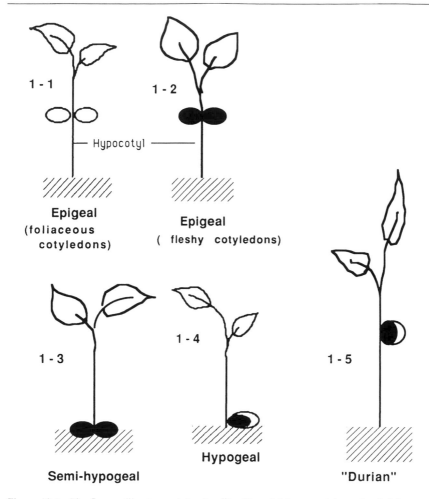

Figure 19.1 The five seedling types: 1-1 = Seedling Type 1 (phanerocotylar, epigeal, foliaceous cotyledons); 1-2 = Seedling Type 2 (phanerocotylar, epigeal, fleshy cotyledons); 1-3 = Seedling Type 3 (phanerocotylar, hypogeal); 1-4 = Seedling Type 4 (cryptocotylar, hypogeal); 1-5 = Seedling Type 5 (cryptocotylar, epigeal)

During the same period, Garwood (1983) pointed out the important difference, within the *phanerocotylar/epigeal* class, between seedlings with leaf-like green cotyledons (Figure 19.1-1) and seedlings with fleshy food-storing cotyledons (Figure 19.1-2). This morphological character can play an obviously important role in seedling establishment.

In order to take account of these different aspects of seedling morphology, Miquel (1985, 1987) recognized five distinctive seedling types, using three basic characters (Figure 19.1):

Seedling Type 1 – *phanerocotylar, epigeal* seedlings with *foliaceous cotyledons*

Seedling Type 2 – *phanerocotylar, epigeal* seedlings with *fleshy cotyledons*

Seedling Type 3 – *phanerocotylar/hypogeal* seedlings

Seedling Type 4 – *cryptocotylar/hypogeal* seedlings

Seedling Type 5 – *cryptocotylar/epigeal* seedlings

SEEDLING TYPES

Data have been obtained for 210 plant species, either trees or woody lianas, the latter accounting for a non-negligible part of the rain forest composition at Makokou (Hladik, 1974, 1978). This is not a random sample. Data were simply collected from seeds and seedlings available during the periods of field work. Complete details on seedling growth experiments are given in Miquel (1987).

Percentages of the five seedling types

Using a comparable system of classification of seedling types, the frequencies of the five seedling types as recorded in the Makokou forest are compared with three other tropical forests in different continents (Figure 19.2). There are no significant differences ($p < 0.05$).

Seedling Type 1 is the most common (39%). Seedling Type 5, the "Durian" type, although rare (5%), is present in all tropical forests. This last type was considered as "suicidal" by Wright (in Ng, 1983), because the cotyledons could be trapped in the coat. In fact, a high humidity during germination is the only limiting factor and, as shown in the comparative data presented in this paper, the "Durian" type is actually quite abundant in the Makokou forest, if measured in terms of tree density. Seedling Type 2 (25%) and Seedling Type 4 (22%) occur in about the same proportion, while Seedling Type 3 (9%) is less frequent.

The persistence of different types can be related to the potential function of the plant species in the early stages. It is likely that each of the five seedling types can cope with various limiting factors in the heterogeneous structure of a rain forest – such as humidity, temperature, light (in quantity and spectral composition), carbon dioxide level – according to photosynthetic threshold, potential resting stages with storage organs, growth rate for aerial system and root system, etc. The five seedling types

	Number of species	epigeal phanerocotylar		hypogeal cryptocot.		epig.
		1	2	3	4	5
African evergreen forest, Gabon (Miquel, 1987)	210	39%	25%	9%	22%	5%
Asian evergreen forest, Malaya (Ng, 1978)	300	64%		10%	18%	8%
American semi-deciduous forest Panama (Garwood,1983)	206	43%	16%	41%		
American tropical forest, Guadeloupe (Rousteau,1983)	102	51%	16%	31%		2%

Figure 19.2 Occurrence of the five seedling types (percentages of observed species) in various tropical rain forests

which we retained have been distinguished on the basis of strictly morphological characters. Several other seedling classes have been described according to the ontogeny of the storage organs, which can be endosperm, perisperm, cotyledons, or hypocotyl (Bokdam, 1977; de Vogel, 1980; Ye, 1983). Among the types considered here, the cotyledons as storage organs decrease during seedling growth, except for Seedling Type 1, with leaf-like cotyledons increasing throughout a long period.

Seedling types, seed size and life forms

Relationships between seed size and seedling morphological types have been examined in 172 species which have been identified (136 trees and 36 woody lianas, listed in Appendix 1). The smallest seeds, Class A in Figure 19.3 (A < 0.5 cm as measured along the largest axis, excluding wing), always belong to Seedling Type 1, both in trees and lianas. This is due to the necessity for small seeds without storage organs to utilize light energy as rapidly as possible, through the foliaceous cotyledons.

A large proportion of relatively small seeds, Class B in Figure 19.3 (0.5 < B < 2 cm), have seedlings with foliaceous cotyledons, but we also find all the other seedling types among the tree species. Among liana species, Seedling Type 4 is more frequent. Our data on the "Durian" type are restricted to nine tree species, though Hallé (1962) described one liana species of Seedling Type 5 among the family Hippocrateaceae in Côte d'Ivoire.

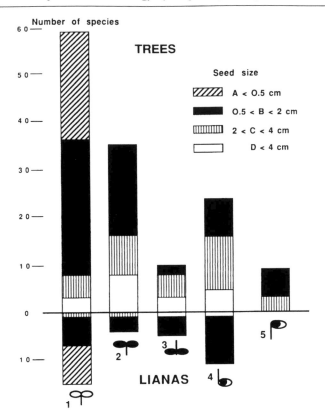

Figure 19.3 Number of species within seed size classes, for each of the five seedling types among tree and woody liana species

Large seeds, Class C and D in Figure 19.3 ($2 < C < 4$ cm and $D > 4$ cm), are found almost exclusively among tree species. Very few woody lianas have large seeds, although they can bear large fruits with strong peduncules. The exception is the seed of the liana *Lavigeria macrocarpa*, which can reach 11 cm, and which is, as far as we know, the largest seed of the African rain forest (Hladik *et al.*, 1984). From these large seeds emerge different seedling types. In the case of seeds exceeding 4 cm, there is no "Durian" type, and one wonders whether these seeds are not too heavy to be supported by a hypocotyl.

Seedling Type 1 (with foliaceous cotyledons) has often been considered as exclusively typical of small seeds and corresponding to pioneers such as *Musanga cecropioides, Harungana madagascariensis, Trema guineensis, Anthocleista* spp., etc... Nevertheless, this type is also found among shade-tolerant species with large seeds, such as *Panda oleosa, Gambeya* spp., *Strombosia* spp., the seedlings of which develop very large cotyledons, particularly efficient in the understorey where light intensity is minimal. The herbaceous monocotyledons, which are not taken into consideration in this comparative study, also have small seeds, but belong mostly to Seedling Type 4.

Thus, to the same seed size class may correspond different seedling types, involving different physiological adaptations to germination. The exception is the smallest seed size class (< 0.5 cm), with only the Seedling Type 1.

Seedling types and seed dispersal

The large diversity of seedling types found in the tropics reflects various adaptive physiological strategies to different micro-environmental conditions. In this respect, dispersal increases the probability that a seed may reach the best place at the best time for seedling establishment and growth (after a successful germination). There should thus be a relationship between seedling type and seed dispersal.

Seedling types and animal dispersal

Studies at Makokou (Gautier-Hion *et al.*, 1985) have revealed that 40 vertebrate species (elephant, six monkey species, seven large bird species, seven ruminant species, two large rodents, nine squirrel species, eight small rodent species) are involved in seed dispersal. Out of 122 plant species on which these animals were observed to feed, there are indications of positive or negative interaction with vertebrates in 100 species (seed dispersal in 85 species when fruits are used by at least one animal species; destruction in 15 species when seeds are exclusively eaten). There is no significant difference ($\chi^2 = 3.41$; $p < 0.05$) in the proportion of the different seedling types between this sample of 100 species (Figure 19.4) and our reference set of 210 analysed species (Figure 19.2).

Plant species with Seedling Type 1 clearly are dispersed by all the different groups of vertebrates. As discussed above, many of these seedlings, with foliaceous cotyledons, belong to light-demanding species. In the mosaic of the forest, tree gaps are unpredictable in their location and, altogether, account for only 5–10% of the total area (Poore, 1968; Oldeman, 1974). Accordingly, seed dispersal appears as a necessity for this category of plant species. The number and variety of dispersal sites are increased by the diversity of foraging patterns of the different groups of animals.

Plant species with Seedling Types 2–5 are dispersed by a more limited number of animals. Moreover, species with Seedling Type 4, corresponding to lianas (in the families Apocynaceae and Dichapetalaceae), are typically dispersed by monkeys.

Seedling types

	1	2	3	4	5	Unknown seedlings
100 species with ripe fruit eaten	44	16	6	20	5	9
85 species with dispersed seeds	39	13	3	18	5	7

Animal dispersers

Elephants	17	5	0	5	3	0
Monkeys	21	9	3	14	3	7
Large birds	17	7	0	6	1	0
Ruminants	9	3	0	0	1	0
Large rodents	6	1	0	4	1	0
Squirrels	4	0	0	3	0	0
Small rodents	11	0	0	2	0	2

Figure 19.4 Number of plant species with seeds dispersed by various vertebrates (after Gautier-Hion *et al.*, 1985), for each of the five seedling types

Seed size and seed dispersal

Seed dispersal by animals is partly related to seed size, as illustrated by a comparison of the seed size spectrum of the five seedling types among elephants, monkeys, and large birds (Figure 19.5). Plant species with large seeds (D) such as *Panda oleosa*, *Parinari excelsa*, *Klainedoxa* spp., *Irvingia* spp., *Drypetes gossweileri*, are dispersed exclusively by elephants (endozoochory), and have been defined as "loxodontochores" by Alexandre (1978). Monkeys and large birds disperse small seeds (A and B), either by endozoochory or epizoochory (Gautier-Hion, 1984).

Plants with large seeds seem to depend on specialized dispersers, as described by Wheelwright (1985) for all fruit species eaten by birds. Although data concerning seed dispersal by the smallest birds or frugivorous bats are incomplete for Makokou forest, the observation that small seeds are dispersed by a large number of animal species supports this claim.

Seedling types and seed dispersal status

Seed dispersal status has been attributed to the 172 plant species of known seedling type (Appendix 1), according to data from Alexandre (1978),

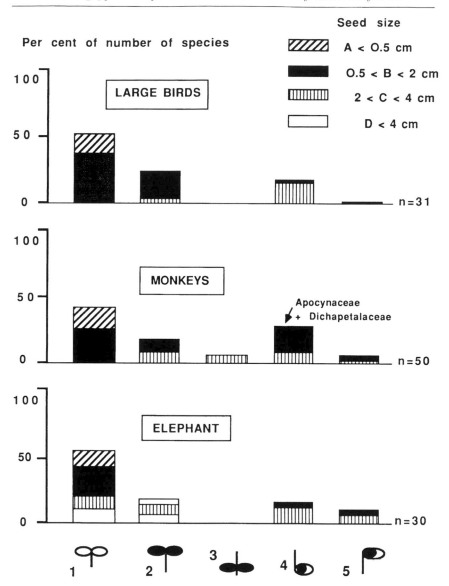

Figure 19.5 Seed size spectrum for three major groups of animal dispersers, according to the five seedling types

Gautier-Hion *et al.* (1985), C.M. Hladik (1973), Tutin and Fernandez (1985) and A. Hladik (unpublished), and from fruit morphological evidence. Among these species from Makokou, only 14% bear winged seeds or fruits, likely to be dispersed by wind. A majority of these anemochorous species have small seeds and Seedling Type 1, while a few have such heavy seeds that dispersal by wind would not be efficient. In this respect, these plants with heavy winged seeds are similar to the so-called "autochorous" species (accounting for 11%), which are not, as far as we know, dispersed by animals.

The autochorous species have Seedlings Types 2 or 3, with fleshy exposed cotyledons that could be considered as potential animal food. In this category, tree species such as *Baphia leptobotrys* and *Pentaclethra* spp. have sufficient energetic reserves for long resting stages. Moreover, they have a large plasticity in light requirements, which may explain, to some extent, a lesser dependency on animal dispersers.

The large majority of the rain forest plants are animal dispersed (70% among the plant species observed for seedling type; probably up to 80% for the total plant community). Many seeds are swallowed and retain a germinative potential, with a possible reduction in delay of germination (Hladik and Hladik, 1967, 1969; Alexandre, 1978). There are also reports of plant species requiring, for germination, a prior "seed clearing" by animals (Ng, 1983). Finally, the environmental conditions necessary for germination (see review by Vázquez-Yanes and Orozco-Segovia, 1984) must occur at the dispersal site.

Seedling types and plant establishment

We also compared the rate of establishment and success of seedlings and saplings of the different plant species. For this purpose, several plots and transects were used, depending on the size of the individual plant: saplings were recorded on 10 quadrats of 16 m^2 (Hladik and Blanc, 1987); the density of trees >5 cm diameter at breast height (dbh) was calculated on two transects each of 10 m by 200 m (Hladik, 1982); and the density of trees >30 cm diameter was averaged per ha, based on measures covering a total area of 19 000 m^2 (Hladik, 1982, and unpublished data). These samples were compared to the seedling stock inside the rain forest. The seedling stock was determined along trails, during a 3-month period, for 51 plant species (710 seedlings at different stages; Miquel, 1985). The seedling stock was also compared to the number of trees established in abandoned fields after shifting cultivation as exclusively produced from seed germination and not from sprouting stems (Miquel, 1985, 1987).

First, the five seedling types have been compared in terms of number of species occurring in the different samples (Figure 19.6-A). Considering the spectrum of occurrence of the different seedling types among our reference set of 210 plant species (Figure 19.6-A-a), the initial seedling stock on the forest floor, calculated on 51 plant species, presents a very similar pattern (Figure 19.6-A-b), with $\chi^2 = 0.704$ ($p < 0.05$). Accordingly, for most species of the different seedling types, dispersal is efficient and all the necessary

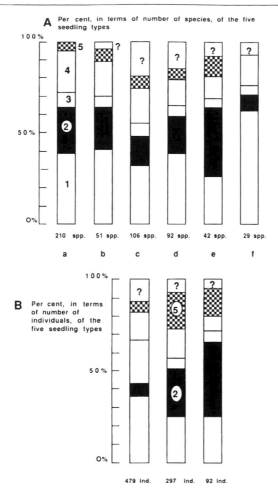

Figure 19.6 Comparison of the occurrence of the five seedling types (indicated by numbers in the first column), in relation to plant density:

A – Percentages, in terms of number of species, among various plant categories in plots of different size:

    (a) total species reference inventory (210 species);

    (b) seedling stock (51 species)

    (c) saplings on 160 m² (106 species)

    (d) trees over 5 cm diameter on 4000 m² (92 species)

    (e) trees over 30 cm diameter on 10 000 m² (42 species)

    (f) trees in natural regeneration after shifting cultivation (29 species)

B – Percentages, in terms of number of individuals:

    (c) saplings on 160 m² (479 individuals)

    (d) trees over 5 cm diameter, on 4000 m² (297 individuals)

    (e) trees over 30 cm diameter, on 10 000 m² (92 individuals)

271

conditions for germination are encountered at the various dispersal sites. Furthermore, the seedling stock appears to fulfil the important role of a resting stage, as already mentioned by Alexandre (1977). In contrast, the seed bank in the soil, which is important for regeneration after forest destruction, is limited to tree species with Seedling Type 1. Comparing saplings below 3 m in height (Figure 19.6-A-c), trees >5 cm diameter (Figure 19.6-A-d), and trees >30 cm (Figure 19.6-A-e), there is an increase in the percentages of species with Seedling Types 2 and 5, which accompanies the increase in areal size of the different plots studied. For trees > 30 cm diameter, the percentage of species with Seedling Type 1 is half that found in our reference set of 210 plant species. Conversely, in the shifting cultivation plot (Figure 19.6-A-f), most of the tree species are of Seedling Type 1 (i.e. pioneer species often originating from the seed bank).

A second comparison is that of the five seedling types in terms of the number of individuals occurring in three of the plots (Figure 19.6-B). The species with foliaceous cotyledons (Type 1), which make up 40% (82 species) of the total number of species in our reference set, account for only 20% of the total number of individuals in the Makokou rain forest. By contrast, plants with Seedling Type 5 (the "Durian" type), which only account for nine known species of the Makokou forest, are abundant among medium-sized trees (20% in the plot of 4000 m^2, Figure 19.6-B-d). Accordingly, this "strange" seedling type which is so difficult to raise in the nursery is quite successful in the Makokou forest, even though only a limited number of these trees actually grow over 30 cm in diameter (Figure 19.6-B-e). Moreover, plant species of Type 2, which account for 25% of the total number of species, are clearly represented by increased numbers of individuals in plots of increasing individual plant size. As a matter of fact, these last-mentioned species belong to the characteristic families of very tall trees of the African forest, such as the Leguminosae (Caesalpiniaceae, Mimosaceae, Papilionaceae), Meliaceae, Irvingiaceae, and Burseraceae.

CONCLUSION

A comparison of different tropical forests throughout the world reveals many similarities about distribution of the different seedling types. The pan-tropical seedling diversity that is related to the heterogeneous forest structure, coupled with the long resting seedling stages, represents an important potential for regeneration in the understorey. Additional studies are required to confirm inter-regional similarities and to elucidate the role of seedling diversity in regeneration under different natural and disturbed conditions.

Seed dispersal is a mechanism for escaping seed predators, and for ensuring that some seeds encounter adequate conditions for germination.

Therefore, a survey of fruit morphological and biochemical characters favouring seed dispersal in plant population studies calls for a supplementary approach of seedling types related to different limiting factors for germination (as studied by Miquel, 1987).

In fact, each seedling type represents a functional adaptation to forest regeneration, with special relationships to seed size class, seed dispersal and plant establishment.

Seedling Type 1, with foliaceous photosynthetic cotyledons, is the most abundant one among tropical rain forest species, and not only in the case of small seeds (< 0.5 cm) for which this type is obligatory (except in the case of the herbaceous monocotyledons). Very large, leaf-like cotyledons are also known to emerge from big seeds. Nevertheless, this type is not so abundant if measured as the number of individuals established inside the rain forest. This type is the most widely dispersed, involving a large variety of animals, in order to encounter conditions on the forest floor with enough light for photosynthetic requirements. The result is a relative low density of individuals compared to a high number of species. Outside the forest, the establishment of the Seedling Type 1 is the most successful.

Seedling Type 2, with a hypocotyl, and the less frequent Type 3 at ground level, both with fleshy exposed cotyledons, are present in plant species with seeds > 0.5 cm. They are dispersed by a more limited number of animals and quite frequently belong to the so-called "autochorous" plant species. When the fleshy cotyledons escape animal predation, they have sufficient energetic reserves for long resting stages in the understorey, and they often become the giant trees of the Makokou forest.

Seedling Type 4, with hidden cotyledons at ground level, is associated generally, but not exclusively, with large seeds. Associated animal-dispersed plant species include notably lianas, preferentially dispersed by monkeys.

Seedling Type 5, the "Durian" type, with hidden cotyledons supported by a hypocotyl, cannot be too heavy, and we have no records of seeds exceeding 4 cm. In spite of difficulties in seed germination and seedling development in the nursery, the densities of these plant species inside the rain forest of Makokou is high.

In terms of management implications, experiments with mixed species cultivated in association at Makokou field station (Miquel, unpub.) show that species with Seedling Type 2 or 3 were particularly successful (over 60% establishment success). On the other hand, the only performant seedlings in the case of natural regeneration after shifting cultivation were those of Type 1. The overall implication is that we have to be attentive to seedling types in any project concerning forest regeneration.

The different plant species can be characterized by a multifactorial "regeneration niche", as supported by Peart's (1985) mathematical approach to hypothetical seed density dispersal functions. Among the dimensions of this niche are the various patterns of plant phenology (flowering time, pollination systems, fruit maturation) and the various systems of dispersal, germination and seedling establishment, the importance of which have been pointed out by Grubb (1977). In addition, the seedling types discussed here have to be taken into consideration in studies of plant population dynamics.

ACKNOWLEDGEMENTS

The field work conducted at the Makokou Research Station in Gabon was supported by funds from the Centre National de la Recherche Scientifique (CNRS, ECOTROP, France) and from the Man and Biosphere Programme of Unesco. The authors are grateful to Malcolm Hadley (Unesco-MAB), Professor Kamal Bawa, and Claude Marcel Hladik for reviewing and commenting helpfully on the manuscript.

REFERENCES

Alexandre, D.-Y. (1977). Régénération naturelle d'un arbre caractéristique de la forêt équatoriale de Côte d'Ivoire: *Turraeanthus africana* Pellgr. *Oecologia Plantarum*, **12**, 241–62

Alexandre, D.-Y. (1978). Le rôle disséminateur des éléphants en forêt de Taï, Côte d'Ivoire. *La Terre et la Vie*, **32**, 47–72

Bokdam, J. (1977). Seedling morphology of some African Sapotaceae and its taxonomical significance. *Mededelingen Landbouwhogeschool Wageningen*, **77**, 1–84

Duke, J.A. (1965). Keys of the identification of seedlings of some prominent woody species in eight forest types in Puerto Rico. *Annals of the Missouri Botanical Garden*, **52**, 314–50

Florence, J. and Hladik, A. (1980). Catalogue des phanérogames du N.-E. du Gabon (sixième liste). *Adansonia*, sér. 2, **20**, 235–53

Garwood, N.C. (1983). Seed germination in a seasonal tropical forest in Panama. A community study. *Ecological Monographs*, **53**, 159–84

Gautier-Hion, A. (1984). La dissémination des graines par les Cercopithécidés forestiers africains. *Revue d'Ecologie (Terre Vie)*, **39**, 159–65

Gautier-Hion, A., Duplantier, J.-M., Quris, R., Feer, F., Sourd, C., Decoux, J.-P., Dubost, G., Emmons, L., Erard, C., Hecketsweiler, P., Moungazi, A., Roussilhon, C. and Thiollay, J.-M. (1985). Fruit characters as a basis of fruit choice and seed dispersal in a tropical forest vertebrate community. *Oecologia*, **65**, 324–37

Gilbert, G. (1939). Observations préliminaires sur la morphologie des plantules forestières au Congo belge. *Institut National pour l'Etude Agronomique du Congo belge*, Série Scientifique, **17**, 1–28

Grubb, P.J. (1977). The maintenance of species-richness in plant communities: The importance of the regeneration niche. *Biological Review*, **52**, 107–45

Hallé, N. (1962). Monographie des Hippocrateacées d'Afrique occidentale. *Mémoires de l'Institut français d'Afrique Noire*, **64**, 1–215

Hallé, N. (1964). Première liste de Phanérogames et de Ptéridophytes des environs de Makokou, Kemboma et Bélinga. *Biologia Gabonica*, **1**, 41–6

Hallé, N. (1965). Seconde liste de Phanérogames et Ptéridophytes du N.-E. Gabon (Makokou, Bélinga et Mékambo). *Biologia Gabonica*, **1**, 337–44

Hallé, N. and Le Thomas, A. (1967). Troisième liste de Phanérogames et Ptéridophytes du N.-E. Gabon. *Biologia Gabonica*, **3**, 113–20

Hallé, N. and Le Thomas, A. (1970). Quatrième liste de Phanérogames et Ptéridophytes du N.-E.

Gabon (Bassin de l'Ivindo). *Biologica Gabonica*, **6**, 131–8

Heckel, E. (1901). Sur le processus germinatif dans le genre *"Onguekoa"* et *Strombosia* de la famille des Olacaceae. *Annales du Musée Colonial de Marseille*, **8**, 17–27

Hladik, C.M. (1973). Alimentation et activité d'un groupe de chimpanzés réintroduits en forêt gabonaise. *La Terre et la Vie*, **27**, 343–413

Hladik, A. (1974). Importance des lianes dans la production foliaire de la forêt équatoriale du Nord-Est du Gabon. *Comptes-rendus de l'Académie des Sciences, Paris, série D*, **278**, 2527–30

Hladik, A. (1978). Phenology of leaf production in rain forest of Gabon: Distribution and composition of food for folivores. In Montgomery, G.G. (ed.), *The Ecology of Arboreal Folivores*, pp. 51–71. (Washington, D.C.: Smithsonian Institution Press)

Hladik, A. (1982). Dynamique d'une forêt equatoriale africaine. Mesures en temps réel et comparaison du potentiel de croissance des différentes espèces. *Acta Oecologica, Oecologia Generalis*, **3**, 373–92

Hladik, A., Bahuchet, S., Ducatillion, C. and Hladik, C.M. (1984). Les plantes á tubercules de la forêt dense d'Afrique centrale. *Revue d'Ecologie (Terre Vie)*, **39**, 249–90

Hladik, A. and Blanc, P. (1987). Croissance des plantes en sous-bois de forêt dense humide africaine (Makokou, Gabon). *Revue d'Ecologie (Terre Vie)*, **42**, 209–34

Hladik, A. and Hallé, N. (1973). Catalogue des Phanérogames du N.-E. du Gabon (5ème liste). *Adansonia, série 2*, **13**, 527–44

Hladik, A. and Hladik, C.M. (1969). Rapports trophiques entre la végétation et les Primates dans la forêt de Barro Colorado (Panama). *La Terre et la Vie*, **23**, 25–117

Hladik, C.M. and Hladik, A. (1967). Observations sur le rôle des Primates dans la dissémination des végétaux de la forêt gabonaise. *Biologia Gabonica*, **3**, 43–58

Léonard, J. (1957). Génèse des Cynometrae, Légumineuses Caesalpiniaceae. Essai de blastogénie appliquée à la systématique. *Mémoires de l'Académie Royale de Belgique, Classe des Sciences*, **30**, 1–312

Mensbruge, G. de la. (1966). *La germination et les plantules de la forêt dense humide de la Côte d'Ivoire*. Centre Technique Forestier Tropical, Nogent sur Marne.

Merlier, H. and Montégut, J. (1982). *Adventices tropicales*. (Paris: ORSTOM-GERDAT-ENSH)

Miquel, S. (1985). Plantules et premiers stades du croissance des espèces forestières du Gabon: Potentialités d'utilisation en agroforesterie. Thèse 3ème cycle. Université Paris VI, Paris.

Miquel, S. (1987). Morphologie fonctionnelle de plantules d'espèces forestières du Gabon. *Bulletin du Muséum National d'Histoire Naturelle, Paris, 4ème série, section B, Adansonia*, **9**, 101–21

Miquel, S. and Hladik, A. (1984). Sur le concept d'agroforesterie: exemple d'expériences en cours dans la région de Makokou (Gabon). *Bulletin d'Ecologie*, **15**, 163–73

Ng, F.S.P. (1978). Strategies of establishment in Malayan forest trees. In Tomlinson, P.B. and Zimmermann, H.M. (eds), *Tropical Trees as Living Systems*, pp. 129–62. (Cambridge: Cambridge University Press)

Ng, F.S.P. (1983). Ecological principles of tropical lowland rain forest conservation. In Sutton, S.L., Whitmore, T.C. and Chadwick, A.C. (eds), *Tropical Rain Forest: Ecology and Management*, pp. 359–75. (Oxford: Blackwell Scientific Publications)

Oldeman, R.A.A. (1974). L'architecture de la forêt guyanaise. *Mémoires ORSTOM*, **73**, 1–204

Peart, D.R. (1985). The quantitative representation of seed and pollen dispersal. *Ecology*, **66**, 1081–3

Poore, M.E.D. (1968). Studies in Malaysian rain forest. I. The forest on Triassic sediments in Jengka reserve. *Journal of Ecology*, **56**, 143–96

Rousteau, A. (1983). 100 plantules d'arbres de Guadeloupe: Aspects morphologiques et écologiques. Thèse 3ème cycle. Université Paris VI, Paris.

Taylor, C.J. (1960). *Synecology and Silviculture in Ghana*. (Edinburgh: Nelson)

Tutin, C.E.G. and Fernandez, M. (1985). Foods consumed by sympatric populations of *Gorilla g. gorilla* and *Pan t. troglodytes* in Gabon: some preliminary data. *International Journal of Primatology*, **6**, 27–43

Vázquez-Yanes, C. and Orozco-Segovia, A. (1984). Ecophysiology of seed germination in the tropical humid forest of the world: a review. In Medina, E., Mooney, H.A. and Vázquez Yanes, C. (eds), *Physiological Ecology of Plants of the Wet Tropics*, pp. 37–49. (The Hague: Dr. W. Junk Publishers)

Vogel, E.F. de (1980). *Seedlings of Dicotyledons*. (Wageningen: Pudoc)

Voorheve, A.G. (1967). *Liberian High Forest Trees*. Belmontia Miscellaneous Publications in Botany 8. (Wageningen: Centrum voor Landbouw Publikaties en Landbouw documentarie)

Wheelwright, N.T. (1985). Fruit size, gape width, and the diets of fruit-eating birds. *Ecology*, **66**, 808–18

Ye, N. (1983). Studies on the seedling type of dicotyledonous plants (Magnoliophyta-Magnoliosida). *Phytologia*, **54**, 161–89

Appendix 1 Checklist of the 172 identified plant species (for authors' name, see the lists of the Phanerogamous plants of Makokou: Hallé, 1964, 1965; Hallé and Le Thomas, 1967, 1970; Hladik and Hallé, 1973; Florence and Hladik, 1980). Additional information about seedling morphology is from Gilbert (1939), Léonard (1957), Taylor (1960), de la Mensbruge (1966) and Voorheve (1967)

	Life form T=Tree L=Liana	Dispersal agent	Seed size A < 0.5 0.5 < B < 2 2 < C < 4 D > 4 cm	Seedling type
AGAVACEAE				
Dracaena arborea	T	Animal	B	4
ANACARDIACEAE				
Antrocaryon klaineanum	T	Animal	C	2
Antrocaryon nannanii	T	Animal	D	2
Pseudospondias longifolia	T	Animal	B	2
Trichoscypha abut	T	Animal	C	3
Trichoscypha acuminata	T	Animal	C	3
Trichoscypha arborea	T	Animal	C	3
ANNONACEAE				
Anonidium mannii	T	Animal	C	5
Friesodielsa enghiana	L	Animal	B	4
Hexalobus sp.	T	Animal	B	1
Isolona letestui	T	Animal	B	5
Pachypodanthium barteri	T	Animal	B	1
Pachypodanthium staudtii	T	Animal	B	1
Polyalthia suaveolens	T	Animal	B	5
Uvariopsis solheidii	T	Animal	B	1
Xylopia aethiopica	T	Animal	A	1
Xylopia hypolampra	T	Animal	B	1
Xylopia quintasii	T	Animal	B	1
Xylopia staudtii	T	Animal	B	1
APOCYNACEAE				
Alstonia boonei	T	Wind	A	1
Aphanostylis mannii	L	Animal	B	4
Cylindropsis parvifolia	L	Animal	B	4
Dictyophleba stipulosa	L	Animal	B	4
Funtumia elastica	T	Wind	A	1
Landolphia owariensis	L	Animal	B	4
Picralima nitida	T	Animal	C	1
Pycnobotrya nitida	L	Wind	C	2
Strophanthus sarmentosus	L	Wind	A	1
BIGNONIACEAE				
Markhamia sessilis	T	Wind	B	1
Spathodea campanulata	T	Wind	B	1
BURSERACEAE				
Aucoumea klaineana	T	Wind	B	1
Canarium schweinfurthianum	T	Animal	C	1

(continued)

277

Life form T=Tree L=Liana	Dispersal agent	Seed size A < 0.5 0.5 < B < 2 2 < C < 4 D > 4 cm	Seedling type
Dacryodes buttneri	T	Animal	C	2
Dacryodes edulis	T	Animal	C	2
Dacryodes klaineana	T	Animal	B	2
Dacryodes macrophylla	T	Animal	C	2
Santiria sp.I	T	Animal	B	2
Santiria sp.II	T	Animal	C	2
CAESALPINIACEAE				
Afzelia bipendensis	T	Auto.	B	2
Amphimas ferrugineus	T	Wind	C	2
Anthonotha macrophylla	T	Auto.	D	4
Berlinia bracteosa	T	Auto.	D	3
Crudia gabonensis	T	Animal	C	4
Detarium macrocarpum	T	Animal	D	2
Dialium dinklagei	T	Animal	B	2
Distemonanthus benthamianus	T	Auto.	B	1
Griffonia physocarpa	L	Auto.	B	3
Guibourtia tessmannii	T	Animal	C	2
Hymenostegia pellegrini	T	Auto.	B	2
Scorodophloeus zenkeri	T	Auto.	C	2
Swartzia fistuloides	T	Animal	B	1
CHRYSOBALANACEAE				
Parinari excelsa	T	Animal	D	4
COMBRETACEAE				
Combretum bipendense	L	Wind	B	1
Combretum sp.	L	Wind	A	1
CONNARACEAE				
Roureopsis obliquifoliolata	L	Animal	B	3
sp.	L	Animal	B	3
CONVOLVULACEAE				
Dipteropeltis poranoides	L	Wind	A	1
Neuropeltis sp.	L	Wind	A	1
DICHAPETALACEAE				
Dichapetalum heudelotii	L	Animal	B	4
Dichapetalum integripetalum	L	Animal	B	4
Dichapetalum mombuttense	L	Animal	B	4
Dichapetalum thollonii	L	Animal	B	4
Dichapetalum unguiculatum	L	Animal	B	4
DILLENIACEAE				
Tetracera alnifolia	L	?	A	1
DIPTEROCARPACEAE				
Marquesia excelsa	T	Wind	B	1
EUPHORBIACEAE				
Bridelia atroviridis	T	Animal	A	1

(continued)

278

	Life form T=Tree L=Liana	Dispersal agent	Seed size A < 0.5 0.5 < B < 2 2 < C < 4 D > 4 cm	Seedling type
Croton oligandrus	T	Animal	B	1
Dichostemma glaucescens	T	Animal	B	5
Discoglypremna caloneura	T	Animal	A	1
Drypetes gossweileri	T	Animal	C	1
Klaineanthus gaboniae	T	?	B	1
Macaranga barteri	T	Animal	A	1
Macaranga spinosa	T	Animal	A	1
Mareyopsis longifolia	T	?	B	5
Margaritaria discoidea	T	?	A	1
Phyllanthus polyanthus	T	?	B	1
Plagiostyles africana	T	Animal	B	5
Ricinodendron heudelotii	T	Animal	B	1
Uapaca sp.	T	Animal	B	1
FLACOURTIACEAE				
Caloncoba welwitschii	T	Animal	A	1
Campostylus mannii	T	Animal	A	1
Lindackeria dentata	T	Animal	A	1
GUTTIFERAE				
Allanblackia klainei	T	?	C	4
Garcinia polyantha	T	Animal	C	4
Garcinia sp.	T	Animal	C	4
Mammea africana	T	Animal	C	4
Pentadesma butyracea	T	Animal	D	4
Symphonia globulifera	T	Animal	C	4
HERNANDIACEAE				
Illigera pentaphylla	L	Wind	A	1
HIPPOCRATEACEAE				
Loeseneriella sp.	L	Wind	B	1
Salacia sp.	L	Animal	C	1
HYPERICACEAE				
Harungana madagascariensis	T	Animal	A	1
ICACINACEAE				
Lavigeria macrocarpa	L	?	D	4
IRVINGIACEAE				
Irvingia gabonensis	T	Animal	D	2
Irvingia grandifolia	T	Animal	D	2
Klainedoxa gabonensis	T	Animal	D	1
LAURACEAE				
Beilschmiedia sp.	T	Animal	C	4
LECYTHIDACEAE				
Petersianthus macrocarpus	T	Wind	B	1
LINACEAE				
Hugonia planchonnii	L	Animal	B	1

(continued)

Life form T=Tree L=Liana	Dispersal agent	Seed size A < 0.5 0.5 < B < 2 2 < C < 4 D > 4 cm	Seedling type
LOGANIACEAE			
Anthocleista schweinfurthii T	Animal	A	1
Strychnos aculeata L	Animal	B	1
Strychnos sp. L	Animal	B	1
MELIACEAE			
Carapa procera T	Animal	D	4
Guarea sp. T	Animal	B	2
Entandrophragma sp. T	Wind	B	2
Lovoa trichilloides T	Wind	B	2
Trichilia gilgiana T	Animal	B	2
Trichilia prieureana T	Animal	B	2
MENISPERMACEAE			
Stephania sp. L	?	A	1
MIMOSACEAE			
Adenanthera sp. L	Auto.	B	2
Albizia sp. T	Auto.	B	2
Cylicodiscus gabunensis T	Wind	D	2
Entada gigas L	Auto.	D	3
Entada scelerata L	Auto.	B	2
Fillaeopsis discophora T	Wind	D	2
Parkia bicolor T	Animal	B	2
Pentaclethra eetveldeana T	Auto.	C	3
Pentaclethra macrophylla T	Auto.	D	3
Piptadeniastrum africanum T	Wind	B	2
Tetrapleura tetraptera T	Animal	B	2
MORACEAE			
Ficus macrosperma T	Animal	A	1
Ficus wildemaniana T	Animal	A	1
Musanga cecropioides T	Animal	A	1
Myrianthus arboreus T	Animal	B	4
Treculia africana T	Animal	B	3
MYRISTICACEAE			
Coelocaryon preussii T	Animal	C	4
Pycnanthus angolensis T	Animal	C	4
Staudtia gabonensis T	Animal	C	4
MYRTACEAE			
Syzygium sp. T	Animal	B	4
OCHNACEAE			
Lophira alata T	Wind	B	4
OLACACEAE			
Coula edulis T	Animal	C	5
Heisteria parvifolia T	Animal	B	1
Ongokea gore T	Animal	C	5

(continued)

Life form T=Tree L=Liana	Dispersal agent	Seed size A < 0.5 0.5 < B < 2 2 < C < 4 D > 4 cm	Seedling type
Strombosia sp.	T	Animal	B	1
PANDACEAE				
Panda oleosa	T	Animal	D	1
PAPILIONACEAE				
Baphia leptobotrys	T	Auto.	B	2
Baphia pubescens	T	Auto.	B	2
Dalbergia hostilis	L	Auto.	B	2
Dalhousiea africana	L	Auto.	B	3
Milletia mannii	T	Auto.	B	2
Pterocarpus soyauxii	T	Wind	B	3
PASSIFLORACEAE				
Paropsis grewioides	T	Animal	A	1
RHAMNACEAE				
Maesopsis eminii	T	Animal	C	1
RUBIACEAE				
Atractogyne gabonii	L	Animal	A	1
Massularia acuminata	T	Animal	A	1
Nauclea diderrichii	T	Animal	A	1
RUTACEAE				
Zanthoxylum gilletii	T	Animal	A	1
SAMYDACEAE				
Casearia barteri	T	Animal	A	1
SAPINDACEAE				
Allophyllus cobbe	T	?	B	1
Blighia welwitschii	T	Animal	C	4
Chytranthus angustifolius	T	Animal	B	4
Eriocoelum macrocarpum	T	Animal	B	4
Pancovia pedicellaris	T	Animal	B	4
SAPOTACEAE				
Baillonella toxisperma	T	?	D	2
Gambeya africana	T	Animal	B	1
Gambeya beguei	T	Animal	B	1
Gambeya lacourtiana	T	Animal	C	1
Omphalocarpum procerum	T	Animal	D	1
Synsepalum letestui	T	Animal	B	3
Tieghemella africana	T	Animal	D	4
SIMAROUBACEAE				
Quassia gabonensis	T	Auto.	D	2
STERCULIACEAE				
Cola acuminata	T	?	C	3
Cola rostrata	T	?	D	3
Sterculia tragacantha	T	Animal	B	1

(continued)

Life form T=Tree L=Liana	Dispersal agent	Seed size A < 0.5 0.5 < B < 2 2 < C < 4 D > 4 cm	Seedling type
STYRACACEAE			
Afrostyrax lepidophyllus	T	?	B	5
TILIACEAE			
Duboscia macrocarpa	T	Animal	A	1
Grewia coriacea	T	Animal	B	1
Grewia sp.	L	Animal	B	1
ULMACEAE			
Celtis tessmannii	T	Animal	B	1
Trema guineensis	T	Animal	A	1
VITACEAE			
Cissus dinklagei	L	Animal	B	1
VOCHYSIACEAE			
Erismadelphus exsul	T	Wind	B	4

Section 6

Regeneration

CHAPTER 20

REGENERATION – COMMENTARY

Richard B. Primack

Reproduction in higher plants may be considered as a series of consecutive phases which affect final reproductive success, such as flowering phenology, pollination and breeding systems, fruit and seed dispersal, seed dormancy and seed germination. The final phase in this process is successful establishment of seedlings and saplings in an environment where they can grow into reproductively mature trees. This successful regeneration allows a species to be maintained over time at a particular forest. Successful regeneration away from the existing population allows each species to extend its range into new habitats. These processes are significant for the understanding of tropical forest and are relevant to the long-term production management of these forests.

In the last few years, several large, long-term primary forest plots have been set up at Barro Colorado Island in Panama, La Selva in Costa Rica, Macambo in Brazil, Pasoh Forest in Peninsular Malaysia, and Sarawak in Malaysian Borneo, to obtain demographic data from tropical tree populations. Such data include the establishment, growth and survival rates of seedlings, saplings and adults of common timber species. The goals of this research are several-fold. First, information will allow estimates to be made of timber and total biomass production in these forests. These data will be valuable for comparison with silvicultural plots that have been treated in order to determine how many years are required for a logged forest to return to the characteristic state of the primary forest. Second, these plots are valuable to determine how stable tropical rain forest tree populations are over time. Since a goal of tropical forestry is to manage forests for a high production of particular timber tree species, this may not be possible if populations are unstable at a local scale. And, third, such plots are relevant to explaining the great species richness of the rain forests.

Two major hypotheses are currently being debated by ecologists to resolve this question, as described in the paper by Hubbell and Foster. A prediction of the equilibrium hypothesis is that species may coexist in the rain forest because each species has a specialized niche in which it is competitively superior to all other species. Coexistence of species in an equilibrium can also occur if density- and frequency-dependent mortality caused by predators, herbivores, and pathogens limits the population size of individual tree species that begin to become common. This limitation will have the effect of preventing any one species or group of species from dominating a forest. The non-equilibrium hypothesis states that the present species composition and abundance of individual species were determined, in part, by random factors such as the types of seeds available when canopy gaps were formed. This hypothesis predicts that there will be a species turnover in forests, with some species going extinct and other species becoming established.

This academic debate about equilibrium and non-equilibrium models has practical significance for conservation biologists and managers of national parks. If a forest is found to be in a non-equilibrium state, many species will be eliminated rapidly from the forest. In isolated conservation areas with no external sources of seeds, such losses of rare species will be permanent and result in the gradual species impoverishment of the forest. Also, in a non-equilibrium state, it may be difficult or even impossible to predict the minimal forest area necessary to preserve particular species.

The largest long-term plot, and the one most completely analysed to date, is the 50 ha plot established in 1980 at Barro Colorado Island in Panama by Steve Hubbell, Robin Foster, other collaborators, and the Smithsonian Institution. The forest is semi-evergreen and seasonal and is at least 500–600 years old. The plot is relatively flat and at an elevation at 150 m. Within the plot are two specialized habitats: 2 ha of swamp and 1 ha of ravines and creeks. In 1980, all trees and saplings with a diameter of 1 cm at breast height were tagged, measured, identified, and mapped. The plot was reassessed in 1985 for tree growth and survival as well as for sapling recruitment. In addition, the vegetation, canopy gaps and terrain were mapped in great detail. The overall conclusion of the 1980 census was that the forest was not in a state of equilibrium. While a few species were habitat specialists, many species appeared to have general habitat requirements. Further, the density of virtually all species appears to be far below the levels at which density-dependent factors would begin to influence populations.

Consequently, the species richness at Barro Colorado Island could best be explained by a non-equilibrium, island biogeography model in which there is continual species turnover. The 1985 re-assessment showed that saplings had higher growth and survival rates when further away from conspecific trees. Such data provide evidence for an equilibrium view of the forest composition controlled by density-dependent effects. However, it is relevant to ask how density-dependent effects can be distinguished from microsuccessional effects. If saplings establish in one type of forest and grow up as trees, the forest environment will be altered and may no longer be suitable for the establishment of seedlings and saplings. Such successional effects on a local scale will be indistinguishable from density-dependent effects. Hubbell and Foster's final conclusions are that there is a complex mixture of equilibrium and non-equilibrium forces affecting the forest at Barro Colorado Island that defies simple explanations and requires considerable long-term study and analysis.

The paper by David Clark discusses the effects of disturbance on patterns of regeneration in neotropical moist forests. The agents of disturbance and tree fall are varied but include wind, floods, fire, landslides, and lightning. The tree falls cause changes in the physical environment of the forest floor which affect seedling and sapling regeneration. The primary effects appear to be the quantity and quality of light, temperature, air and soil humidity, and soil nutrients and physical structure. Some trees uproot and fall over, some trees snap off at the trunk, while other trees die standing. Trees that die standing will shed their branches over a period of years, creating many small canopy gaps with little or no exposure of soil. In contrast, trees that fall over and uproot typically cause large canopy gaps and expose a lot of soil. The ecological significance of these modes of tree death remains relatively unexplored but they almost certainly have major consequences on the size and type of canopy gap formed.

Advances in the equipment used to measure light and other physical variables have allowed great advances to take place in understanding the impact of gaps on the physiological ecology of seedlings and saplings. At the centre of large gaps, light levels are 10–19% of full sunlight, while light is about 4% at the gap edge and only 1% in the forest understorey. Low light levels often appear to limit seedling growth, showing that adult trees pre-empt a key resource required by seedlings. In an intensive review over 20 years ago, Dawkins showed that competition models can only explain a small fraction of the variation among adult trees in growth rates. Combining modern competition models with the physiological methods described by Clark potentially could explain a greater proportion of this unexplained variation.

Some of the characteristic features and problems associated with long-term management of Malaysian dipterocarp forests using selective logging systems were examined by Manokaran, Wan Razali and Kochummen in a presentation to the Bangi workshop (paper published elsewhere than in this volume). One dominant factor affecting regeneration is the irregular fruiting cycle, with a good fruit crop produced only every 3 to 6 years. The net result is that the number of seedlings on the ground fluctuates dramatically over time. Consequently, it is not practical to restrict logging activities only to forests where there is an abundance of seedlings on the ground, since this situation may only occur infrequently. Silvicultural management systems in Malaysia have generally been altered to encourage advanced regeneration in the form of saplings and of small trees because these are more consistently present than seedlings. In the Selective Management System as practised in Sarawak, the assumption is that advanced regeneration (i.e. small trees of desirable timber species) will grow rapidly once the canopy is opened up around them. In fact, many trees of the main timber species fail to respond to silvicultural treatment, with a few trees growing rapidly and most trees growing slowly or not at all. Some of these rapidly growing trees show this enhanced growth for only the first 2 years following logging, and then they also begin to slow down.

A critical need is to determine why only certain trees respond to silvicultural treatment with this enhanced growth. Perhaps modern methods of plant physiological ecology and light detection, as described by David Clark, could identify the critical parameters of light, water, and nutrient availability which allow this dramatic growth response. However, the complexity of changes in the physical environment near a particular tree following disturbance, including decreased root and canopy competition with neighbours and physical damage to the roots, trunk and branches by logging, make this a difficult goal. A further promising avenue will be investigations to determine if there is a genetic basis to the superior growth rates of certain trees. This problem could potentially be investigated by searching for an isozyme marker associated with fast-growing trees. A more likely method would involve a quantitative genetic comparison of the growth rates of either seedlings or leaf cuttings from fast- and slow-growing trees. The technology of clonal propagation using leaves, described in recent articles in the Malaysian Forester, may be one key to conducting such experiments and eventually instituting extensive planting schemes in the absence of seeds and seedlings.

Work is needed to determine how logging affects the reproductive stock of the forest. During logging, large trees of desirable species and of good form are removed selectively, leaving behind the smaller trees and the poor-quality trees. If there is a genetic basis to growth rates and tree form, which seems very probable, then the seeds produced by the remaining trees may be of poor genetic stock. Silvicultural treatments must be undertaken to remove these poor-quality trees after selective logging operations.

The species composition of seedlings and adult trees in the forest changes dramatically following logging, and the goal of producing a dense stand of a particular, desirable timber species is often not achieved. In many ways such changes in species composition are not unexpected. The physical and biological environment is so altered by logging that the original common species present on the site may no longer be able to grow there. Also, if the forest is not in an equilibrium stage, a timber species might become extinct on the local scale due to random factors, such as weather patterns or the availability of seeds from other species.

In conclusion, these studies have provided a broad review of the factors affecting tree regeneration. The key points are that fruiting is often irregular, resulting in great fluctuations in seedling density on the ground. Further, seedlings and saplings of certain common tree species may be virtually absent in primary forest or after logging, indicating that the species is no longer regenerating on a local scale. Such considerations have implications both to the forester who is interested in encouraging further regeneration of the common, valuable timber species as well as to the park manager interested in the preservation of tree species diversity. The forester, in trying to encourage the regeneration of a common and/or desirable tree species through silvicultural treatment, may be bringing the density of the species above a certain critical level, resulting in an explosive growth of destructive pests. Major advances in measuring light, canopy structure, and the photosynthetic responses of seedlings are also described. Such research might serve to identify the best environment for seedling growth. In addition, genetic variation for photosynthetic rates and consequent growth rate also needs to be investigated. The importance of genetic variability in explaining the great phenotypic variation in growth rates is a critical area in which immediate research is required.

As a final point, ecologists either prefer, or are required, to work at pristine, isolated field stations where the impact of humans is minimized. Yet the majority of tropical forests have been greatly altered by humans, or soon will be. Forest managers need research results which are relevant to the disturbed forests that they are required to manage. Instead of speculating as to how human activities such as logging or hunting might affect the reproductive ecology of tropical plants, ecologists need to conduct explicit comparisons of primary forest and disturbed forest. For example, if field studies in a protected forest show that a particular vertebrate is the key fruit disperser of a tree species, what actually happens in nearby unprotected forest when that vertebrate has been hunted out? Does that tree species fail to regenerate? Or does another animal take over the role of principal disperser? As another example, how does the decrease in density of particular timber species following logging affect pollinator flight distances, fruit set, and the rate of outcrossing? While forest managers need to learn more about the ecology of the primary forests, ecologists must also be willing to spend their time investigating managed forests that have been affected by human activities.

CHAPTER 21

THE ROLE OF DISTURBANCE IN THE REGENERATION OF NEOTROPICAL MOIST FORESTS

David B. Clark

ABSTRACT

Information on the incidence, extent and effects of different agents and types of disturbance in neotropical moist forests is reviewed. Disturbance in terrestrial ecosystems is defined as a relatively discrete event causing changes in the physical structure of the environment. Disturbance affects forests by altering the structure of vegetation or exposing soil. An important descriptor of a forest's disturbance patterns is the mortality regime (modes of death) of the canopy trees. Different modes of death affect regeneration in quite different ways. Assessing the effects of disturbance other than outright mortality requires longitudinal studies of individual plants. Recent data on the physiological impacts of disturbance indicate that small changes in light levels can significantly affect regeneration. New technologies for measuring these small changes are discussed. Methodological problems in measuring and comparing disturbance regimes are evaluated, and some solutions are proposed. More research is needed to quantify different types of disturbances in tropical forests and to evaluate their ecophysiological effects. Disturbance regimes are important controllers of community structure and composition; they should be studied in order to understand how pristine communities function, and to facilitate management of altered systems.

INTRODUCTION

The structural matrix of tropical rain forest is formed by the trees which make up the canopy. Each element of this matrix has a finite lifespan. After some period, each individual dies and is replaced, on the average, by another individual. The study of the processes associated with these tree-by-tree

Table 21.1 Some principal neotropical research sites for the study of forest regeneration and disturbance

Site	Location	Annual rainfall
Los Tuxtlas Biology Station Instituto de Biología Universidad Nacional Autónoma de México	Mexico	4900 mm (Estrada *et al.*, 1985)
La Selva Biological Station, Organization for Tropical Studies	Costa Rica	4000 mm (OTS data)
San Carlos de Rio Negro International Amazon Project, co ordinated by Instituto Venezolano de Investigaciones Científicas	Venezuela	3500 mm (Uhl, 1982)
Barro Colorado Island Smithsonian Tropical Research Institute	Panama	2600 mm (Dietrich *et al.*, 1982)

replacement events has occupied a central place in the study of tropical forest ecology for the last 2 decades. The general topic of regeneration in wet tropical forests was recently covered in a review by Clark (1986). My purpose here is to examine the interaction of disturbance and forest regeneration in moist tropical forests of the neotropics. First, I review the different types of disturbance and discuss their ecological consequences for tree regeneration. I then summarize data from the neotropics on disturbance and regeneration and point out areas where more research is needed. In conclusion, I suggest some scientific and practical reasons why disturbance in forest ecosystems merits more attention.

The forests discussed in this paper occur in areas with ≥ 1800 mm of rainfall annually and a mean annual biotemperature of $\geq 19\,°C$, and are classified as tropical moist, wet, and rain forests in the Holdridge life zone system (Holdridge, 1979). The drier end of this spectrum encompasses forests which some authors call seasonal or monsoon forests. In the neotropics – Mexico, Central America, parts of South America, and the Caribbean – there are around 5 million km² of tropical wet and rain forests (*sensu* Holdridge), plus a substantial area of tropical moist forest (Myers, 1980). I will refer to all of these forests as "tropical moist forests" for convenience. As will become apparent in this review, research on regeneration in these forests has been concentrated at a few sites (Table 21.1). While there are several other active research sites in moist tropical forests (such as

Manú National Park in Peru, Corcovado National Park in Costa Rica, and the World Wildlife Minimum Critical Size of Ecosystems project in Brazil), the majority of publications to date on ecological aspects of regeneration and disturbance have come from the four sites listed in Table 21.1.

THE DEFINITION OF DISTURBANCE

The factors affecting tree regeneration can be divided conceptually into two categories. One group comprises biotic agents affecting recruitment. Examples of these processes include pollinators, seed dispersers and predators, herbivores and pathogens. Another set of factors affect regeneration by altering the structure of the physical environment. These factors I will group under the general heading of "disturbance".

Bazzaz (1983) defined disturbance as "a sudden change in the resource base of a unit of the landscape that is expressed as a readily detectable change in population response". This definition makes explicit the population biologist's ultimate interest in disturbance, that is, its effects on individuals and populations. As a practical definition, however, it is difficult to apply. Because disturbance is defined by population responses, one cannot talk about disturbance in general, but only in reference to particular populations' responses. White and Pickett (1985) proposed an alternative definition: "a disturbance is any relatively discrete event in time that disrupts ecosystem, community, or population structure and changes resources, substrate availability, or the physical environment." This definition incorporates two of Bazzaz's key points, a relative sudden change in resource availability and a resultant change in population structure. It also includes the idea of disturbance as a disruption of the "normal" state of an ecosystem, community, or population. In common usage, to disrupt means "to throw into confusion" (Davies, 1969). Given that many species depend on periodic disturbance for their existence, the idea of disturbance as a disruption of the normal does not seem appropriate.

Building on these definitions, I define disturbance in terrestrial forest ecosystems as "a relatively discrete event causing a change in the physical structure of the environment." Here, I use "physical structure" to mean the vegetation and surface soil of a site. In focusing on discrete events, I exclude longer term phenomena which change physical structure, such as the change of seasons in temperate forest inducing leaf flush. By focusing on physical structure, not populations of particular organisms, this definition lends itself to inter-community comparisons. The biological effects of a given disturbance will vary with the organisms of interest. In the following sections, I attempt to show that investigation of disturbance and its effects on plants is leading to a new understanding of neotropical forest regeneration.

Table 21.2   Disturbances in neotropical moist forests

Type of disturbance	Approximate area of effect (m^2)
Hurricane Flood Volcanism Fire	$10^7 - 10^5$
Regional wind storm Multiple tree fall	$10^5 - 10^3$
Lightning Single tree fall	$10^3 - 10^2$
Branch fall	$10^2 - 10^1$
Leaf fall (especially palms) Animal effects (nests, rootings, burrows)	$10^2 - 10^{-1}$

ECOLOGICAL AND PHYSIOLOGICAL CONSEQUENCES OF DISTURBANCE

The disturbances listed in Table 21.2 vary by roughly nine orders of magnitude in terms of the area each affects. Regardless of the size of the disturbance, however, each affects forest structure either by destroying vegetation or by exposing bare soil. These two types of environmental changes can be produced in several different ways, either separately or in combination. These different outcomes have profoundly different consequences for tree regeneration.

Tree death and disturbance in forest ecosystems

In forest ecosystems, many small-scale and all large-scale disturbances involve the death of trees. To discuss, compare, and analyse disturbance and its ecological effects, it is useful to consider the ways in which trees can die. These modes of death can be classified as follows:

(a)   Alive until instant of falling
    (a.1) Uprooted
    (a.2) Snapped off

(b)   Died standing, then fell
    (b.1) Dropped branches before falling
        b.1.1. Tipped up
        b.1.2. Snapped off
    (b.2) Branches intact at falling
        b.2.1. Tipped up
        b.2.2. Snapped off

While these are the principal modes of tree death, intermediate situations clearly occur. A tree may die standing, yet carry a considerable quantity of live liana canopy when it falls. A tree's condition may decline over a period of years, so that when it finally falls its crown may be much smaller than in its prime. In a given forest, every species of tree has a characteristic distribution of modes of death, which depends on factors such as longevity, degree of buttressing, average size, and wood density. The sum of these distributions, weighted by species abundance, is a forest-level description of the average way in which tree death occurs at a particular site. This pattern, which I call the mortality regime, is an important descriptor of the disturbance characteristics of any forest ecosystem. Tree regeneration biology is closely tied to the mortality regime. In forests where trees tend to die standing and decompose before falling, nurse logs may be unimportant. If there are no tip-ups, the species which specialize on this microhabitat (cf. Riera, 1985; Brandani *et al.*, 1988) will be rare or absent. Forests where trees tend to fall while still alive will experience proportionately greater understorey disturbance.

Ecological effects of modes of tree death

The canopy gap formed by the tip-up of a living tree has been described as having three zones: a crown zone, a bole zone, and a root zone (Orians, 1982). These zones have been hypothesized to provide different micro-environments which serve as the basis for microhabitat specialization by tropical trees (Orians, 1982). In this conceptual model, the crown zone represents the zone of maximum damage to existing vegetation, maximum organic matter input, and little soil disturbance. The bole zone is characterized by high light, little organic matter input or soil disturbance, and intermediate damage to existing vegetation. The root zone, where nutrient-poor sub-surface soil is lifted to the surface, has high soil and vegetation disturbance and no organic matter input.

This simple model does not apply to at least four of the six modes of canopy tree death. Root zones are absent in all gaps formed by trees which snap off, as well as in disturbances caused by branch and large leaf-falls. The crown zone and root zone may be coincident if a tree dies standing and sheds its branches before tipping up. The amount of light penetrating to the forest floor is also affected by the mode of death. The sudden death of an intact tree, whether by tip-up or snapping off, will in general open a larger hole than the gradual disintegration of a standing dead tree.

Ecological effects of soil disturbance

Soil disturbances vary tremendously in area and in the nature of their effect. Both floods and volcanic action are large-scale disturbances that, in general, increase the availability of soil nutrients. Earthquakes and resultant landslides are large-scale disturbances that expose large quantities of generally nutrient-poor subsoil (Garwood *et al.*, 1979). Mammal rooting and burrowing are small-scale disturbances that resemble root zones of gaps in that nutrient-poor soil is brought to the surface; they differ in that light levels remain low. Certain insect-caused disturbances such as termitaria (Salick *et al.*, 1983) and leaf-cutting ant nests (Haines, 1978) can cause nutrient enrichment in the disturbed area. The combined actions of tip-ups, tree falls, and animal disturbances to the soil all act to create heterogeneity in soil nutrient and physical properties on relatively short time scales. On a longer time scale, and at a larger spatial scale, soil disturbance events act to retard or inhibit formation of soil horizons. At this scale, soil disturbance also acts to minimize the effects of individual plant species on soil properties.

Physiological consequences of disturbance

The tree mortality regime and the level of variation in soil properties are system-level characteristics of a forest. The ultimate effects of disturbance on regeneration, however, are expressed at the level of individual plants. Much of the theoretical literature concerning the effects of gaps has concentrated on this aspect. Field data, however, have been biased towards system-level properties such as gap formation rates.

The agents of disturbance and their potential effects change throughout the life of an individual. Seedlings can be affected by essentially all types of disturbances (even lightning, pers. obs.), whereas pole-sized saplings and adults are relatively unaffected by most animal agents of soil disturbance. Disturbances which increase understorey light from 1% of full sunlight to 2% may be critical for the survival of seedlings, but may be irrelevant to larger juveniles if these are not light-limited.

Assessing immediate mortality due to disturbance is relatively straightforward, at least in longitudinal studies. Measuring effects other than mortality is much more difficult, as it involves measuring both the disturbance and the individual's response. For example, many vegetation disturbances increase understorey light levels, which in turn increase growth rates of established seedlings. Documentation of this effect requires pre- and post-disturbance measurement of both light levels and growth rates. Similarly, physical injury is a frequent consequence of disturbance. While it is easy to measure levels of damage at a given time, the long-term effects of this damage are generally unknown (and certainly vary interspecifically). A full understanding of the effects of disturbance on regeneration will depend on concurrent investigation of the demography, physiology, and microsite of individuals affected by disturbance (cf. Oberbauer *et al.*, 1988). This is a fertile field for collaboration between forest ecologists and plant ecophysiologists.

DISTURBANCE IN NEOTROPICAL MOIST FORESTS

Hurricanes and smaller storms

Hurricanes are important agents of disturbance in tropical moist forests on islands in the Caribbean, but have little effect on the bulk of neotropical forests, which occur in Central and South America. On the island of Puerto Rico, hurricanes occur with sufficient frequency to mould a characteristic flat-surfaced canopy with no emergents (Brown *et al.*, 1983). In a Puerto Rican lower montane rain forest, four hurricanes occurred in 48 years at the site of a long-term study of forest regeneration (Crow, 1980). In this study, significant changes in species abundances were still occurring at the end of 33 years of observation; notably, a tolerant palm increased dramatically in density. In contrast to forests which undergo primarily small-scale disturbance, these hurricane forests probably never approach a species equilibrium at the stand level. The scale of disturbance, both in time and space, is sufficient to maintain large areas in successional states.

While hurricanes do not affect substantial areas of neotropical continental forest, strong windstorms can be important local disturbances. At Barro Colorado Island (BCI), one storm blew down several hectares (Foster and Brokaw, 1982). At La Selva, strong winds cause frequent tree falls, occasionally opening large multiple-tree gaps. The Los Tuxtlas Station in Mexico also receives regular strong (c. 80 km hr^{-1}) wind storms (Estrada *et al.*, 1985). Strong storms also occur in South American Amazonian rain forests. Uhl (1982) reports one storm at San Carlos de Rio Negro in Venezuela that flattened 49 trees ≥ 20 cm diameter at breast height (dbh) in a 0.41 ha area. Riera (1983) mentions a several-hectare area in French Guyana which was levelled by a strong local storm.

Floods

Large areas of Amazonia, perhaps more than 100 000 km^2, flood annually (Goulding, 1988). The effect of this flooding varies depending on whether nutrient-rich sediments are deposited (*varzea* forests), or if the flooding is caused by nutrient-poor black or clear-water rivers (*igapo* forest) (Pires and Prance, 1985). In addition to these annual floods, rivers also cause extensive disturbance by meander erosion. Salo *et al.* (1986) showed that 12% of the Amazonian lowlands of Peru are in successional stages along rivers. They further suggest that at a regional level this process serves to enhance between-habitat diversity. This study is unique in providing a regional estimate of the importance of flooding on forest regeneration.

Landslides

Landslides can affect large areas of neotropical rain forest. For example, two earthquakes in Panama caused slides which denuded approximately 12% of the 450 km^2 affected area (Garwood *et al.*, 1979). Approximately 14% of neotropical moist forests occur in zones of high seismic activity (Garwood *et al.*, 1979). In addition, substantial areas occur on steep slopes. The data of Garwood *et al.* are apparently the only regional-level data on the importance of this type of disturbance.

Fire

The importance of anthropogenic fire as part of the cycle of slash-and-burn agriculture has long been appreciated. Apart from these impacts, however, fire has not generally been considered to be a major disturbance of neotropical moist forests. New evidence is forcing a re-evaluation of this view. Sanford *et al.* (1985) found evidence of burning over large areas of forest in the Rio Negro basin. These fires appeared too extensive to be related solely to agriculture. In a considerably wetter, apparently primary forest at La Selva, R. Sanford (pers. comm.) has also found widespread evidence of charcoal in the upper 50 cm of soil. Preliminary carbon-dating of this charcoal gave a date of 2400 years BP. The origin of the charcoal in these forests is not clear. Neither the Amazonian nor Costa Rican rain forests currently experience large-scale fire. This assessment may be due in part to our short-term human perspective. Is a rain forest dry enough to burn during the 100-year or 1000-year drought? More than 4 000 000 ha of supposedly non-inflammable tropical forest burned in Borneo in 1983 during an El Niño drought (Beaman *et al.*, 1985). Careful excavations and radiodatings will be necessary to determine the extent and frequency of large-scale fires in moist forests of the neotropics.

Tree-fall gaps

The general subject of tree-fall gaps and tropical forest regeneration has been reviewed extensively in recent years (Orians, 1982; Pickett, 1983; Brokaw, 1985; Denslow, 1987). Here, I will summarize the data on neotropical gap formation rates, point out deficiencies in the current data, and suggest lines of future research.

The gap formation rates listed in Table 21.3 at first glance suggest a remarkable similarity between forests in Panama, Costa Rica, Mexico, and Venezuela. In fact, however, the rates were calculated using a variety of methods and assumptions; the comparability of the different methods has not yet been established. One method of obtaining gap occurrence rates is to assess the per cent area in gaps at one time and divide this per cent by the number of years required to pass from the gap to non-gap category. This approach depends critically on knowing gap/non-gap transition rates. These rates clearly vary between large and small gaps; large gaps probably close by vertical growth and small gaps by lateral ingrowth (Denslow, 1987). The amount of gap area lost by lateral ingrowth has never to my knowledge been measured in the neotropics, but the data in Hubbell and Foster (1986; their Table 1, row 1) suggest that the figure may be as much as 30% or more. None of the estimates in Table 21.3 based on one-time gap surveys (Uhl and Murphy, 1981; Sanford *et al.*, 1986) present data on gap closure rates, so the validity of the proposed gap turnover rates cannot be assessed.

Another problem with comparing turnover rates is that different authors have used different minimum sizes for the smallest gap category. Because the smaller gap categories are always the most frequent (Brokaw, 1982a; Martínez-Ramos and Alvarez-Buylla, 1986; Sanford *et al.*, 1986; Hubbell and Foster, 1986), this can cause major differences between estimates. Brokaw (1982a) lucidly explained the importance of using a standard gap definition, and proposed the following: "a gap is a 'hole' in the forest extending through all levels down to an average height of 2 m above ground". Popma *et al.* (1988) found that gap area measured by Brokaw's definition underestimated by 44–515% the area colonized by pioneer species. They point out that the influence of gaps of any given size will vary as a function of forest height. They also recognize, however, the necessity of a structural definition for inter-forest comparison. Brokaw's definition is both objective and practical, and at the moment seems to be the preferable method for inter-forest structural comparisons. The very interesting question of the relation between gap physical structure and gap biological effect remains a fruitful area for research.

Table 21.3 Gap formation rates in mature neotropical rain forests. See text for explanation of limitations on comparability of estimates

Location	Gap formation rate (% area/yr)	Gap definition used	Remarks	Citation
Costa Rica (La Selva)	1.05	"Vertical opening greater than 40 m² in mature forest canopy, with vegetation less than appoximately 5 m tall"	Underestimates true rate due to (1) effects of lateral closure of gaps, and (2) omission of gaps whose vegetation reaches >5 m in <6 yrs	Sanford *et al.*, 1986
Costa Rica (La Selva)	0.7–1.2	None given		Hartshorn 1978
Panama (Barro Colorado Island)	0.8	"Vertical 'hole' in the forest extending through the canopy to within 2 m of the forest floor." Minimum size 20 m²		Brokaw 1982*b*
Panama (Barro Colorado Island)	1.0	Vegetation < 2m tall. Resolution c. 25 m²	Rate calculated from transition probabilities (Table 1) assuming 1 yr between censuses	Hubbell and Foster 1986 (Table 1)
Venezuela (San Carlos)	1.0	"Any opening in the forest canopy with seedlings and saplings not exceeding 2.7 m tall." No minimum size given	Assumes that "gap phase" (vegetation < 2.7 m) persists for "about 5 years". Many gaps fill faster (cf. Uhl. 1982. Fig. 1)	Uhl and Murphy 1981

(continued)

Table 21.3 continued

Location	Gap formation rate	Gap definition used	Citation
Mexico (Los Tuxtlas)	1.5	No definition given	Martinez-Ramos and Alvarez-Buylla 1986
Mexico (Los Tuxtlas)	2.1 (mean) 2.9 (med.)	Brokaw (1982a) combined with physical damage to understorey palms. Minimum size 25 m^2	Martinez-Ramos *et al.* 1988
Mexico (Los Tuxtlas)	0.7	Brokaw (1982a)	Bongers *et al.* 1988
French Guyana (St. Elie)	1.1	Brokaw (1982a) (?)	Riera 1983

I propose the following standards for gap turnover calculations:

(1)   The definition of gap used must be given in sufficient detail to be replicable by other investigators – future studies should follow Brokaw's (1982*a*) definition;

(2)   The minimum size of gap detected should be stated;

(3)   If gap closure rates are used, the data on which they are based should be published;

(4)   The size/frequency distribution of gaps should be presented so that data can be recalculated for comparability with other studies; and,

(5)   Statistics of replicability should be given.

When the studies listed in Table 21.3 are examined in this light, none meet all of these criteria. To date, there are no published data on the precision of any of these estimates; if an area were to be surveyed twice in succession with the same methods, how close would the estimates of gap area be? It is encouraging that the estimates from the two studies at Barro Colorado Island, which used completely different techniques to measure gaps and which did not overlap temporally, are extremely similar. The current gap formation rate on Barro Colorado Island is probably very close to 1% yr^{-1}. At Los Tuxtlas, however, the estimates of Bongers *et al.* (1988) and Martínez-Ramos *et al.* (1988) differ by a factor of three. Bongers *et al.* suggest that the difference may be due to differences in the slope and exposure of the two study areas. It is equally possible that the difference is a product of two very different methods of measuring gap area.

Branch-fall and leaf-fall gaps

Gaps caused by falling branches and large leaves should be comparable to the crown zone of tree-fall gaps. Light levels increase, there is input of organic matter, existing vegetation may be damaged, but there is no exposure of mineral soil. I have been unable to find any published studies which document the importance of total branch-fall and leaf-fall gaps relative to tree-fall gaps. Martínez-Ramos *et al.* (1988) found that 22% of the fallen stems >10 cm diameter at Los Tuxtlas were branches. Aide (1987) showed that limb falls accounted for an average of 34% of the mortality of juveniles of a tolerant liana in Panama. Palm leaves are likely to be important agents of disturbance in neotropical forests, where up to 26% of the stems ≥ 10 cm diameter may be palms (cf. Lieberman *et al.*, 1985*a*). At La Selva, fronds of subcanopy palms regularly exceed 6 m in length (Rich, 1986), and falling fronds are an important source of seedling mortality (Vandermeer, 1977).

Soil disturbances

The data of Garwood *et al.* (1979) and Salo *et al.* (1986) on landslides and river erosion respectively are the best available on neotropical large-scale disturbances causing exposure of bare soil. There are numerous agents of small-scale soil disturbance. Unfortunately, there are only isolated data to assess their relative importance, and comparative studies are yet to be done.

Tip-up mounds from uprooted trees may be important agents of soil disturbance in neotropical forests. The only quantitative data I am aware of are from Barro Colorado Island, where Putz (1983) reported that 0.09% of the forest floor is occupied by pit and mound topography. These pits filled at the rate of 8 cm yr[-1].

A wide variety of neotropical animals cause soil disturbance by burrowing, nesting, or foraging. These animals include leaf-cutting ants (*Atta*), termites, agoutis (*Dasyprocta agouti*), pacas (*Agouti paca*), armadillos (*Dasypus novemcinctus*), and wild pigs (*Tayassu*). Both *Atta* nests and termitaria can form local nutrient hot spots in neotropical rain forest understorey (Haines, 1975, 1978, 1983; Salick *et al.*, 1983). On the refuse dumps of *Atta* nests, these nutrient differences are sufficient to affect plant growth (Haines, 1975). Baseline data on termite and *Atta* nests in mature neotropical moist forest are scarce. Haines reports 0.02 *Atta* colonies ha[-1] in mature forest on Barro Colorado Island (Haines, 1978), and <0.04 colonies ha[-1] at Rio Negro (Haines, 1983). There appear to be no data on nest turnover times or the effect of nests on mature forest regeneration. Salick *et al.* (1983) encountered 1500 termite mounds ha[-1] on San Carlos lateritic soils; the nest turnover rate was 165 ha[-1] yr[-1], involving 0.8 tons of soil ha[-1] yr[-1]. The authors noted vigorous seedling growth in abandoned termitaria. They hypothesized that abandoned termitaria increase microhabitat heterogeneity and may be important establishment sites for tree seedlings.

In temperate ecosystems, the influence of mammals as agents of soil disturbance has been the object of intensive study (cf. Platt, 1975; Bratton, 1975; Kalisz and Stone, 1984; Gutterman, 1987). For neotropical forests, in contrast, I can find no references to the effects of mammal nests and burrowings. There are apparently no data on the extent or turnover of these disturbances, nor information on their effects on plant regeneration. The only information I am aware of is a study at La Selva on the rates of damage to artificial seedlings in primary forest (Clark and Clark, 1989). During 1 year of observation, 21% of these model seedlings were knocked over or destroyed by vertebrates. The "mortality" was caused by diggings, trampling, animal trails, and occasionally "herbivory", and much of the mortality appeared to be associated with soil disturbance.

I have not discussed earthworms as agents of soil disturbance, but have concentrated on visually conspicuous disturbances. This reflects only the bias of an ecologist who tends not to notice events at scales smaller than a tree seedling. It may well be that future studies will show that the most important organisms depositing bare soil on the soil surface, at least in terms of mass of soil, are earthworms. Earthworms are quite common in the soils of at least some neotropical moist forests (R. Sanford, pers. comm.), and certainly merit further study.

Forest mortality regimes

There are large differences between neotropical forests in the tree mortality regime. The most complete data come from Barro Colorado Island. Putz and Milton (1982) studied mortality of trees > 60 cm girth in five plots in old second growth and two plots in mature phase forest. Of 115 deaths in all plots, 52% snapped off, 17% uprooted, 14% died standing, and 9% died from unknown causes. Brokaw (1982*b*) studied gaps on Barro Colorado Island in a 28 ha plot that was approximately half old second growth and half primary forest. Of the 52 deaths of gap-making trees, 69% were snapped off and 31% were tipped up. At Los Tuxtlas, 69% of trees died by snapping off (Martínez-Ramos *et al.*, 1988). Riera (1983) found that in French Guyana 53% of gaps were due to snap-offs, while 33% were caused by tip-ups and 13% by mixed causes or trees losing tops. At La Selva, Hartshorn (1980) estimated that "90% of the gaps are caused by uprooted trees" but presented no data. Lieberman *et al.* (1985*b*) found that, for 1302 trees and lianas > 10 cm dbh which died over a 13-year period, "26% died standing, 31% had fallen, 7% were found buried under tree-falls, and 37% had decomposed entirely, leaving no trace."

Forest disturbance regimes

To understand the differences between tropical rain forests, measures of structure and process which are comparable between sites need to be developed. Describing the rate of large-scale disturbance can be difficult because important events may occur very infrequently on a human time-scale. On the other hand, these large-scale events can now be assessed by remote sensing (cf. Garwood *et al.*, 1979; Beaman *et al.*, 1985; Salo *et al.*, 1986), so, at a regional level, estimates will improve steadily.

There are substantial areas of tropical forest which are not subject to regional disturbances for periods equal to at least several generations of the canopy trees. In these systems, local disturbances will be the primary factors affecting regeneration in ecological time. In the neotropics, there are still no sites which are adequately characterized with respect to rates of local disturbance. Based on the data and arguments outlined above, I propose that the following data should be gathered at least at the few tropical moist forest sites where ecological research is concentrated.

Gap formation rates

The most basic measures of forest disturbance are the rate at which gaps form, and the size/frequency distribution of these gaps. Because these rates may vary seasonally and within habitats (cf. Brokaw, 1982*b*; Riera, 1983; Martínez-Ramos and Alvarez-Buylla, 1986; Clark and Clark, 1989), temporal and spatial variation should be assessed. A key system-level characteristic which has never, to my knowledge, been measured is the frequency distribution of agents causing gaps (multiple tree, single tree, branch-fall, etc.).

Soil disturbance rates

How much mineral soil is turned up at the surface every year? What is the size and half-life of each disturbance? What are the agents of disturbance? A complete picture of soil disturbance is not available for any moist tropical forest site.

Canopy tree mortality regime

How do trees on the average die? A quantitative description of the tree mortality regime is important to interpret the relation between rates of gap formation and soil disturbance.

ECOPHYSIOLOGICAL EFFECTS OF DISTURBANCE IN NEOTROPICAL MOIST FORESTS

Physical damage and forest regeneration

There are few data on forest-level rates of physical damage to trees and the effects of this damage on regeneration. Uhl (1982) found that 38% of the

305

mortality to tree saplings 1–10 cm diameter at San Carlos was due to falling trees or branches. Uhl also found that, at a given time, 4% of the small trees were bent to the forest floor; some of these plants subsequently produced new roots and sprouts along the horizontal stem. Gartner (1989) looked at mechanical damage to 16 species of shrubs of the genus *Piper* at La Selva. She found that an average of 51% of the individuals showed evidence of past physical damage. D.A. Clark and I have obtained similar preliminary data for trees. We annually score individuals in our long-term demographic study (D.A. Clark, 1986; Clark and Clark, 1987a) for gross stem damage, defined as an abrupt change in stem diameter ≥25% of the uninjured stem. This criterion is both arbitrary and highly conservative, as it does not measure smaller traumas or old ones which have healed. Nevertheless, by this definition, 23% of the stems of six tree species in mature forest were damaged (N=597 individuals >50 cm tall and ≤20 cm diameter). Putz and Brokaw (1989), using the same criterion, found an almost identical rate of damage (27%) for trees >10 cm diameter on BCI. They also found that 16% of 165 trees which suffered trunk breakage between 1976 and 1980 resprouted and survived at least 7 years. Clearly, mechanical damage and its effects are important factors in the regeneration of neotropical trees.

Soil nutrients and disturbance

A primary effect of soil disturbance is to increase microhabitat heterogeneity. Various authors have predicted that systematic differences in soil properties occur within tree-fall gaps (Hartshorn, 1978; Whitmore, 1978; Orians, 1982). Vitousek and Denslow (1986) studied nutrient levels in different zones of gaps at La Selva. As predicted, nitrogen mineralization rates and phosphorus concentrations were lower in the root zone than in the crown and bole zones. However, neither nutrient increased significantly in the crown zone. The authors speculate that increases in light availability are likely to be more important than increases in nutrients as factors affecting regeneration at La Selva. Uhl *et al.* (1988) found no increases in soil fertility or nutrient concentration in leachate water in six gaps they studied at San Carlos.

The impact of variance in nutrient availability on forest regeneration has not yet been evaluated. Nevertheless, several studies provide suggestive evidence that such effects may be important. At La Selva, Brandani *et al.* (1988) examined species distributions of young trees in different zones of 51 gaps. Root, bole, and crown zones of different gaps were more similar to each other in species composition than they were to the other zones in the same gaps. Root zones were dominated by fewer species than crown or bole zones. Barton (1984) also found within-gap zonal preferences for several tree species at La Selva. At Los Tuxtlas, Núñez-Farfán and Dirzo (1988) found differences in species composition and diversity between the root zone of one gap and and the crown zone of another gap. In French Guyana, Riera (1985) found that most *Cecropia obtusa* established on root tip-ups, while other common gap species were usually found in the non-tip-up area of gaps. All of these studies were based on observing the distribution of naturally established individuals. Because not only soil nutrients but also temperature, soil moisture, and light vary among gap zones, the exact factors resulting in non-random species distributions within gaps are not known.

Light levels

The extensive literature on gaps and forest regeneration (reviewed in Denslow, 1987) has focused on light as a key controlling factor. There are excellent reasons for this emphasis. Several studies have documented very low light levels in the neotropical rain forest understorey. At La Selva, Chazdon and Fetcher (1984) showed that understorey light levels averaged 1–2% of incident sunlight; light in a moderate-sized gap averaged 20–35% of full sun. Chazdon (1986) measured light at La Selva in gap centres, edges, and adjacent understorey of three gaps measuring $150 \, m^2$, $200 \, m^2$, and $400 \, m^2$. Median light levels were 2.9–5.8 moles m^{-2} day^{-1} at the centre of the gaps, 1.1–1.5 moles m^{-2} day^{-1} at gap edges, and 0.26–0.33 moles m^{-2} day^{-1} in adjacent understorey. Full sun irradiance averages *c.* 30 moles m^{-2} day^{-1} at La Selva (Oberbauer *et al.*, 1988), so Chazdon's data could be expressed as *c.* 1% sunlight in shaded understorey, 4% at gap edges, and 10–19% in gap centres.

Oberbauer *et al.* (1988, 1989) measured light levels for 7 days in the crowns of ten 1–3 m tall saplings in each of four species of trees in primary forest at La Selva. The species were selected to represent a range of understorey tolerance; individuals were selected to sample the species' range of observed understorey growth. All species averaged between 0.31 and 0.44 moles m^{-2} day^{-1}, or < 2% of full sunlight. Daily total photosynthetically active radiation (PAR) for the 40 saplings ranged from 0.3–13.8% of full sun.

Physiological effects of light

All tropical tree species examined to date have higher photosynthetic rates at light levels above the 1–2% gloom of the forest understorey. For example, Fetcher *et al.* (1987) examined photosynthetic response of eight La Selva tree species selected to represent a range of regeneration patterns. All showed a steep linear photosynthetic response at understorey light levels. Even very shade-adapted species which show depression of photosynthesis at 26% of full sun respond positively to small increases in light (Chazdon,

1986). These laboratory studies suggest that very small differences in light level can significantly affect performance of seedlings and saplings in the field. Several recent studies confirm this prediction.

In *Zamia skinneri*, a very tolerant understorey cycad at La Selva, leaf production and reproduction were positively correlated with light (Clark and Clark, 1987*b*, 1988). Females occurred in more open sites than males, which in turn were better-lit than non-reproductives. While the differences were highly significant statistically, and consistent in two different reproductive episodes, more interesting were the small absolute differences between the groups. Canopy openness (measured with a spherical densiometer; Lemmon, 1956) above females, males, and non-reproductives at the two reproductive episodes averaged 1.8%, 1.4%, and 1.1%; the total range of canopy openness over all 172 plants was only 0% to 4.4%.

Oberbauer *et al.* (1988, 1989) found significant correlations between annual height growth and the amount of irradiance received by primary forest saplings of four species of rain forest emergent trees at La Selva. These correlations are notable because they are based on only 10 individuals per species, on short-term (one week) light measurements only, and because the range of weekly means was small (0.9–6.1% of full sunlight for *Dipteryx panamensis*, 0.6–3.9% for *Lecythis ampla*).

Chazdon (1986) studied the response of different tolerant palms to understorey and gap light levels at La Selva. Light levels comparable to the centre of a large gap (26% of full sun) significantly lowered rates of light-saturated photosynthesis and quantum yield. In contrast, potted seedlings planted in gap edge sites grew taller and produced more and larger leaves than seedlings grown in understorey closed-canopy sites. In this case, the optimum light level for these species lies somewhere between 1% and 26% of full sun (approximately 0.3–8 moles m^{-2} day^{-1})

Barton (1984) found indirect evidence of partitioning of gap environments by tree species at La Selva, possibly related to differences in light levels. All six pioneer species occurred at higher densities in large gaps than in small ones; only one of five tolerant species showed a significant distributional difference between large and small gaps (biased towards small gaps). Some evidence of intra-gap specialization was also found. Two of six pioneer species were significantly more common in gap centres than in gap edges, whereas two of five primary species were significantly more common in gap edges. The factors responsible for the preferences for different zones were not measured, but response to different light regimes is an obvious possibility.

Much of the theoretical literature on gap effects implicitly has considered only the effects of vertical direct light. The studies cited above suggest that, in fact, the total quantity of light received may be the variable of interest, and that this light may be received as vertical direct light, lateral direct light, or indirect light. In a study of saplings of six La Selva tree species, Clark and

Clark (1987a) showed that the mean condition for all species was to receive direct light only from the sides. Nevertheless, annual diameter growth was positively correlated with an index of crown lighting for 8 of 17 size class/species combinations. The number of significant correlations was noteworthy because sample sizes were small in many cases, the crown lighting index could assume only a few values, and the annual diameter growths were mainly ≤ 2 mm.

In addition to the total quantity of light intercepted, the temporal pattern of light availability is important. This pattern is related to forest structure, which is related in turn to disturbance. As the degree of local disturbance increases, canopy unevenness will increase, as will the effect of shafts of light penetrating laterally into the forest. Plants in the tropical moist forest understorey that receive 1–2% of full sunlight usually receive that energy in short, high-energy episodes. The 40 saplings of four tree species studied by Oberbauer *et al.* (1988, 1989) spent more than 90% of the time at irradiances of < 25 μmol m^{-2} sec^{-1}; all species received an average of only 11–33 minutes day^{-1} of sunflecks. Similarly, at Los Tuxtlas four species of understorey *Piper* received 4–22 minutes of sunflecks per day (Chazdon *et al.*, 1988). At La Selva, sunflecks may contribute 10–78% of the daily PAR on clear days (Chazdon, 1986). Sunflecks cause photosynthetic induction of understorey leaves such that response to subsequent sunflecks is more efficient (Chazdon and Pearcy, 1986; Chazdon, 1988).

Recent advances in measurement of light environments

These studies emphasize the importance of measuring light environments at the level of individuals. Several new techniques have recently been developed which have greatly increased our capacity to rapidly and precisely measure light environments of plants in the field.

Hemispherical photography using fisheye lenses has long been used to measure light conditions (Anderson, 1964). The techniques are simple and relatively inexpensive, and the resulting estimates of light environment have been shown to be correlated with physiological and ecological performance of individual plants (Pearcy, 1983; Ustin *et al.*, 1984; Chazdon and Field, 1987a). The biggest drawback to widespread application of the technique has been the high labour cost of analysing the photographs by hand. Recently, three parallel projects in the neotropics have developed image processing systems based on microcomputers to perform this analysis electronically. One of these systems (Chazdon and Field, 1987b) is based on an Apple Macintosh. The other two systems (Rich, 1988; Becker *et al.*, 1989) were developed for use on IBM and compatible microcomputers. All three systems use an electronic digitizer to convert a photographic negative to a digital image. A computer programme then analyses the photograph and calculates the amount of direct and indirect light that could reach a plant crown, based on the structure of the surrounding canopy and the calculated track of the sun.

These systems show tremendous potential as a means to evaluate quickly and inexpensively the light environments of individual plants in a repeatable, documentable manner. To date, the technique has not proven sensitive enough to detect small differences in tropical understorey light environments (Chazdon and Field, 1987*b*). Research is underway to determine the limits of resolution of these systems (cf. Clark *et al.*, 1988), and to examine the correlation between canopy photographic indices of light and demographic performance of individual plants. Another active line of research is based on the availability of low-cost PAR sensors. Kitajima and Augspurger (1989) used packets of photosensitive ozalid paper to estimate total daily PAR. The number of sheets of paper exposed was highly correlated with $\log^{10}$ (PAR) measured with a PAR sensor ($r^2 = 0.98$, n = 21 packets).

Another approach has taken advantage of the introduction of low-cost electronic PAR sensors. Gutschick *et al.* (1985) demonstrated that small gallium-arsenide phosphide photodiodes have light-response characteristics very similar to conventional PAR sensors. These photodiodes cost <$10 each (compared to several hundred dollars for a conventional quantum sensor), and are only a few millimetres wide. A second advance has been the evolution of field computers to record and manage the data from these sensors. The Campbell 21X datalogger, for example, weighs only a few kilos and can handle dozens of input channels. Output from photodiodes can also be integrated with inexpensive coulometers (Benzing *et al.*, 1988). Because of their low cost, no moving parts, and compact size, coulometers appear to be very promising as medium- to long-term light integrators for ecophysiological studies.

With inexpensive sensors and increased data handling capacity, it is now feasible to mount arrays of sensors instead of single point sensors. These techniques are currently being used to examine intracrown, interindividual, and interspecific differences in neotropical forest light environments (Chazdon and Field, 1987*a*; Oberbauer *et al.*, 1988, 1989).

A new perspective on tropical moist forest light environments and their effects on regeneration

The studies listed above lead to a new appreciation of understorey light environments in neotropical moist forests. Direct vertical illumination in gaps is undoubtedly important for those plants exposed to it. However, the influence of a given gap will be much greater than its vertical projection (Popma *et al.*, 1988). A given gap will cause increased illumination at gap edge and adjacent understorey sites. Shafts of light may penetrate tens of metres into the forest, greatly increasing the radius of effect of the gap. The temporal pattern of these sunflecks will be crucial in determining their physiological effect. Although the total increased irradiance caused by the gap may not be large numerically, increases of a few per cent of full sunlight will have major physiological impacts on plants.

In the next decade, there will be an exponential increase in our understanding of the relationships between disturbance, light environments, and plant performance. Part of this advance will stem from technological progress in measuring light environments and plant performance in the field. The other key factor will be integration of these studies into existing demographic investigations. Such integrated research will have a profound impact on our understanding of tropical forest regeneration. A typological approach to species' regeneration patterns (gap species versus shade tolerant, for example), will be replaced by objective measures of demographic and physiological performance. With these approaches, it will be possible to evaluate quantitatively the existence of regeneration guilds and their relation to disturbance. The answers obtained will be critical to understanding the origin and maintenance of tropical tree species diversity.

CONCLUSIONS

It is clear from this review that there are insufficient data available to assess total disturbance regimes in neotropical moist forests. Furthermore, research analysing the effects of disturbance at the leaf and crown level (cf. Chazdon and Field, 1987*a*; Oberbauer *et al.*, 1988) is proceeding faster than system-level studies. Why should this be? One reason, I think, is the incorrect perception that approximate disturbance rates are now known. At only one neotropical site, Barro Colorado Island, are there good data on gap formation rates as well as information on the canopy tree mortality regime. At no site is there an adequate description of soil disturbances. Soil disturbances may have been neglected because they are difficult to quantify. Both observational demographic studies and experimentation will be required to document the effects of soil disturbance on plant performance.

Why study disturbance? I believe that understanding disturbance processes is critical for understanding the suite of regeneration pathways followed by trees in moist tropical forests. In addition, local and regional biodiversity is intimately related to disturbance (Pickett and Thompson, 1978). It is already clear that disturbance regimes are changing in remnant and successional tropical forests (cf. Foster and Brokaw, 1982). It is important to document disturbance regimes in at least a few sites in order to understand the baseline condition of the world's most complex terrestrial communities.

ACKNOWLEDGEMENTS

Many of the ideas presented here were developed jointly with Deborah A. Clark; I thank her for critical input to the organization and writing of this chapter. I also thank an anonymous reviewer, Paul M. Rich, Mauricio Quesada, the researchers at La Selva for their suggestions. These ideas were developed in the course of work supported by a grant from the United States National Science Foundation to D.A. Clark, D.B. Clark, and S.F. Oberbauer (BSR 85-16371).

REFERENCES

Aide, T.M. (1987). Limbfalls: a major cause of sapling mortality for tropical forest plants. *Biotropica*, **19**, 284–5

Anderson, M.C. (1964). Studies of the woodland light climate. I. The photographic computation of light conditions. *Journal of Ecology*, **52**, 27–41

Barton, A.M. (1984). Neotropical pioneer and shade-tolerant tree species: do they partition tree-fall gaps? *Tropical Ecology*, **25**, 196–202

Bazzaz, F.A. (1983). Characteristics of populations in relation to disturbance in natural and man-modified ecosystems. In Mooney, H.A. and Godron, M. (eds.) *Disturbance and Ecosystems: Components of Response*, pp. 259–75. (Berlin: Springer-Verlag)

Beaman, R.S., Beaman, J.H., Marsh, C.W. and Woods, P.V. (1985). Drought and forest fires in Sabah in 1983. *Sabah Society Journal*, **8**, 10–30

Becker, P.F., Erhardt, D. and Smith, A.P. (1989). Analysis of forest light environments: I. Computerized estimation of solar radiation from hemispherical canopy photographs. *Agriculture and Forest Meteorology*, **44**, 217–32

Benzing, D.H., Renfrow, W.B., Jones, P. and Haldeman, J. (1988). Inexpensive coulometers for integration of sensor output: measurement of solar radiation. *Biotropica*, **20**, 84–8

Bongers, F., Popma, J., Meave del Castillo, J. and Carabias, J. (1988). Structure and floristic composition of the lowland rain forest of Los Tuxtlas, Mexico. *Vegetatio*, **74**, 55–80

Brandani, A., Hartshorn, G.S. and Orians, G.H. (1988). Internal heterogeneity of gaps and richness in Costa Rican tropical wet forest. *Journal of Tropical Ecology*, **4**, 99–119

Bratton, S.P. (1975). The effect of the European wild boar, *Sus scrofa*, on grey beech forest in the Great Smokey Mountains. *Ecology*, **56**, 1356–66

Brokaw, N.V.L. (1982*a*). The definition of treefall gap and its effect on measures of forest dynamics. *Biotropica*, **14**, 158–60

Brokaw, N.V.L. (1982*b*). Treefalls: frequency, timing and consequences. In Leigh, E.G., Jr., Rand, A.S. and Windsor, D.M. (eds), *The Ecology of a Tropical Forest: Seasonal Rhythms and Long-term Changes*, pp. 101–8. (Washington, D.C.: Smithsonian Institution Press)

Brokaw, N.V.L. (1985). Treefalls, regrowth, and community structure in tropical forests. In Pickett, S.T.A. and White, P.S. (eds), *The Ecology of Natural Disturbance and Patch Dynamics*, pp. 53–69. (New York: Academic Press)

Brown, S., Lugo, A.E., Silander, S. and Liegel, L. (1983). *Research History and Opportunities in the Luquillo Experimental Forest*. United States Department of Agriculture, Southern Forest Experimental Station, General Technical Report SO-44. USDA Forest Service, New Orleans.

Chazdon, R.L. (1986). Light variation and carbon gain in rain forest understorey palms. *Journal of Ecology*, **74**, 995–1012

Chazdon, R.L. (1988). Sunflecks and their importance to forest understorey plants. *Advances in Ecological Research*, **18**, 1–63

Chazdon, R.L. and Fetcher, N. (1984). Photosynthetic light environments in a lowland tropical rain forest in Costa Rica. *Journal of Ecology*, **72**, 553–64

Chazdon, R.L. and Field, C.B. (1987*a*) Determinants of photosynthetic capacity in six rain forest

Piper species. *Oecologia*, **73**, 222–30

Chazdon, R.L. and Field, C.B. (1987*b*). Photographic estimation of photosynthetically active radiation: evaluation of a computerized technique. *Oecologia*, **73**, 525–32

Chazdon, R.L. and Pearcy, R.W. (1986). Photosynthetic responses to light variation in rain forest species. I. Induction under constant and fluctuating light conditions. *Oecologia*, **69**, 517–23

Chazdon, R.L., Williams, K. and Field, C.B. (1989). Interactions between crown structure and light environments in five rain forest *Piper* species. *American Journal of Botany*, **75**, 1459–71

Clark, D.A. (1986). Regeneration of canopy trees in tropical wet forests. *Trends in Ecology and Evolution*, **1**, 150–4

Clark, D.A. and Clark, D.B. (1987*a*). Análisis de la regeneración de árboles del dosel en bosque muy húmedo tropical: aspectos teóricos y prácticos. *Revista de Biología Tropical*, **35**, (Supplement 1), 41–54

Clark, D.A. and Clark, D.B. (1987*b*). Temporal and environmental patterns of reproduction in *Zamia skinneri*, a tropical rain forest cycad. *Journal of Ecology*, **75**, 135–49

Clark, D.B. and Clark, D.A. (1988). Leaf production and the cost of reproduction in the tropical rain forest cycad, *Zamia skinneri. Journal of Ecology*, **76**, 1153–63

Clark, D.B. and Clark, D.A. (1989). The role of physical damage in the seedling mortality regime of a neotropical rain forest. *Oikos*, **55**, 225–30

Clark, D.A., Rich, P.M., Clark, D.B. and Oberbauer, S.F. (1988). Long-term monitoring of photosynthetically active radiation in tropical wet forest understory. *Bulletin of the Ecological Society of America*, **69**, 99 (abstract).

Crow, T.R. (1980). A rain forest chronicle: a 30-year record change in structure and composition at El Verde, Puerto Rico. *Biotropica*, **12**, 43–55

Davies, P. (ed.) (1969). *The American Heritage Dictionary of the English Language*. (New York: Dell Publishing)

Denslow, J.S. (1987). Tropical rain forest gaps and tree species diversity. *Annual Review of Ecology and Systematics*, **18**, 432–51

Dietrich, W.E., Windsor, D.M. and Dunne, T. (1982). Geology, climate, and hydrology of Barro Colorado Island. In Leigh, E.G., Jr., Rand, A.S. and Windsor, D.M. (eds), *The Ecology of a Tropical Forest: Seasonal Rhythms and Long-term Changes*, pp. 21–46. (Washington., D.C.: Smithsonian Institution Press)

Estrada, A., Coates-Estrada, R. and Martínez-Ramos, M. (1985). La estación de biología tropical Los Tuxtlas: un recurso para el estudio y conservación de las selvas del trópico húmedo. In Gómez-Pompa, A. and del Amo R., S. (eds) *Investigaciones sobre la Regeneracion de Selvas altas en Veracruz, Mexico*, pp. 379–93. (Mexico City: Editorial Alhambra Mexicana)

Fetcher, N., Oberbauer, S.F., Rojas, G. and Strain, B. (1987). Efectos del régimen de luz en la fotosintesis y el crecimiento de plantulas de árboles en bosque muy húmedo tropical. *Revista de Biologia Tropical*, **35**, (Supplement 1), 97–110

Foster, R.B. and Brokaw, N.V.L. (1982). Structure and history of the vegetation of Barro Colorado Island. In Leigh, E.G., Jr., Rand, A.S. and Windsor, D.M. (eds), *The Ecology of a Tropical Forest: Seasonal Rhythms and Long-term Changes*, pp. 67–81. (Washington D.C.: Smithsonian Institution Press)

Gartner, B.L. (1989). Breakage and regrowth of *Piper* species in rain forest understorey. *Biotropica*, **21**, 303–7

Garwood, N.C., Janos, D.P. and Brokaw, N. (1979). Earthquake-caused landslides: a major disturbance to tropical forests. *Science*, **205**, 997–9

Goulding, M. (1988). Ecology and management of migratory food fishes of the Amazon Basin. In Almeda, F.A. and Pringle, C.M. (eds), *Tropical Rainforests: Diversity and Conservation*, pp. 71–85 (San Francisco: California Academy of Sciences)

Gutshick, V.P., Barron, M.H., Waechter, D.A. and Wolf, M.A. (1985). Portable monitor for solar radiation that accumulates irradiance histograms for 32 leaf-mounted sensors. *Agriculture and Forest Meteorology*, **33**, 281–90

Gutterman, Y. (1987). Dynamics of porcupine (*Hystrix indica* Kerr) diggings: their role in the survival and renewal of geophytes and hemicryptophytes in the Negev desert highlands. *Israel Journal of Botany*, **36**, 133–43

Haines, B. (1975). Impact of leaf-cutting ants on vegetation development at Barro Colorado Island. In Golley, F.B. and Medina, E. (eds), *Tropical Ecological Systems: Trends in Terrestrial and*

Aquatic Research, pp. 99 111. (New York: Springer-Verlag)

Haines, B.L. (1978). Element and energy flows through colonies of the leaf-cutting ant, *Atta colombica*, in Panama. *Biotropica*, **10**, 270 7

Haines, B.L. (1983). Leaf-cutting ants bleed mineral elements out of rain forest in southern Venezuela. *Tropical Ecology*, **24**, 85–93

Hartshorn, G.S. (1978). Tree falls and tropical forest dynamics. In Tomlinson, P.B. and Zimmermann, M.H. (eds), *Tropical Trees as Living Systems*, pp. 617–38. (London: Cambridge University Press)

Hartshorn, G.S. (1980). Neotropical forest dynamics. In Ewel, J. (ed.), *Tropical Succession*. *Biotropica*, **12** (Suppl.), 23 30

Holdridge, L.R. (1979). *Ecologia Basada en Zonas de Vida*. (San José, Costa Rica: Editorial IICA)

Hubbell, S.P. and Foster, R.B. (1986). Canopy gaps and the dynamics of a neotropical forest. In Crawley, M.J. (ed.) *Plant Ecology*, pp. 77–96. (Oxford: Blackwell Scientific Publications)

Kalisz, P.J. and Stone, E.L. (1984). Soil mixing by scarab beetles and pocket gophers in North-Central Florida. *Soil Science Society of America Journal*, **48**, 169–72

Kitajima, K. and Augspurger, C. (1989). Seed and seedling ecology of a monocarpic tropical tree, *Tachigali versicolor*. *Ecology*, **70**, 1102 14

Lemmon, P.E. (1956). A spherical densiometer for estimating forest overstorey density. *Forest Science*, **2**, 314 20

Lieberman, M., Lieberman, D., Hartshorn, G.S. and Peralta, R. (1985*a*). Small-scale altitudinal variation in lowland wet tropical forest vegetation. *Journal of Ecology*, **73**, 505 16

Lieberman, D., Lieberman, M., Peralta, R. and Hartshorn, G.S. (1985*b*). Mortality patterns and stand turnover rates in a wet tropical forest in Costa Rica. *Journal of Ecology*, **73**, 915–24

Martínez-Ramos, M. and Alvarez-Buylla, E. (1986). Seed dispersal, gap dynamics and tree recruitment: the case of *Cecropia obtusifolia* at Los Tuxtlas, Mexico. In Estrada, E. and Fleming, T. (eds), *Frugivores and Seed Dispersal*, pp. 333–46 (Dordrecht: Dr. W. Junk Publishers)

Martínez-Ramos, M., Alvarez-Buylla, E., Sarukhán, J. and Piñero, D. (1988). Treefall age determination and gap dynamics in a tropical forest. *Journal of Ecology*, **76**, 700–16

Myers, N. (1980). *Conversion of Tropical Moist Forests*. (Washington, D.C.: National Academy of Sciences)

Nũnez-Farfán, J. and Dirzo, R. (1988). Within-gap spatial heterogeneity and seedling performance in a Mexical tropical forest. *Oikos*, **51**, 274–84

Oberbauer, S.F., Clark, D.B., Clark, D.A. and Quesada, M. (1988). Crown light environments of saplings of two species of rain forest emergent trees. *Oecologia*, **75**, 207–12

Oberbauer, S.F., Clark, D.A., Clark, D.B. and Quesada, M. (1989). Comparative analysis of photosynthetic light environments within the crowns of juvenile rain forest trees. *Tree Physiology*, **5**, 13–23

Orians, G.H. (1982). The influence of tree-falls on tropical forest species richness. *Tropical Ecology*, **23**, 255–79

Pearcy, R.W. (1983). The light environment and growth of C_3 and C_4 tree species in the understorey of a Hawaiian forest. *Oecologia*, **58**, 19 25

Pickett, S.T.A. (1983). Differential adaptation of tropical tree species to canopy gaps and its role in community dynamics. *Tropical Ecology*, **24**, 68–84

Pickett, S.T.A. and Thompson, J.N. (1978). Patch dynamics and the design of nature reserves. *Biological Conservation*, **13**, 27–37

Pires, J.M. and Prance, G.T. (1985). The vegetation types of the Brazilian Amazon. In Prance, G.T. and Lovejoy, T.E. (eds), *Amazonia*, pp. 109–45 (New York: Pergamon Press)

Platt, W.J. (1975). The colonization and formation of equilibrium plant speces associations on badger disturbances in a tall-grass prairie. *Ecological Monographs*, **45**, 285–305

Popma, J., Bongers, F., Martínez-Ramos, M. and Veneklaas, E. (1988). Pioneer species distribution in treefall gaps in neotropical rain forest: a gap definition and its consequences. *Journal of Tropical Ecology*, **4**, 77 88

Putz, F.E. (1983). Treefall pits and mounds, buried seeds, and the importance of soil disturbance to pioneer trees on Barro Colorado Island, Panama. *Ecology*, **64**, 1069–74

Putz, F.E. and Milton, K. (1982). Tree mortality rates on Barro Colorado Island. In Leigh, E.G., Jr., Rand, A.S., Windsor, D.M. (eds), *The Ecology of a Tropical Forest: Seasonal Rhythms and Long-term Changes*, pp. 95 100. (Washington D.C.: Smithsonian Institution Press)

Putz, F.E. and Brokaw, N.V.L. (1988). Sprouting of broken trees on Barro Colorado Island. *Ecology*, **70**, 508–12

Rich, P.M. (1986). Mechanical architecture of arborescent rain forest palms. *Principes*, **30**, 117–31

Rich, P.M. (1988). Video image analysis of hemispherical canopy photographs. In Mausel, P.W. (ed.) *First Special Workshop* on *Videography*, Terra Haute, Indiana, May 19–20 1988. American Society of Photogrammetry and Remote Sensing.

Riera, B. (1981). Chablis et cicatrisation en foret Guyanaise. Ph.D. dissertation, Université Paul Sabatier de Toulouse, Toulouse.

Riera, B. (1985). Importance des buttes de deracinement dans la regeneration forestiere en Guyane francaise. *Revue d'Ecologie (Terre Vie)*, **40**, 321–9

Salick, J., Herrera, R. and Jordan, C.F. (1983). Termitaria: nutrient patchiness in nutrient-deficient rain forests. *Biotropica*, **15**, 1–7

Salo, J., Kalliola, R., Häkkinen, I., Mäkinen, Y., Niemelä, P., Puhakka, M. and Coley, P.D. (1986). River dynamics and the diversity of Amazon lowland forest. *Nature*, **322**, 254–8

Sanford, R.L., Saldarriaga, J., Clark, K.E., Uhl, C. and Herrera, R. (1985). Amazon rain forest fires. *Science*, **227**, 53–5

Sanford, R.L., Braker, H.E. and Hartshorn, G.S. (1986). Canopy openings in a primary neotropical lowland forest. *Journal of Tropical Ecology*, **2**, 277–82

Uhl, C. (1982). Recovery following disturbances of different intensities in the Amazon rain forest of Venezuela. *Interciencia*, **7**, 19–24

Uhl, C. and Murphy, P.G. (1981). Composition, structure, and regeneration of a tierra firme forest in the Amazon Basin of Venezuela. *Tropical Ecology*, **22**, 219–37

Uhl, C., Clark, K., Dezzeo, N. and Maquirino, P. (1988). Vegetation dynamics in Amazonian treefall gaps. *Ecology*, **69**, 751–63

Ustin, S.L., Woodward, R.A., Barbour, M.G. and Hatfield, J.L. (1984). Relationships between sunfleck dynamics and red fir seedling distribution. *Ecology*, **65**, 1420–8

Vandermeer, J.H. (1977). Notes on density dependence in *Welfia georgii* Wendl. ex Burrett (Palmae), a lowland rain forest species in Costa Rica. *Brenesia*, **10/11**, 9–15

Vitousek, P.M. and Denslow, J.S. (1986). Nitrogen and phosphorus availability in treefall gaps of a lowland tropical rain forest. *Journal of Ecology*, **74**, 1167–78

White, P.S. and Pickett, S.T.A. (1985). Natural disturbance and patch dynamics: an introduction. In Pickett, S.T.A. and White, P.S. (eds), *The Ecology of Natural Disturbance and Patch Dynamics*, pp. 3–13 (New York: Academic Press)

Whitmore, T.C. (1978). Gaps in the forest canopy. In Tomlinson, P.B. and Zimmermann, M.H. (eds), *Tropical Trees as Living Systems*, pp. 639–55 (London: Cambridge University Press)

CHAPTER 22

THE FATE OF JUVENILE TREES IN A NEOTROPICAL FOREST: IMPLICATIONS FOR THE NATURAL MAINTENANCE OF TROPICAL TREE DIVERSITY

Stephen P. Hubbell and Robin B. Foster

ABSTRACT

A long-term demographic study of woody plants is underway in a 50-hectare plot of lowland tropical forest on Barro Colorado Island in Panama. Results and inferences from a primary census completed in 1982 are re-evaluated in the light of a recensus carried out in 1985, which provided the first data on patterns of tree recruitment, growth and mortality in the plot. In 1985, the data base contained records on a total of 269 021 plants, of which 242 218 were alive and had a stem diameter of 1 cm dbh or larger. There was an increase of 2.7% in individuals compared to the primary census, but this increase masked considerable turnover of plants in the plot. The patterns of recruitment, growth and mortality are considered in relation to the question as to whether natural regulating factors, operating especially on the juvenile stages of tropical trees, are sufficient to maintain an equilibrium tree species richness without management intervention. The recensus results suggest that negative conspecific effects could be important in the maintenance of tree species richness in tropical forests, but they fall short of proof. Taken altogether, the evidence for equilibrating forces might seem very strong, but caution is needed and it is premature to evaluate the quantitative importance of these effects in stabilizing the species assemblage. The results contribute to the academic debate on equilibrium versus non-equilibrium hypotheses in community dynamics, and also have practical implications for conservation biologist and forest managers.

317

INTRODUCTION

Research on the reproductive biology of tropical forest plants has traditionally focused on the relatively brief reproductive processes that occur early in the life of the plant, such as pollination, seed dispersal, and the fate of seeds and seedlings. If the ultimate focus is on reproductive success, however, then the fate of established, juvenile plants is equally important, especially in the case of long-lived organisms such as trees. Sapling demography is important to reproductive success because the juveniles must persist in the forest, often for several decades, before reaching reproductive maturity.

Because of the difficulties of studying long-term processes, we know much more about early events in the reproduction of tropical trees than we know about the demography of the post-seedling juvenile stages. The longest running studies of tropical forest plots are about 40 years old – less than the generation time of all but the shortest-lived tree species. These studies have unfortunately not included saplings, or any trees smaller than 10 cm diameter at breast height (dbh) (Manokaran and Kochummen, 1987; Wyatt-Smith, 1966). Also, most standard forestry plots in species-rich tropical rain forest have been small – from one to a few hectares (Synnott, 1979) – samples too small for analysing size- and species-specific survival and growth in any species except those that are extremely abundant.

Recently, several large forest plot studies have been established or are in the planning stage that can provide missing demographic data on juvenile tropical trees in the New and Old World tropics. Understanding the natural regeneration of tropical trees through their complete life cycle, and in their natural community setting, is one of the principal objectives of these studies (Hubbell, 1984).

One such study, the Barro Colorado Island Forest Dynamics Project, is an ongoing study of a 50-hectare plot of lowland tropical forest in Panama. We review here some of the demographic patterns found among the tree species in this plot over a 3-year interval, and discuss their implications for the maintenance of tree species richness in naturally regenerated tropical forests. We consider patterns of sapling distribution, recruitment, growth, and mortality in relation to the question as to whether natural regulating factors, operating especially on the juvenile stages of tropical trees, are sufficient to maintain an equilibrium tree species richness without management intervention.

The question of the equilibrium or non-equilibrium status of tropical forests is of practical interest to forest conservationists and managers. If local stands of tropical forest represent biotically stabilized communities of particular tree taxa at or near equilibrium, then the conservation and management of these forests will be a different and far easier task than if they are non-equilibrium species assemblages, drifting continually in floristic composition in space and time. By equilibrium, we mean that a particular assemblage of tree species has some collective stability, such that the mixture of species tends to return to its original composition if disturbed. This generally implies that the population growth of each species is limited by density- and frequency-dependent factors that prevent it from assuming complete dominance. By non-equilibrium, we refer to a community that lacks such stability, and in which changes in species composition are the result of unique historical and biogeographical events. We discuss the conservation and management implications of the equilibrium and non-equilibrium status of tropical forests in greater detail elsewhere (Hubbell and Foster, 1986b).

Distinguishing between these alternatives requires that we understand the processes that determine relative tree species abundances in the forests. If the forests are near equilibrium, then we would expect species abundances to remain relatively constant. If currently common species show lower per-capita reproductive success than less common species, then we would have evidence for the stabilizing density- and frequency-dependence required by the equilibrium hypotheses. If the community is disturbed, the succession of species following the disturbance should be reasonably predictable, and the original species mix should be nearly, if not exactly, restored. Evaluating the evidence for equilibrium is not trivial. If the forest is continually disturbed, then it will rarely, if ever, be "at rest", even if strong equilibrating forces are at work. There are always density-independent factors causing population fluctuations in organisms. The challenge therefore is to evaluate the quantitative effectiveness of density-dependent factors in stabilizing populations and community structure. As mentioned, a practical problem is the longevity of trees, which makes the study of their community dynamics an exercise in patience or else extrapolative inference based on short-term data.

Hypotheses to explain the maintenance of large numbers of tree species in close sympatry in the richest tropical forests have been reviewed extensively (e.g. Ashton, 1969; Janzen, 1970; Connell, 1971, 1978; Hubbell and Foster, 1986b). One classical hypothesis, that tropical trees coexist at an equilibrium resulting from resource partitioning, seemingly ran into difficulties because plants compete for relatively few limiting resources (Hubbell, 1980). In classical equilibrium competitive models, species can outnumber limiting resources only under special circumstances. Although it later turned out that this problem could be solved easily if there was spatial heterogeneity in resource supply rates (Tilman, 1982), it stimulated a search for other biotic mechanisms that could explain high species richness in tropical tree communities.

Nearly 2 decades ago, it was recognized that equilibrium coexistence of arbitrary numbers of tropical tree species, even species with identical resource requirements, could be achieved if their populations were limited by density-dependent and frequency-dependent mortality from predators, herbivores, or pathogens. Janzen (1970) and Connell (1971) suggested that the critical density-dependent mortality occurred very early in the life history of tropical trees, in the seed and seedling stages. Heavy early mortality is also found in trees in species-poor temperate forests, however, so the causal connection between species richness and density-dependent seed or seedling mortality is far from proven. Indeed, there is nothing particularly remarkable about heavy early mortality in tree species, a virtual geometrical necessity for any species which produces many seeds and grows to a large size when mature (Silvertown, 1982).

More recently, it has been shown that tree species identical in resource requirements can persist in a non-equilibrium community in drifting relative abundance for long time periods, sufficient even for speciation, in the absence of any stabilizing density- or frequency-dependence or refuge from competition (Hubbell, 1979; Hubbell and Foster, 1986*b*). In these model communities, the loss of any reasonably abundant species takes a very long time because each species exhibits the same or similar per-capita birth and death rates. This model is the neutral version of "lottery" models of coexistence, in which long-lived, polycarpic species can coexist in stochastic equilibrium via frequency-dependent fluctuations in recruitment (Chesson and Warner, 1981). In the latter models, coexistence arises when each species has a period in which its per-capita recruitment success exceeds the mean per-capita recruitment success of all other species in the community.

The problem now is not that there are too few, but that there are too many explanations for the coexistence of tropical tree species. Unfortunately, many of the theoretically possible explanations are not mutually exclusive. This means that the persistent coexistence of tropical tree species is probably mediated by several factors, including niche specialization and resource partitioning, the density-dependent action of enemies, and the historical biogeography of each species as well. The current challenge is to find clever ways to assess the quantitative contribution of each possible mechanism to the maintenance of tropical tree diversity in nature.

This contribution provides a historical review of the Barro Colorado Island (BCI) Forest Dynamics Project. We begin with an outline of the results from the primary census of the BCI plot completed in 1982, and discuss the inferences that were drawn about the dynamics of the BCI forest based only on the static primary census information. Then, we re-evaluate these inferences in the light of the results from the recensus of 1985, which provided the first data on patterns of tree recruitment, growth, and mortality in the plot. The accuracy of inferences about forest dynamics from single-census "snapshot" studies is of both practical and theoretical interest. From a theoretical viewpoint, it is important to know whether accurate predictions of forest dynamics can be made from static data. Practically, we need to know if long-term studies are necessary to understand the dynamic ecosystem. To follow large forest plots for an appreciable time is expensive and justifiable only if such studies are critical to understanding forest dynamics.

STUDY SITE AND METHODS

In 1980, a 50 ha permanent plot was established in old-growth forest on Barro Colorado Island (BCI), a 15 km^2 biological reserve on a former hilltop in artificial Lake Gatun, part of the Panama Canal. The forest is very well known botanically (Croat, 1978), and is semi-evergreen and seasonal, with a pronounced dry season from January through to mid-April (Leigh *et al.*, 1982). The plot is located on relatively flat terrain at a mean elevation of about 150 m above mean sea level (msl); 25 ha are on grades less than 3%; 13 ha have slopes of 3–10%, and the remaining 12 ha are steeper, with slopes between 10–20% (Hubbell and Foster, 1986c). The site has no permanent water and the soils are similar to Amazonian Ultisols (Stallard, pers. comm.). Spatially restricted habitats include a seasonal swamp with acidic, anaerobic soils that occupies about 2 ha in the centre of the plot, and several steep ravines bordering seasonal creeks that make up less than 1 ha in total area. Extensive palaeoecological research in the plot has shown that the forest has been undisturbed by man essentially for the last 500 years, and that the site has never been used for maize agriculture (Piperno, pers. comm.). Radio-carbon dating of charcoal indicates that small occasional clearings were made in the forest between 500 A.D. and 1500 A.D., and associated artefacts indicate that the clearings were used as seasonal camps by hunters and gatherers.

All freestanding woody plants in the plot with a stem diameter of at least 1 cm dbh were tagged, measured, and mapped in the primary census (1980–82) (Hubbell and Foster, 1983). The plot was completely recensused in 1985, and all tree growth, survival, and new sapling recruitment into the 1 cm dbh class was recorded. The plot is scheduled for recensus again in 1990, and at regular 5-year intervals thereafter. Annual measurements of canopy height, vegetation layering, and new gaps have been taken since 1983 at all points of a 5 m grid covering the entire plot. From these data, it has been possible to reconstruct the recent gap disturbance regime of the BCI forest (Hubbell and Foster, 1986a). The detailed gap and canopy data also permitted us to classify the regeneration strategies of the tree species into categories based on the distribution of their small saplings in recent gaps, older regenerating gaps, and the understorey of mature-phase forest (Hubbell and Foster, 1986a, 1986c, 1987a).

PRIMARY CENSUS RESULTS

In 1982, after the primary census was completed, there were 235 895 individually tagged plants in the plot; the mean number of stems per ha of all size classes over 1 cm dbh was 4711. A total of 303 species, from shrubs to overstorey trees, was recorded in the census. Species abundances varied over 4.6 orders of magnitude, from 21 species represented by single plants, to one shrub species, *Hybanthus prunifolius* (Violaceae), with 39 911 individuals (Hubbell and Foster, 1990*a*). The rarest third of all species contributed less than 1% of the total number of plants in the plot. Only 22 canopy tree species achieved mean stand densities of more than one adult tree per ha; but the commonest canopy tree, *Trichilia tuberculata* (Meliaceae) accounted for nearly 1 out of 8 trees over 20 cm dbh (Hubbell and Foster, 1983, 1986*c*). Although some species were nearly randomly distributed through the plot, many populations were quite patchily distributed. A community analysis of species composition showed that considerable local differentiation of stands occurred within the 50 ha. Spatial autocorrelation analyses of relative species abundance revealed that the scale of stand differentiation in species composition was very local. Community similarity decreased by about half within just 200 m from any given point (Hubbell and Foster, 1983).

The most obvious evidence from niche specialization among the freestanding woody plants was growth form: 58 species achieved adult size below 4 m in height; 60 species were small understorey treelets (adults between 4 and 10 m tall); 71 species were subcanopy trees (adults between 10 and 20 m tall); and 114 species were overstorey trees (canopy plus emergent trees, and strangler figs, adults over 20 m tall).

The distribution of saplings also provided strong circumstantial evidence of habitat niche specialization in some species. By "habitat niche" we refer to microsite conditions that are relatively fixed in space and time, such as conditions of drainage and soil fertility controlled by topography and edaphic characteristics. Of 239 species evaluated, half of the species (120) exhibited statistically significant habitat correlations with identified topographic or edaphic features in the plot (Table 22.1). For the remaining half, however, it was impossible to reject the null hypothesis of indifference to the habitat variation occurring in the plot. For many of the common species, the hypothesis of habitat indifference had very high statistical power ($p > 0.8$). We could also show that commonness *per se* did not force the statistical conclusion of habitat indifference in any species (Hubbell and Foster, 1986*c*).

By "regeneration niche" we refer to relatively ephemeral microsite conditions, such as gap conditions created by tree falls. Evidence for regeneration niche specialization was based on the distribution of saplings into recent or old gaps, or in the understorey of mature-phase forest. We were able to discriminate a range of species responses, from species that were apparently extremely light-demanding and shade-intolerant, through species that were gap-avoiders and highly shade-tolerant. We were surprised to find, however, that there was a large group of species whose saplings were distributed indistinguishably from random with respect to new and old gaps, and the mature-phase understorey. Of the 239 species tested, 98 (41%) showed indifference to gap or shaded understorey conditions (Table 22.1). Once again, the indifferent group included the commonest species. For most of these species, we could accept the generalist (null) hypothesis with high statistical power. Although these species are all shade-tolerant they also evidently do well if they are in a light gap.

In summary, from this sapling analysis, a correlation was apparent between commonness and being a generalist species, and rarity and being a habitat or regeneration niche specialist (Tables 22.2 and 22.3). It is unlikely that the common generalist species are really completely indifferent to the variation in habitat and regeneration niche conditions in BCI forest. More important, however, is the conclusion that they are much less specialized in their regeneration niche requirements than are the rarer species. There are no very common habitat or regeneration niche specialists, but some rare generalist species do exist. We therefore concluded that a necessary but insufficient condition for becoming a common species in the BCI forest is having the physiological and dispersal capacity to recruit, survive, and grow under the prevailing (i.e. spatially and temporally common) habitat and regeneration conditions in the BCI forest. This is insufficient to guarantee commonness, however, because other factors such as predators or pathogens may also limit the abundance of a tree species.

It is reasonable to expect that specialization should generally lead to reduced abundance. Consider, for example, the gap size requirements of BCI tree species as one component of their regeneration niche specialization. Most of the plot is covered by mature-phase forest having a deeply shaded understorey; and the gap disturbance regime on BCI produces mostly small gaps ($<25\,m^2$) and large gaps (e.g. $>400\,m^2$) only rarely

Table 22.1 Classification of 239 woody plant species in relation to habitat and regeneration niche specialization in the 50 ha plot on Barro Colorado Island (after Hubbell and Foster 1986c)

Habitat niche category	Regeneration niche category			
	Sun	*Partial sun*	*Indifferent*	*Shade*
Slope	0	10	19	11
Indifferent	27	28	53	11
Flat terrain	9	31	24	6
Swamp/Ravine	3	4	2	1

Table 22.2 Numbers of very common species (> 1000 individuals/species) and rare species (10–49 individuals/species) in relation to gap regeneration niche categories

| | Regeneration niche category | | | |
	Sun	*Partial sun*	*Indifferent*	*Shade*
Common species	0	4	18	19
Rare species	17	19	8	3

Table 22.3 Numbers of very common species (> 1000 individuals/species) and rare species (10–49 individuals/species) in relation to habitat niche categories

| | Habitat niche category | | | |
	Slope	*Indifferent*	*Flat terrain*	*Swamp/ravine*
Common species	5	26	10	0
Rare species	3	27	9	8

(Hubbell and Foster, 1986*a*). We therefore would expect that heliophilic species that require a large gap to regenerate, should be less common than shade-tolerant or indifferent species. Note that members of the "sun" group of species were never common (Table 22.2). On the other hand, the "indifferent" group of species, which presumably persist nearly as well as in the shade as in gaps, contains the largest number of common species.

INTERPRETING THE PRIMARY CENSUS RESULTS

Niche differentiation

It is reasonable to expect that niche differentiation, whether by growth form or habitat or regeneration niche specialization, should promote equilibrial coexistence. These differences presumably reflect morphological and physiological adaptations to different subsets of the environmental conditions found in the BCI forest, and reflect genuine life history trade-offs. However, with the possible exception of growth-form differences, we still lack the mechanistic proof that these trade-offs are real, and if so, that these differences are actually necessary or sufficient to explain coexistence. If these niche differences are important to coexistence, then they must be shown to generate density- and frequency-dependence to have dynamic sufficiency. Otherwise, these differences are irrelevant to coexistence, and their origins through competitive niche displacement are called into question.

What is more remarkable is not that such niche differences sometimes exist, but that so many common species appear to be functionally identical or nearly identical generalists in terms of habitat and regeneration niche requirements. It is probable, of course, that we have not discovered all the critical niche differences between the tree species; but, in any event, the number of species which are ecologically very similar in the BCI forest is large, and it includes many of the commonest species. How are we to explain the coexistence of all these nearly identical species?

Enemies

One possibility is that the coexistence of such species is maintained not by resource partitioning and niche differentiation, but by the density-dependent action of enemies or through frequency-dependent recruitment success. Seed and seedling predation is often heavy immediately beneath the parent tree, and Janzen (1970) and Connell (1971) suggested that only seeds and seedlings that had been dispersed some distance from the parent would escape death. According to this model, no tree species can monopolize a given space permanently because no species can replace itself in the same place in the next generation. Such an inability of self-replacement would open up habitat to other species. The spacing of adult trees should therefore appear more regular than if new adults were drawn at random from the seed shadows of parent trees.

As noted above, the relative size of seedlings and adult trees makes a random draw of seed shadows impossible anyway; but the dense clumping of some tropical tree species led Hubbell (1979, 1980) to question the efficacy of the Janzen-Connell mechanism for maintaining tree species diversity in tropical forests. Saplings of many common species are often, in fact, located beneath conspecific adults, and sometimes at higher densities there than elsewhere in the forest. Of course, unless mortality beneath adults is nearly total, this result is not too surprising since so many more seeds fall nearby than are transported away from the parent (Hubbell, 1980). The critical issue is whether the differential mortality beneath conspecifics is sufficient to give seedlings of heterospecific trees a clear, collective competitive advantage.

Pairwise species replacements

The BCI plot study provides no information on seeds and seedlings, but we can look for effects in the sapling stages above 1 cm dbh. From the frequencies of saplings of species i beneath the adults of species j, we calculated the pairwise tree species replacement probabilities for the 10 most common BCI overstorey, mature-phase species. The remaining species were pooled into a "black-box" category. The calculations for sapling size classes 1–2 cm, 2–4 cm, and 4–8 cm dbh all indicated that the mean probability of conspecific replacement was not significantly lower than the mean probability of heterospecific replacement (Hubbell and Foster, 1987b). This result suggested that any of these common species could replace itself by any other, and that self-limiting factors among these habitat and regeneration niche generalists were weak or absent.

Because of the high species richness, however, the probabilities of self-replacement, or of replacement by any other species, were low. We therefore predicted that the sequence of mature-phase tree species occupying any given site in the forest through time would be highly variable, rather than a relatively predictable succession of species (Hubbell and Foster, 1987b). We concluded that the low rate of self-replacement of tropical trees was simply the passive result of high species richness in the forest, and that it was not due to an incapacity of species to replace themselves.

Density-dependence and frequency-dependence

We also looked at the static primary-census data for any phenomenological evidence for density- and frequency-dependence. For density-dependence, we searched for evidence that the per-capita production of juveniles by adults dropped as adult density increased. For frequency-dependence, we looked for evidence that rare species had a per-capita reproductive advantage over common species, suggesting that they would increase in the future at the expense of the common species.

Inferences of this sort are subject to many assumptions (Hubbell and Foster, 1986b); but, with two exceptions, there was no evidence for quantitatively important density-dependence among the tree species. This conclusion was based on regressions of the number of small saplings per hectare on the number of conspecific adults per hectare (Hubbell and Foster, 1986b). Tests at smaller and larger spatial scales gave the same result. The two exceptions were the two most common overstorey tree species, *Trichilia tuberculata* (Meliaceae) and *Alseis blackiana* (Rubiaceae). In these species, the sapling distributions showed statistically significant avoidance of conspecific large trees (Hubbell and Foster, 1987b). We concluded therefore that most of the tree species in the BCI forest were far below the abundance at which negative density-dependence could be detected readily or would be effective at limiting population growth. We also concluded that, if any species did manage to increase in abundance to very high levels (such as *Trichilia* and *Alseis*), it would be prevented from assuming complete dominance by strong intraspecific density-dependence. What the actual mechanisms behind this density-dependence might be cannot presently be determined.

The evidence for frequency-dependence, or rare-species competitive advantage, was similarly weak. Rare tree species did not appear to exhibit a higher per-capita rate of sapling production than common species (Hubbell and Foster, 1986*b*). If anything, rare species tended to exhibit a greater percentage of adults, suggesting infrequent or episodic reproduction, or species on their way to local extinction.

Coevolutionary convergence and coexistence

Another possibility for the maintenance of species richness is simply species persistence in the absence of strongly stabilizing biotic factors. There may be no destabilizing factors either, so that the tree species are drifting slowly in non-equilibrium abundance in the BCI forest. This drift would be possible if the species were competitively similar generalists. A few years ago, Connell (1980) suggested that coevolution and character displacement among competing species should occur at much slower rates in species-rich communities. This is likely, he argued, because the identity of one's competitors will be more unpredictable in ecological and evolutionary time in species-rich communities. Recent theoretical investigations of rates of coevolution in model tree communities lend some support to Connell's argument, at least for certain reasonable selection regimes (Hoffmaster and Hubbell, 1990). Uncertainty of one's competitors is potentially very high for trees in species-rich tropical forests. We calculated the tree diversities in neighbourhoods of individual trees in the BCI forest. We found that, in the two 20-tree immediate neighbourhoods of two conspecific adult trees, only about seven trees could be matched to species (Hubbell and Foster, 1986*b*).

We extended Connell's argument and hypothesized that tree species may not diverge in species-rich tropical forests, but may often converge (Hubbell and Foster, 1986*b*). Coevolutionary convergence might be expected whenever a large suite of diffuse competitors creates a relatively constant biotic environment, stabilized by the law of large numbers over evolutionary time. As one example, mature-phase tropical tree species should evolve generalized shade tolerance to the average shade cast by the suite of their competitors, rather than to specialize on the shade of a particular intensity and spectral quality cast by any one competitor. This tolerance may partially explain the high degree of morphological similarity of the juvenile stages among many shade-tolerant rain forest tree species. Character convergence among competing species also arises in models of sessile organisms (e.g. plants) that monopolize local resource patches once established, and exhibit local density-dependence (S. Pacala, pers. comm.). Once evolved, there would be few forces besides drifting abundance to eliminate such species from the community. As mentioned earlier, theoretical studies have shown that the mean times to extinction by drift for ecologically equivalent competitors are very long indeed (Hubbell and Foster, 1986*b*).

Evidence for non-equilibrium

If convergent evolution has produced large suites of common generalist tree species, coexisting in tropical forests in drifting relative abundance, then there should be evidence that the BCI tree community is not in equilibrium. The evidence after the primary census was circumstantial but quite strong: many common as well as rare species exhibited great variation in their sapling and adult diameter class distributions from one part of the plot to another, suggesting that the populations are far from stationary (Hubbell and Foster, 1987*a*). An equilibrium argument could explain this result, but then one must assume that the demography and life history of each species is radically different at equilibrium from one part of the plot to the next. Non-equilibrium was also suggested by the spatial distributions of saplings and adults in some species, indicating that they were probably invading the forest (Hubbell and Foster, 1986*b*,1986*c*).

Primary census conclusion

Based on the results and interpretations of the static population patterns from the primary census, we concluded that the BCI tree community was not in taxonomic equilibrium. The evidence for niche specialization was strong, but there also appeared to be many generalist tree species in each major life history mode, broadly overlapping in terms of growth form, and habitat and regeneration requirements. We found little evidence for intraspecific density-dependence or intraspecific frequency-dependence. Evidence of this sort would be expected from the niche differentiation model. Except for the two most abundant canopy species, most tree species appeared to be far below the abundances at which strong density-dependence would be observed.

On this evidence, we criticized the general equilibrium view of tropical forest tree communities in a series of recent papers (Hubbell, 1984; Hubbell and Foster, 1986*b*, 1986*c*, 1987*a*, 1987*b*). We concluded that, although there was clear evidence for strong diffuse competition among BCI trees, this competition was probably not effective in stabilizing a particular taxonomic assemblage of tree species, in causing competitive exclusion, or in preventing invasion of additional tree species. At the conclusion of the primary census, our model for the dynamics of the BCI tree community was similar to the theory of island biogeography (MacArthur and Wilson, 1967). Although this theory predicts a steady-state species richness at the balance point between species immigration and local extinction, it also describes a community in taxonomic non-equilibrium with continual species turnover. We concluded that the present BCI tree species richness probably has more to do with the historical biogeography of the Isthmus of Panama and Central America than with the community ecology and biotic interactions among the current tree species residents on BCI.

THE RECENSUS RESULTS

In 1985, the 50 ha plot at BCI was completely recensused. After the new sapling recruits found during the recensus were incorporated, the database contained records on a total of 269 021 plants, of which 242 218 were alive and had a stem diameter of 1 cm dbh or larger (Hubbell and Foster, 1990*b*). The net increase of only 2.7% masked considerable turnover of plants in the plot. There were 20 705 (8.78%) deaths and 33 126 (14.04%) new recruits over the 1982–85 interval. The difference in number of recruits and deaths exceeded the net increase in number of living stems over 1 cm dbh in 1985 because 6098 (2.6%) of the primary-census plants lost their main stem and survived in 1985 as a stem of less than 1 cm dbh. There was a net increase of one species in the plot; two species went locally extinct, and three new species, all rare, were encountered for the first time in 1985 (Hubbell and Foster, 1990*a*, 1990*b*).

The question now arises, do the data on temporal and spatial patterns of sapling recruitment, growth, and mortality support the non-equilibrium conclusion drawn from the static data of the primary census? The answer in brief is that the recensus data revealed equilibrating forces that were stronger than had been anticipated. However, it would be premature to claim that these forces are sufficient to maintain tree diversity in the BCI forest; and the evidence for non-equilibrium of the BCI forest remains strong. The bulk of the evidence for the equilibrating forces is presented elsewhere (Hubbell and Foster, 1990*c*), but some salient results can be outlined here.

Recensus evidence for non-equilibrium

The recensus data show that there is a significant difference between common and rare tree species in the trend of their population change. If "common" species are taken for discussion to be those having a mean density of at least one individual ha⁻¹, then 191 species are common, and 118 are rare. As a group, rare species showed fewer increases (39%) than common species (77%), a highly significant difference statistically ($p < 0.1$, Table 22.4). Additional evidence comes from the high proportion of species that showed dramatic population fluctuations during the intercensus interval. Forty per cent of all species, from shrubs to trees, exhibited changes in total abundance that exceeded 10% – a large change given the short intercensus interval of 3 years. Many of these changes could be attributed to effects of the severe El Niño drought of 1982–83 (Hubbell and Foster, 1990*a*). This drought was the most intense on record in the 60 years of weather record keeping on BCI. During and immediately following the dry season of 1983, many species suffered severe losses, and 15% of all canopy trees above 32 cm dbh died (Hubbell and Foster, 1990*b*). For example, the common canopy tree, *Poulsenia armata* (Moraceae), declined by 23%.

The recensus sapling growth and survival data also support the primary census conclusion that a large number of common species are very similar in their regeneration requirements. This result is evidence for the non-equilibrium view insofar as it does not support the niche differentiation model of coexistence. Two extreme patterns are illustrated. *Trichilia tuberculata* (Meliaceae) is representative of the pattern seen in many common "generalist" tree species, namely those that were predicted from the primary census to be indifferent to gap versus shaded understorey conditions. The recensus data generally confirm these predictions. Figure 22.1 shows the growth and survival of 1–2 cm dbh saplings of *Trichilia*. In the figure, the "new gap" category refers to sites where a light gap occurred after 1983. The "old gap" category refers to gaps present in 1983, the first year of the canopy measurements. The "non-gap" category refers to sites that were continuously beneath a canopy higher than 10 m throughout the interval between censuses. *Trichilia* showed very high and nearly equal survivorship (90–95%) over the census interval in all three types of

Table 22.4 Non-equilibrium population trends in common species (> 50 individuals/species)versus rare species (< 50 individuals/species) in the BCI plot over the recensus interval 1982–1985

	Population trend	
	Decline or no change	*Increase*
Common species	43 *(71.1)*	148 *(119.9)*
Rare species	72 *(43.9)*	46 *(74.1)*

The expected number of species (in parenthesis) is based on the null hypothesis of independence of population trend and species abundance. The χ^2 value was 46.3, so the null hypothesis is rejected at $p < 0.001$. Rare species were non-increasing, and common species increasing, more often than expected by chance

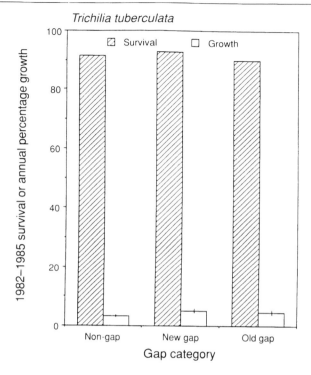

Figure 22.1 Annual per cent growth and 1982-1985 per cent survival in saplings 1.0–1.9 cm diameter at breast height (dbh) of *Trichilia tuberculata* (Meliaceae). Shaded bars are survival; open bars are annual growth. Vertical lines on growth bars are ± 2 standard deviations. "New gaps" are gaps created after 1983; "old gaps" were created before 1983. The "non-gap" areas were shaded understorey sites with canopy heights of 20 m or more

regeneration sites. Growth rates were also similar and very low in all sites, with only a slight absolute increase in response to being in a new or old gap.

In contrast, the light-demanding gap specialist, *Cecropia insignis* (Moraceae), exhibited low (35–40%) survivorship in the 1–2 cm dbh size class in all sites (Figure 22.2). For saplings that survived, growth rates were very high, and the absolute increase in growth rate in old gap sites was large. *Cecropia* showed a growth-rate response to old gaps and not to new gaps. This unexpected result was probably due to the fact that the new gaps that *Cecropia* occupied happened to be created shortly before the 1985 recensus. We were surprised that *Cecropia* survival was as poor in gaps as in the shade. This pattern was found in most other gap pioneer species, so it must reflect in part the intense shade and root competition that occurs in gaps during gap closure.

When the growth and survival data for a large number of canopy and midstorey trees are examined together, the evidence that most of the BCI

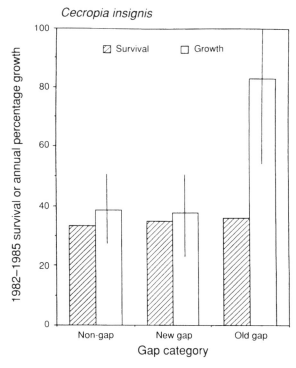

Figure 22.2 Annual per cent growth and 1982–1985 per cent survival in saplings 1.0–1.9 cm diameter at breast height (dbh) of *Cecropia insignis* (Moraceae). Shaded bars are survival; open bars are annual growth. Vertical lines on growth bars are ± 2 standard deviations. "New gaps" are gaps created after 1983; "old gaps" were created before 1983. The "non-gap" areas were shaded understorey sites with canopy heights of 20 m or more

tree species are shade-tolerant, indifferent species is quite strong. The distribution of mean annual growth rates is strongly skewed toward low values, with only a few heliophilic species at higher values (Figure 22.3). Survival rates are likewise strongly skewed, but toward high values (Figure 22.4), as one would also have predicted from the primary-census data.

When survival was analysed by size class, the common generalist species – those that are shade-tolerant and indifferent to gaps or shade – showed a quite remarkable pattern: survival was high and virtually constant over all size classes, from 1 cm dbh right through to the adult size classes (Hubbell and Foster, 1990a). In contrast, survival in gap species like *Cecropia* increases rapidly with size class, approaching the survival rates of the shade-tolerant species when they reached adult sizes and entered the high-light environment of the canopy layer. Across species, log survival and log growth rate were negatively correlated ($r = -0.81$, Hubbell and Foster, 1990a). This correlation is strong demographic evidence for genuine

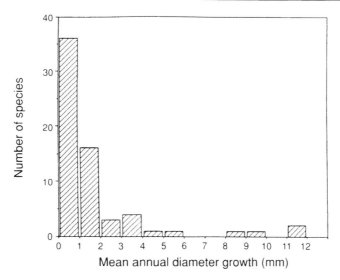

Figure 22.3 Distribution of mean annual growth rate for 1.0–1.9 cm diameter at breast height (dbh) saplings over the 1982–1985 interval for 65 Barro Colorado Island overstorey and midstorey tree species with at least 10 saplings in the 1.0–1.9 cm dbh size class

physiological and life history trade-offs in tropical trees that are specialized either for high- or low-light environments. Although such trade-offs are expected physiologically, these demographic correlates have not been demonstrated in tropical trees before.

Recensus evidence for equilibrating forces

The principal new evidence for equilibrating forces comes from the comparative growth and survival performance of saplings beneath conspecific or beneath heterospecific large trees (Hubbell and Foster, 1990a, 1990c). From the primary census, we were able to detect the density dependence and sapling "avoidance" of large conspecific trees in only two tree species, *Trichilia* and *Alseis*. The recensus data show that negative effects of large trees on conspecific saplings were common in the tree community, and that the effects were quantitatively large in many cases.

Growth and survival in saplings are both negatively affected by being beneath a conspecific adult > 20 cm dbh. In Tables 22.5 and 22.6, "common" species are a set of 11 overstorey species with a minimum of 10 saplings in each of three size classes (1–2, 2–4, and 4–8 cm dbh) beneath a conspecific large tree. "Rare" species are all other species. In saplings 1–4 cm dbh, 3-year survival was 5.8% lower in the common species, and by

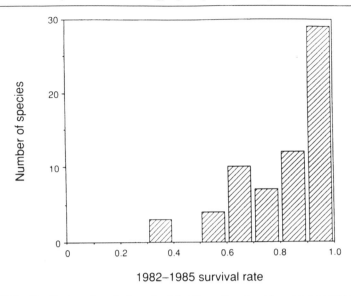

Figure 22.4 Distribution of survival rate for 1.0–1.9 cm diameter at breast height (dbh) saplings over the 1982–1985 interval for 65 Barro Colorado Island overstorey and midstorey tree species with at least 10 saplings in the 1.0–1.9 cm dbh size class

9.6% in the rare species, beneath conspecific versus beneath heterospecific large trees (Table 22.5). Growth rates of 1–4 cm dbh saplings were depressed by 17.1% in the common species, and by 22.3% in the rare species (Table 22.6). The growth rate reduction in the rare species by themselves was not significant, probably due to small sample sizes. Numbers of saplings beneath conspecific and heterospecific trees were very unequal because there are many more heterospecific than conspecific large trees. However, if samples are equalized by a statistical jack-knifing procedure of randomly subsampling heterospecific saplings, the significance of all differences remains unaltered. The analyses reported here are on overstorey trees only, but the same patterns were also found in midstorey tree species. Negative conspecific effects were less prevalent among understorey treelets and shrubs (Hubbell and Foster, 1990c).

These analyses can also be performed on individual common species (Hubbell and Foster, 1990c). There was a clear negative effect on growth among 11 common overstorey species. Although only two species were significant at the $p = 0.05$ level, 1–4 cm dbh saplings showed poorer growth beneath conspecifics than beneath heterospecifics in all but one of the 11 species, a result that was highly significant ($p < 0.005$, sign test). The two species with individually significant reductions in growth were *Tetragastris panamensis* (Burseraceae) and *Poulsenia armata* (Moraceae). In *Alseis*, there were marginally significant reductions in growth ($p < 0.06$); but in *Trichilia*, although growth was depressed, the effect was not significant ($p < 0.27$). The results on survival in common species were more mixed. Only two species, *Trichilia* and *Alseis*, showed significant reduction in survival of 1–4 cm dbh saplings beneath conspecific large trees – the same two species in which density dependence was detected in the primary census data. There were no obvious trends in the survival of the remaining nine species of the set. Four species showed poorer survival beneath conspecifics, whereas five showed poorer survival beneath heterospecifics. In a few common species, there were no clear conspecific effects. For example, in *Quararaibea asterolepis* (Bombacaceae), differences in both growth and survival were non-significant beneath conspecifics and heterospecifics, and the trends were in opposite directions for growth and survival. Growth was slightly better beneath heterospecifics, whereas survival was slightly better beneath conspecifics.

It is important to show that these results are truly conspecific effects and not due to correlated variables (Hubbell and Foster, 1990c). For example, if conspecific large trees happen on average to be larger or closer to the focal saplings than are heterospecific large trees, then a "large-tree" or "close-tree" effect could be mistaken for a "conspecific" effect. Alternative hypotheses for why saplings might survive and grow less well beneath large conspecific trees are:

(1)   Conspecific trees are larger;

(2)   Conspecific trees are closer to the focal saplings;

(3)   Saplings beneath conspecifics are in shadier sites; and,

(4)   Saplings beneath conspecifics are in areas of higher plant density.

Table 22.5   Proportion of 1.0–3.9 cm dbh saplings of BCI overstorey tree species that survived beneath a conspecific or a heterospecific large tree (> 20 cm dbh), over the 1982–1985 census interval

| | Survival rate beneath: | | | 2-sided |
Data set	Conspecific tree		Heterospecific tree	p-value
All species	0.857 (*1040*)	<	0.905 (*32254*)	0.0001
Common species	0.861 (*985*)	<	0.919 (*21212*)	0.0001
Rare species	0.782 (*55*)	<	0.878 (*11042*)	0.037

The "common" species for this analysis were a set of 11 overstorey species in which at least 10 saplings were present in the conspecific category. "Rare" species were all other canopy species. Sample sizes are in parenthesis. A two-sided test was applied because the results could have gone either way

Table 22.6 Mean annual growth rate (cm dbh/yr) of 1.0–3.9 cm dbh saplings of BCI overstorey tree species that were beneath a conspecific or a heterospecific large tree (> 20 cm dbh), over the 1982–1985 census interval

| | Growth rate beneath | | | Two-sided |
Data set	Conspecific tree		Heterospecific tree	p-value
All species	0.065 (*839*)	<	0.085 (*27834*)	0.0003
Common species	0.064 (*798*)	<	0.076 (*18636*)	0.0327
Rare species	0.080 (*41*)	<	0.103 (*9198*)	0.3608

The "common" species for this analysis were a set of 11 overstorey species in which at least 10 saplings were present in the conspecific category. "Rare" species were all other canopy species. Sample sizes are given in parenthesis. A two-sided test was applied because the results could have gone either way

All of these alternative hypotheses could be definitively rejected. The negative conspecific effect was occasionally strong enough to be significant in spite of opposing tendencies in these alternative variables. For example, in some species, negative conspecific effects were significant even though the conspecific adults were significantly smaller or farther from the test saplings than their heterospecific counterparts (Hubbell and Foster, 1990*c*).

INTERPRETING THE RECENSUS RESULTS

The recensus results have confirmed some inferences from the primary census, and caused a reassessment of others. On the one hand, the data on sapling growth, survival, and recruitment generally supported the conclusions about life-history specialization and generalization among the BCI trees. For example, the inference that many common BCI species were habitat and regeneration niche generalists overall was supported by the sapling survival and growth data. Many of these species showed little variation in growth and survival performance, whether they were in new or old gaps or in the shaded understorey. Species interpreted as specialized for high-light environments or restricted habitats on the basis of the static, first-census data were also found to have been correctly identified.

On the other hand, the recensus data strengthens the case for biotic equilibrating forces in the BCI forest. The spatial data from the first census did not prepare us for the pervasiveness and strength of the negative conspecific effects found in both growth and survival, especially among canopy and midstorey tree species. Tree replacement probabilities calculated from the static data did not pick up these effects (Hubbell and Foster, 1987*b*). Spatial patterns of new recruitment in relation to conspecifics may also show similar effects, but they have not yet been analysed.

How strong are these conspecific effects? For canopy tree species, mean annual sapling growth rates are reduced by 10–20%, and 3-year survival rates are 5–10% less beneath large trees of the same species than beneath large trees of different species. For each of 11 common species, we estimated the likelihood of a sapling surviving the time required to grow from 1 to 4 cm dbh beneath a large tree (> 20 cm dbh) of the same species or beneath another species. We took the size-specific growth and survival data for 1–4 cm dbh saplings, and projected the number of survivors expected to remain out of an initial cohort of 1000 saplings starting at 1 cm dbh beneath conspecific and heterospecific large trees. Because saplings usually tended to grow more slowly beneath conspecific large trees, they would be expected to spend a longer time at risk to mortality in the immature stages than saplings beneath heterospecific trees. Thus, even if survival per unit time is no lower in saplings beneath conspecifics, there will be fewer survivors to maturity from the initial cohort beneath conspecifics than from the cohort beneath heterospecifics. If survival and growth are both lower beneath conspecifics, then the difference in number of survivors would be expected to be even greater.

We calculated the ratio of expected numbers of survivors over the 1 to 4 cm dbh growth interval out of cohorts beneath heterospecific and conspecific large trees (Hubbell and Foster, 1990c). For the 11 common species that were analysed individually, these ratios varied from about unity for *Quararaibea*, which showed no conspecific effect, to about nine for *Trichilia*, and about 65 for *Alseis*. Since only a small fraction of the total juvenile growth period is included in this analysis, we can conclude that a 1 cm dbh *Trichilia* sapling has less than one 9th the chance, and an *Alseis* sapling less than one 65th the chance of surviving to maturity beneath a large tree of the same species rather than beneath some other species.

What is the mechanistic basis of these conspecific effects? In our case, it is obviously not seed or seedling predation because we examined the fate of only saplings over 1 cm dbh. It could be species-specific competitive effects between saplings and conspecific adult trees. Possibly the adult alters local soil nutrient ratios away from optimal values for nearby conspecific saplings. It could be pathogens (e.g. fungi attacking cambial tissue) that cause chronic, non-lethal infections in the adult, but that are sublethal or lethal in saplings. Perhaps Connell's (1971) hypothesis is correct, namely that large trees attract host-specific herbivores which then discover, defoliate, and reduce the growth and survival rates of nearby seedlings and saplings. Whatever the case, the effects also depend on the size of the neighbouring large tree (Hubbell and Foster, 1990c). Significant negative conspecific effects are much less common if the lower size cut-off for large trees is reduced from 20 to 10 cm dbh.

Wong *et al.* (1989) had an opportunity to test Connell's hypothesis. In 1985, an outbreak of a lepidopteran occurred on *Quararibea* on BCI. Some trees in the plot were severely defoliated, but others escaped with very light damage. The dominant variable predicting damage level was the phenological state of the plant. If the tree was flushing new leaves, it was more likely to be heavily defoliated; whereas, if it had mature leaves, it was lightly defoliated. Saplings in the vicinity of heavily and lightly defoliated adults were assessed for damage as a function of distance from the adult tree. There was no correlation between damage level to the adult and damage to nearby saplings. However, if plants with mature leaves were excluded from the sample, then there was a proximity effect. Saplings flushing new leaves were more likely to be defoliated if they were near an adult that was also flushing new leaves at the same time.

Quararibea was a species that showed no negative conspecific effects of adults on sapling growth or survival. The results of Wong *et al.* (1990) suggest the following hypothesis: Tree species that have asynchronous leaf flush are less likely to show negative conspecific effects due to insect herbivory. Tropical herbivorous insects probably commonly time their emergence to coincide with new leaf production, a well-known phenomenon in the temperate zone. It is known that new leaves are preferred to mature leaves by some tropical herbivorous insects (e.g. leaf-cutting ants, Hubbell *et al.*, 1984). If this hypothesis is correct, then we should expect species like *Trichilia* which exhibit strong negative conspecific effects to have more synchronous leaf flush than *Quararibea*.

CONCLUSIONS AFTER THE FIRST RECENSUS

The recensus results suggest that negative conspecific effects could be important in the maintenance of tree species richness in tropical forests, but they fall short of proof. At least among canopy and midstorey tree species, intraspecific, density-dependent effects particularly on sapling growth appear to be much larger and more widespread in the tree community than the primary census information suggested. Although we can only guess at present at the true contribution of these effects to coexistence of BCI tree species, there is no doubt that the direction of these forces is toward community equilibrium, not disequilibrium. Similar results have been reported from other rain forest tree communities (e.g. Connell *et al.*, 1984). Recent observational and experimental work on the fate of seeds and seedlings of BCI trees also supports the equilibrium view. With few exceptions, the usual outcome of these experiments is that it is worse for seeds and seedlings to be beneath a conspecific than a heterospecific tree (e.g. Augspurger, 1983*a*, 1983*b*; Howe *et al.*, 1985; Hamill 1986). For some species it is better for seeds to be dispersed into a light gap (Augspurger, 1983*a*; Hamill, 1986); but for other species, it is worse (Schupp, 1987). However, the situation is probably more complicated in some species. For example, in the understorey tree, *Faramea occidentalis* (Rubiaceae), the second most abundant species in the plot (25 118 individuals in 1985), seed survival is actually higher in areas of high adult density, suggesting predator satiation (Schupp, 1987).

Taken altogether, the evidence for equilibrating forces might seem very strong, but caution is needed. We cannot yet evaluate their quantitative importance in stabilizing the species assemblage. Moreover, we have just described how one recensus can alter the apparent conclusions, so we must not be too surprised if changes occur again after additional recensuses are taken. After all, so far we have recorded only a tiny fraction of the total lifespan of most of the plants in the plot. There is real reason for caution because the evidence for non-equilibrium – directional successional change in the forest, and El Niño effects – is also very strong (Hubbell and Foster, 1990*a*, 1990*b*). If there is one lesson from this study, it is that long-term studies are absolutely essential for understanding the dynamics of communities of long-lived organisms.

ACKNOWLEDGMENTS

We thank the National Science Foundation, the Scholarly Studies Program of the Smithsonian Institution, the World Wildlife Fund, the Center for Field Studies, the Geraldine R. Dodge Foundation, and many other private donors who have supported this research. We also thank many students from around the world who have donated time and effort to assist with the field work. Special thanks are due to Dr. Jeff Klahn and to Susan Williams, who served as field project co-ordinators and managed the field crews through the recensus effort. We also thank Brit Minor and Dr. Steven Hewett, who managed the data bases and provided many of the key data summaries and analyses. Finally, we thank the Smithsonian Tropical Research Institute for considerable moral, financial, and logistical support.

REFERENCES

Ashton, P.S. (1969). Speciation among tropical forest trees: some deductions in light of recent evidence. *Biological Journal of the Linnean Society*, **1**, 155–96
Augspurger, C.K. (1983*a*). Offspring recruitment around tropical trees: changes in cohort distance with time. *Oikos*, **20**, 189–96
Augspurger, C. (1983*b*). Seed dispersal of the tropical tree, *Platypodium elegans*, and the escape of its seedlings from fungal pathogens. *Journal of Ecology*, **71**, 759–71
Chesson, R. and Warner, R.R. (1981). Environmental variability promotes coexistence in lottery

competitive systems. *American Naturalist*, **117**, 923–43

Connell, J.H. (1971). On the role of natural enemies in preventing competitive exclusion in some marine animals and rain forest trees. In den Boer, P.J. and Gradwell, G.R. (eds), *Dynamics of Populations*, pp. 298–312 (Wageningen: Centre for Agricultural Publishing and Documentation)

Connell, J.H. (1978). Diversity in tropical rain forests and coral reefs. *Science*, **199**, 1302–10

Connell, J.H. (1980). Diversity and the coevolution of competitors, or the ghost of competition past. *Oikos*, **35**, 131–8

Connell, J.H., Tracey, J.G. and Webb, L.J. (1984). Compensatory recruitment, growth, and mortality as factors maintaining rain forest tree diversity. *Ecological Monographs*, **54**, 141–64

Croat, T. (1978). *The Flora of Barro Colorado Island*. (Palo Alto, California: Stanford University Press)

Hamill, D.N. (1986). Selected studies on dispersion, distibution, and species diversity. Ph.D. Thesis. University of Iowa, Iowa City.

Hoffmaster, D. and Hubbell, S.P. (1990). Increased species richness slows the coevolutionary divergence of competitors, manuscript

Howe, H.F., Schupp, E.W. and Westley, L.C. (1985). Early consequences of seed dispersal for a neotropical tree (*Virola surinamensis*). *Ecology*, **66**, 781–91

Hubbell, S.P. (1979). Tree dispersion, abundance, and diversity in a tropical dry forest. *Science*, **203**, 1299–1309

Hubbell, S.P. (1980). Seed predation and the coexistence of tree species in tropical forests. *Oikos*, **35**, 214–29

Hubbell, S.P. (1984). Methodologies for the study of the origin and maintenance of tree diversity in tropical rain forest. In Maury-Lechon, G., Hadley, M. and Younes, T. (eds), *The Significance of Species Diversity in Tropical Forest Ecosystems*, pp. 8–13. Biology International Special Issue 6. (Paris: IUBS)

Hubbell, S.P. and Foster, R.B. (1983). Diversity of canopy trees in a neotropical forest and implications for the conservation of tropical trees. In Sutton, S.J., Whitmore, T.C. and Chadwick, A.C. (eds), *Tropical Rain Forest: Ecology and Management*, pp. 25–41. (Oxford: Blackwell Scientific Publications)

Hubbell, S.P. and Foster, R.B. (1986a). Canopy gaps and the dynamics of a neotropical forest. In Crawley, M. (ed.) *Plant Ecology*, pp. 77–95. (Oxford: Blackwell Scientific Publications)

Hubbell, S.P. and Foster, R.B. (1986b). Biology, chance and history and the structure of tropical rain forest tree communities. In Diamond, J. and Case, T.J. (eds), *Community Ecology*, pp. 314–29. (New York: Harper and Row)

Hubbell, S.P. and Foster, R.B. (1986c). Commonness and rarity in a neotropical forest: implications for tropical tree conservation. In Soulé, M. (ed.). *Conservation Biology: Science of Scarcity and Diversity*, pp. 205–31. (Sunderland, Massachusetts: Sinauer Associates)

Hubbell, S.P. and Foster, R.B. (1987a). La estructura en gran escala de un bosque neotropical. *Revista de Biologia Tropical*, **35**, (Suppl. 1) 7–22

Hubbell, S.P. and Foster, R.B. (1987b). The spatial context of regeneration in a neotropical forest. In Crawley, M., Gray, A. and Edwards, P.J. (eds), *Colonization, Succession, and Stability*, pp. 395–412. (Oxford: Blackwell Scientific Publications)

Hubbell, S.P. and Foster, R.B. (1990a). Short-term dynamics of a neotropical forest: El Niño effects and successional change. *Ecology*, (in press)

Hubbell, S.P. and Foster, R.B. (1990b). Structure, dynamics, and equilibrium status of old-growth forest on Barro Colorado Island. In Gentry, A. (ed.) *Four Neotropical Forests, pp. 520–41*. (New Haven: Yale University Press)

Hubbell, S.P. and Foster, R.B. (1990c). Short-term dynamics of a neotropical forest: new evidence for equilibrating forces, manuscript

Hubbell, S.P., Howard, J.J. and Wiemer, D.F. (1984). Chemical leaf repellency to an attine ant; seasonal distribution among potential host plant species. *Ecology*, **65**, 1067–76

Janzen, D.H. (1970). Herbivores and the number of trees in tropical forest. *American Naturalist*, **104**, 501–28

Leigh, E.G., Jr., Rand, A.S. and Windsor, D.M. (eds), (1982). *The Ecology of a Tropical Forest: Seasonal Rhythms and Long-Term Changes*. (Washington, D.C.: Smithsonian Institution Press)

MacArthur, R.H. and Wilson, E.O. (1967). *The Theory of Island Biogeography*. (Princeton, New Jersey: Princeton University Press)

Section 7

Reproductive biology and
tree improvement programmes

CHAPTER 23

REPRODUCTIVE BIOLOGY AND TREE IMPROVEMENT PROGRAMMES – COMMENTARY

Peter S. Ashton and Kamaljit S. Bawa

Any discussion of tree improvement must take account of future demand for tree products. This is no easy task, for the working life of a rubber or oil palm tree can extend to 30 years, and tree improvement programmes such as the celebrated one for rubber can continue to increase tree yield for at least double that time, even when conducted, as with *Hevea* rubber trees, on a single provenance. Tree biologists who do turn to resource economists for advice are rarely satisfied. Alan Grainger's (not dated) predictions of tropical hardwood supply and demand well into the next century provide a rare exception. His predictions are inevitably regarded as controversial.

All the same, World Health Organization predictions of human demographic trends have little margin for error (barring cataclysm). It is therefore certain that, whereas the population of the industrialized, temperate nations may increase by 15% over the next 30 years, those in the nations of the tropics will double. In other words, the relative proportion of the world's population resident in the former will be halved, while that in the latter will double, to well over half the world's population. This fact has profound implications to forestry and for tree improvement.

We must anticipate, if standards of living world-wide are to be maintained, let alone improved, that the relative demand for tropical products from the traditional consumers in the currently industrialized nations will steadily decline. This means that, unless there is substantial further industrialization, which currently does not seem likely, it is probable that industrial products such as rubber latex will go into over-production. Others, such as vegetable oil, which are produced by temperate as well as tropical plants, are already beginning to experience over-production. Yet the need to create new employment in the tropics, and particularly in the rural sector, is obvious and urgent. Already, natural forest in Asia is almost entirely confined to lands unsuited to permanent herbaceous crops. Within 30 years, this will be the

345

case everywhere except for parts of the Amazon valley and, perhaps, Zaire. The *Hevea* rubber tree remains the only plantation species which can be grown profitably on much of that land in the Old World.

The rapid expansion of human populations in the tropics will itself provide major opportunities for market expansion. Consequent lower transportation costs, plus greater familiarity with regional products among local consumers, provide opportunities for product diversification, and hence for development of new tree crops for marginal lands. This development is doubly fortunate. It not only allows greater opportunities for creating employment through development of a diversified tree-based economy on marginal lands, but it offers much needed opportunities for diversified (hence lower risk) land-use and multiple-species plantations in a climate where pest control in long-lived crops is already expensive and will certainly become more so in the future.

What kind of products will be in demand? Besides industrial products such as rubber latex, it seems likely that the specialized chemical products for which tropical plants are so celebrated, and which are, or have been, used world-wide as pharmaceuticals, flavourings or cosmetics, are at risk of being replaced by manufactured alternatives. Increasingly, pharmaceuticals are manufactured through culture of micro-organisms as well as from fossil fuel by-products. Vanilla can now be synthesized in tissue culture. On the other hand, the relative price of hardwoods has increased more rapidly than those of any other tropical commodity over the last 40 years, and Grainger (not dated) is probably right in assuming that demand, world-wide, will continue to be high. The spectacular growth of the world trade in rattan in the last decade indicates the promise of forest products which form the basis for craft-based industries whose products are in universal demand. Food crops, notably tropical fruit, are assured of a strong home demand and also, with improved marketing and storage technology, international demand.

In many cases, such as rattan and brazil nuts, most or all harvesting is still conducted in natural forest. Increasingly, culture must be intensified, though, as in the case of these two commodities, plantation cultivation will only be possible in mixture with other species.

In summary, therefore, we can fairly reasonably predict the kind of forest product we should be concentrating research on and therefore the species we should select, the attributes which should be enhanced through breeding, and the cultural ground rules which can best ensure high production, low risk and sustainability. Above all, cultivation in the coming decades will increasingly require exploitation of tree diversity in nature.

The four papers in this section are concerned with some selected aspects of reproductive biology that have implications for success in tree improvement programmes and plantation management. Griffin examines the effects of inbreeding on the growth of trees in plantations. Most forest tree species are

346

highly outbred. Inbreeding can have several deleterious consequences, including reduction in growth. As Griffin argues, seeds collected for plantations often have a narrow genetic base. Inbreeding is therefore common in such plantations. Plantations raised from inbred seed may further erode the genetic base of the population, and trees in such plantations may continue to show decline in survival and growth rate. Griffin calls for greater care in the selection of seed during domestication of tree crops.

Brune also touches briefly upon the consequences of casual collection of seed for plantations. He cites the example of *Eucalyptus deglupta* where seeds were collected from plants that tended to flower in the first year. Plants grown from such seeds began to flower earlier and showed poor growth because vegetative growth was not vigorous, and the trees branched precociously as the primary meristems turned into the reproductive shoots. Brune also discusses other problems in reproduction, such as altered flowering seasons and lack of fruit set in the absence of pollinators when plants are introduced into new habitats.

Kageyama reviews patterns of genetic variation in natural populations of rain forest trees. As stated earlier by Bawa, Ashton and Salleh in their introductory chapter to this volume, information about genetic architecture of populations is critical to the success of tree breeding programmes. Direct observations of population genetic structure in tropical forest trees are scanty, but recent studies by Hamrick and his associates in Panama, Bawa and co-workers in Costa Rica, Ashton and his colleagues in South-east Asia, and Kageyama in Brazil, have shown that most species contain considerable variation within populations. Less is known about the extent of genetic differentiation between populations; limited data in a few species show little divergence among widely separated populations (Buckley *et al.*, 1988; Bawa and Krugman, 1990). Kageyama emphasizes that patterns of genetic differentiation are likely to be influenced by modes of pollen and seed dispersal as well as by successional status of the species, as is the case for other angiosperms (Loveless and Hamrick, 1986). Because the various tropical rain forest tree species differ with respect to their pollen and seed dispersal features, as well as their successional status, we might expect a wide variety of population genetic structures to exist.

Longman, Manurung and Leakey describe how flowering can be induced at a convenient height in clonal and grafted forest trees. This is an important area of research because flowers of most canopy species are not easily accessible for controlled breeding work. Methods to induce flowering also need to be developed for those species that flower at very long intervals (e.g. *Shorea*) and for many other dipterocarps in the prehumid regions of the far eastern tropics.

In conclusion, research in many areas of reproductive biology is needed for better management of tree breeding programmes and plantations. However, in the context of future likely market needs, almost no appropriate research has been done. We must identify more precisely what kind of tree commodity is likely to be in increasing demand. If trees are to continue to be grown in the tropics for production of goods other than timber, it is certain that new species will be introduced into cultivation and that some of these will merit genetic improvement. Thus, interdisciplinary collaboration is crucial between tree biologists, including silviculturalists, tree breeders and geneticists, tree nutritionists and pathologists, and economists, market analysts, planners and policy-makers. Upon their recommendations, exploration and introduction must be intensified dramatically. Tree improvement must go hand in hand with new approaches in plantation design with the aim of reducing maintenance costs and risk without detriment to production. Both must be advanced hand in hand with better co-ordinated, and more consistently, if not better, funded research into the reproductive biology, demography and population genetics of tropical trees growing in nature; and into the structure and predictability of natural mixed species plant communities.

REFERENCES

Bawa, K.S. and Krugman, S.L. (1990). Reproductive biology and genetics of tropical trees in relation to conservation and management. In Gómez-Pompa, A., Whitmore, T.C. and Hadley, M. (eds.) *Rain Forest Regeneration and Management*, Man and the Biosphere Series vol. 6, pp. 119–136. (Paris: Unesco and Carnforth: Parthenon Publishing)

Buckley, D.P., O'Malley, D.M., Apsit, V., Prance, G.T. and Bawa, K.S. (1988). Genetics of Brazil nut (*Bertholletia excelsa* Humb. and Bonpl.; Lecythidaceae). 1. Genetic variation in natural populations. *Theoretical and Applied Genetics*, **76**, 923–8

Grainger, A. (not dated). Tropform: A model of future tropical hardwood supplies. Resources for the Future, Washington, D.C.

Loveless, M.D. and Hamrick, J.L. (1984). Ecological determinants of genetic structure in plant populations. *Annual Review of Ecology and Systematics*, **15**, 65–95

CHAPTER 24

REPRODUCTIVE BIOLOGY AND TROPICAL PLANTATION FORESTRY

Arno Brune

ABSTRACT

Some examples are given of problems encountered in the reproductive behaviour of tropical trees planted on a large scale in the tropics, specifically the unintentional selection for early flowering in Eucalyptus deglupta; *problems with seed set of* Pinus caribaea var. hondurensis *in the Amazonian region; interaction between* Bertholletia excelsa *and its pollinator; and extended flowering of several eucalypts.*

INTRODUCTION

The methods of reproduction of forest trees in the tropics are essentially the same as those in non-tropical areas. The differences are largely in the greater diversity of pollination mechanisms, breeding systems and seed dispersal features of tropical trees. In the tropics, species diversity is greater than in the temperate zone, and entomophilous species predominate over anemophilous ones. The greater species diversity leads to a more complex interdependence between plants and animals, rendering plantations (commonly pure plantations of one species) vulnerable to natural influences. Problems in plantation forestry also arise when the phenology is altered following the movement of plant material from one region to another. Thus, the reproductive processes can be disrupted in plantations in the tropics, as illustrated by examples drawn from the author's field experience in Brazil.

349

PROBLEMS IN REPRODUCTION

Early flowering

Precocious flowering often occurs in species transferred to tropical environments, or from one tropical environment to another. Man's interference can further hasten flowering at an early age, as happened with *Eucalyptus deglupta* planted at Monte Dourado, formerly known as the Jari project (see, for example, Palmer, 1990). A previous search of suitable plantation species indicated that the tropical *Eucalyptus deglupta* from the Philippines and Papua New Guinea could be used for pulp production in the area. *Eucalyptus deglupta* tends to flower early, often at an age of 2 years in its native habitat. Seed is small, with little reserve material and loses viability fast. Imported seed led to few seedlings the 1st year. Some seedlings flowered after 1 year, and many in the 2nd year. In order to avoid importing more seed of low viability, seeds originating from the first flowerings were used extensively. Germination of these seeds was very good, and new plantations grew fast. Flowering then started in the whole population at the age of 1 year. Flowers developed on the whole crown of the tree, especially on the top leader. Thus, just about every vegetative bud turned into a generative one. A large portion of energy must be diverted from vegetative growth into reproduction at the expense of the former. Although the amount of energy has not been determined in this case, energy certainly is wasted from the silvicultural view-point. In addition to losing energy, the form of the tree is severely affected, as a new side branch has to take over as leader when height growth resumes. This replacement results in a tree which is not straight.

In this case, the forester inadvertently selected for early flowering by choosing to collect seed from trees which flowered at an early age. Since late-flowering trees had not yet reached reproductive age, selection was very effective: only precocious trees were able to cross with each other, and time of onset of flowering is commonly a highly heritable character. Continuous flowering can be seen as an advantage by those who want honey production, which was not a primary factor in this case.

Seed set

Reproduction is sometimes very poor, resulting in a low or irregular seed set. Although precocious *Eucalyptus deglupta* at Monte Dourado actually produces viable and abundant seed, the same is not true of *Pinus caribaea* var. *hondurensis* planted in that same region. In the case of *Eucalyptus deglupta*, the transfer was from one equatorial region to another; in the case of the tropical pine *Pinus caribaea* var. *hondurensis*,the species originates from Central America around latitude 15°N, where summer rainfall prevails, and it was transferred to Monte Dourado, around 1°S, where rainfall is regularly distributed for most of the year. This pine has considerable growth increments in the sandy soils of parts of the area, making it an excellent candidate for certain kinds of pulp production. It was thought, therefore, that a selection programme would result in still greater increments.

Pinus caribaea var. *hondurensis* is commonly quite variable in its phenotype, including many individuals exhibiting the indefinite growth form known as "fox tail" and considered by many foresters as undesirable due to its wind susceptibility. Selection was easily accomplished in the field. It was also easy to graft trees to start a seed orchard for production of superior seed. However, the trees hardly flowered, and when they did, scarcely ever set sound seed. Several factors may be responsible, solely or in combination, for the absence of seed formation. It may be argued that trees grow continuously and do not "pause" to develop sexual buds. Thus, the absence of a resting season (dry season or winter, however mild) may be the reason for the continuous vegetative development. One could similarly blame the absence of photoperiodic variation around the equator during the year. In addition, it could be that such pollen as is actually produced is not shed properly because of constant high air humidity, or is carried to the ground by the rains, instead of being blown to the pistillate strobili by dry winds, as happens in areas where pines are native. Non-synchronous ripening of male and female strobili may also be involved.

At any rate, no seed of any importance is produced in *Pinus caribaea* var. *hondurensis* in that area. The problem was solved easily by selecting trees in the area, taking scion material to a "cerrado" area in the state of Minas Gerais, where that Caribbean pine does not grow nearly as fast, but flowers and sets seed in abundance. Because of the continental size of Brazil, it was easy to transfer branches 2000 km to the south, simply in order to "entice" them into seed production. The seed, of course, is flown back to the area for which it was selected, in the Amazonian region.

Pollinator–tree interactions

Interesting pollinator–tree relationships are revealed by man's interference. An example is provided by the large Amazonian tree *Bertholletia excelsa* (Lecythidaceae). This tree has yellow flowers, in panicles, with a distinctive shape, an orchid-like labellum, covered with thick short bristles. Pollination is effected by Amazonian Apidae, most probably Anthophorinae, Bombinae or Euglossinae, locally called "mamangabas". This tree produces large pixidia, in which Brazil-nuts are carried. The operculum is small, and most often does not open by itself, so seeds have to be taken out by rodents (agouti), which bury some seeds and eat others. Buried seed may develop into new trees.

Brazil-nuts are a natural crop, and collection is carried out by migrant workers, often also in charge of wild rubber collection. Only in recent years have plantations been attempted. Often the land is cleared for pasture or other extensive plantations. In this case, Brazil-nut trees are left standing, because of their value as nut producers, and due to legal protection. The burning of the land around the standing trees eliminates both the Apidae needed for pollination as well as their native habitat. For this reason, it soon became apparent that isolated trees, far from native vegetation, did not bear fruit.

Interest in the Brazil-nut as an important export product has risen in recent years. Nowadays, grafted orchards are established, but 500 m wide bands of native vegetation are left between strips of planted trees (Adalgisa, 1985). This practice, in part, resolved the problems of pollination. To supplement forage for the pollinators, plantings of *Bixa orellana* and *Passiflora* spp. are established, because these species flower throughout the year, and are also visited by the same Apidae (Adalgisa, 1985). Most probably, the pollinators could be further helped in their reproduction if pieces of bamboo, pierced on one side, were hung around the area. This method is practised in north-eastern Brazil, where *Passiflora* plantations are also pollinated by "mamangabas", which nest in the hollow bamboos. This example underlines the fragile relationship between plants and insects, interference with one endangering the existence of the others.

Extended flowering

Timing of flowering is also crucial for reproduction. Eucalypts native to Australia, after having been transferred to Brazil, changed their reproductive behaviour. Those trees which had 1 or 2 months of flowering period in Australia, started to flower at younger ages in Brazil, and did so for quite an extended period. In trials established early this century by Edmundo Navarro de Andrade in Rio Claro, plots of different species flowered profusely for long periods, allowing widespread hybridization to occur. It was over half a century later that it became clear what was happening, and that plantations were isolated when hybridization had to be avoided. New species were also tried, leading to a renewed interest in eucalypt species for plantations. Seed orchards have to be especially isolated from other hybridizing species, as do seed production areas and seed collection stands. Peaks of flowering differ between clones and this fact has also to be taken into account in creating seed orchards, where flower synchronization is important to assure pollen flow between the predominantly allogamous trees.

It is interesting in this case that the same species which may occur in Australia sympatrically, and do not cross due to temporal separation of flowering times, do so freely in Brazil. Which climatic or edaphic conditions lead to extended flowering? There are eucalypt species like *E. citriodora*, which carries seed crops only every 3–5 years in Australia, but seeds abundantly in Brazil year after year for most of the year, often with mature capsules and buds occurring together on the same tree. The geographic latitude of plantations in Brazil largely coincides with the latitude of natural populations in Australia: in fact, a matching of latitudes is invariably attempted in choosing provenances. Although it may affect growth, advantage is now commonly taken of the extended flowering period by placing beehives for honey production in plantations and seed orchards, *Eucalyptus citriodora* honey being much appreciated in certain areas.

CONCLUSIONS

Plantation forestry is a technique which, due to the ease of management, is commonly preferred over methods which tend to conserve large portions of species diversity. Generally, pure species forests only occur naturally in extreme situations of the world, as in very cold, dry, or salty areas. The warmer and the more humid the climate, the larger the forest species diversity. This diversity may be changed to a uniform forest, with the loss of biological diversity. However, the alternatives to the maintenance of biological diversity in the tropics have not yet received much attention. When a native forest species like *Bertholletia excelsa* is planted, the need to keep at least some of the other organisms which are part of its ecological niche becomes evident. Experience with *Cedrella* and *Hevea* support this notion.

The solution of planting exotics is surely a temporary one, since eventually pests will adapt to pure plantations of these species as well. At the present time, it is probably wisest to conserve as much of the natural forest as we can in the tropics, until suitable non-destructive management techniques have been mastered and perceived as being economically and socially desirable. The adoption of such techniques may still be relatively far off in time, at least over extensive areas in most tropical regions. In the meantime, a combination of parts of an area being used in pure plantation forests, and the rest left as it is, or very carefully exploited, could provide a temporary solution.

An alternative has to be found to prevent large-scale destruction of the forest, be it for pasture, farming or pure species plantation. At the same time, large-scale interests have to be accommodated, as well as the social pressures resulting from a growing, impoverished rural population. Shifting agriculture is a solution for a sparsely distributed population over a large area. Agroforestry systems on a permanent basis may be a better solution for small farmers, but not for large enterprises, at least not yet. The biological challenge is greatest in humid equatorial forests, but cannot be examined without taking into account local social and even political situations. It is a task which foresters, biologists and agronomists can no longer face alone.

REFERENCES

Adalgisa, F. (1985). Castanheira anã, um gigante da Amazônia. *Globo Rural*, **1(3)**, 36 41
Palmer, J. (1990). Jari: lessons for land managers in the tropics. In Gómez-Pompa, A.,, Whitmore, T.C. and Hadley, M. (eds.) *Rain Forest Regeneration and Management*, Man and the Biosphere Series, Vol. 6, pp. 419 429. (Paris: Unesco and Carnforth: Parthenon Publishing)

CHAPTER 25

EFFECTS OF INBREEDING ON GROWTH OF FOREST TREES AND IMPLICATIONS FOR MANAGEMENT OF SEED SUPPLIES FOR PLANTATION PROGRAMMES

A. Rod Griffin

ABSTRACT

Growth rate of tree species is generally reduced following inbreeding. Under natural regeneration inbred plants will tend to be eliminated as the stand self-thins and they will have minimal effect on final crop yield. By contrast, it is desirable that every tree established in plantations has the genetic potential to produce a final crop tree. Experimental results with conifers indicate a linear decline in early height growth of about 5% per 0.1 increase in inbreeding coefficient (F). Effects are similar for several eucalypt and acacia species. Where species introductions are made with a small sample of germ plasm, inbreeding problems are particularly likely to develop. Examples of such introductions are discussed, the best documented of which is the case of Acacia mangium *in Sabah, Malaysia. Inbreeding problems can be minimized by prompt management action at the start of the domestication programme. Problems and opportunities (with respect to maximizing outcrossed seed production) are discussed for natural populations, unpedigreed plantations, and seed stands or orchards.*

INTRODUCTION

As natural forest resources decline throughout the world, there is an increasing need to utilize planted trees – for industrial wood production, soil conservation, fuelwood and agroforestry. Evans (1982) estimated that about 11 million ha of trees were planted in the tropics in the 15 years to 1980, with a projected doubling of annual establishment rates in the succeeding 5 years. Approximately one third of this area was planted with eucalypts, one

355

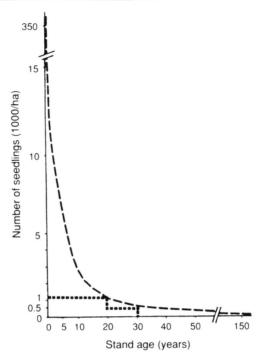

Figure 25.1 Mortality curve for *Eucalyptus regnans* following natural regeneration (− −) (after Cunningham, 1960), compared with stocking rate (....) for a pulpwood plantation of the same species in S.E. Australia (APM Forests, pers. comm.)

third with pines and the rest mostly with other hardwoods, the foremost of which was teak. One consequence of this move to increasing domestication of trees is that the phenomenon of inbreeding depression – reduced vigour and viability of plants derived by selfing or mating of related individuals – is no longer of interest solely to geneticists and population biologists. If the plantation manager lacks understanding of the consequences of using inbred seed, he risks substantial losses of productivity.

In a plantation, it is highly desirable that every tree established has the potential to survive and contribute to the harvested crop. In contrast, self-thinning is a characteristic of successful natural regeneration, with only a small fraction of the trees developing to harvestable size. As an example, the natural mortality curve for *Eucalyptus regnans* F. Muell., an important timber species from south-eastern Australia which regenerates from seed after fire, is shown in Figure 25.1. This mortality is contrasted with the stand density of a plantation of the same species established at 1100 stems ha⁻¹, subject to one commercial thinning at age 20 years and clear-felled at age 30 years. Although the time scale may vary, such a management regime would be realistic for many industrial plantations of forest trees in the tropics.

In the natural stand of *E. regnans*, all but 1 in 300 of the original germinants will have become suppressed and will have died by the age of 20 years, even if every one of the other 299 had had the genetic potential to produce a final crop tree. Productivity of the stand will therefore be unaffected even if a large proportion of germinants are actually of poor genetic quality. In contrast, the ideal situation for the manager of the plantation in Figure 25.1 is that each one of the trees established produces a usable stem by the time of thinning at age 20 years. Admittedly biomass production per hectare is, within limits, independent of stocking rate, but mortality or presence of moribund individuals leads to undesirable variation in tree size. At lower initial stocking rates, which might for example be employed in agroforestry, returns will be more closely proportional to the yield of individual trees, and it is even more vital that every one is harvestable.

Achievement of optimal yield from planted trees requires both good management practice and that each plant has the genetic potential to be productive. Thus, attention must be paid not only to the choice of species, provenance and trees within provenance as sources of seed (see e.g. Burley and Wood, 1976), but also to the need for appropriate "packaging" of the genes in the planting stock. Even a highly desirable genotype may produce poor quality progeny if these are derived by inbreeding. The latter point is well recognized by those prescribing tree introduction and improvement strategies, but there is a strong circumstantial (and in a few cases documentary) evidence that much of the seed actually used for plantation establishment in the tropics is inbred to an unacceptable extent.

The aims of this contribution are: first, to illustrate the effects of inbreeding, with particular reference to experiments with pines (Pinaceae), eucalypts (Myrtaceae) and acacias (Mimosoideae); second, to draw attention to the extent to which inbreeding presents a real or potential problem to managers of planted tree crops (particularly in the early stages of domestication of species as exotics); and, third, to suggest some ways of minimizing these effects.

THE PHENOMENON OF INBREEDING

In order to appreciate how seed production management practices may have adverse effects on genetic quality of planting stock, it is necessary to review some basics of quantitative genetics (see e.g. Falconer, 1970; Simmonds, 1979). In a large randomly mating population, the probability that two alleles carried by an individual are identical by descent approaches zero and the population is fully outbred (inbreeding coefficient F=0). Under several conditions, F will depart from 0:

(1)   If the population is small so that there is a likelihood of mating between relatives. The actual population of interbreeding individuals is of concern here. This is frequently less than the census number of trees in a stand because it excludes those which do not flower at all, or are spatially isolated;

(2)   If there is assortative mating between like phenotypes. It is clear, for example, that trees which flower at identical times will have a greater probability of mating than ones which are not synchronized. Any preference of animal pollinators for particular morphological or chemical phenotypes would also encourage assortative mating; and,

(3)   If there is any tendency to assortative mating between relatives, that is sib-mating, or, in the extreme case, selfing. In natural populations, seed and pollen dispersal characteristics tend to increase the probability that adjacent individuals will be related. Because of their proximity, such trees are also more likely to interbreed than those which are spatially isolated. Assortative mating is further encouraged to the extent that flowering time is under genetic control and relatives thus tend to flower in synchrony.

All of these conditions affect the rate of increase in inbreeding (ΔF) from one generation to the next, through their impact on effective population size (N_e). N_e is defined as the number of individuals which would give rise to the rate of inbreeding under consideration if they bred at random (Falconer, 1970). In general, $\Delta F = 1/2N_e$.

If we assume that open-pollinated seed from a tree is fathered by a number of different individuals, then we are applying different sampling intensities to the populations of male and female gametes during seed collection. The two sexes, whatever their number, contribute equally to the genes in the next generation. N_e is twice the harmonic mean of the numbers of the two sexes and $\Delta F = (1/8N_m) + (1/8 N_f)$. The rate of inbreeding is thus largely dependent upon the numbers of the less numerous sex – a point of great importance to development of seed procurement strategy.

Whatever the cause of inbreeding, the result is an increase in homozygosity and decrease in heterozygosity, with an observed phenotypic effect (at least in predominantly outbreeding organisms) of the reduced vigour, size and fertility, known as inbreeding depression. The mode of gene action responsible for inbreeding depression or its converse, heterosis, is still a matter for debate – the major issue being whether heterozygosity *per se* is beneficial (overdominance argument) or whether the more homozygous inbreds are merely expressing a higher proportion of deleterious recessive alleles (dominance argument). While this distinction may influence choice of advanced generation breeding strategy, it is of no great importance to the present discussion.

BREEDING SYSTEMS OF FOREST TREES

The breeding system, and in particular the facility to produce viable seed after self-fertilization, is an important determinant of the potential rate of increase of F over successive generations. Botanists and plant breeders have long distinguished between cross-pollinated and self-pollinated species, but it is increasingly recognized that either extreme is rare when one looks at the population as opposed to individual genotype level (Schemske and Lande, 1985; Stern and Roche, 1974). Studies using isoenzyme techniques commonly estimate outcrossing rates (t) around 0.9 for conifers (Schemske and Lande, 1985). Fewer estimates are available for hardwoods, but evidence is accumulating that they are also predominantly outbreeders. Thus, t values for 10 subtropical and temperate eucalypt species ranged from 0.69–0.86 and averaged 0.78 (Moran and Bell, 1983).

Such outcrossing estimates derived from natural populations are not necessarily good indicators of the breeding system under cultivation. For example, *Eucalyptus regnans* shows the preferential outcrossing typical of the genus when pollinated with mixed self and outcross pollen (Griffin *et al.*, 1987), but all trees tested have proven self-fertile to some extent after controlled self-pollination (Eldridge and Griffin, 1983), as did 44 of 45 trees of *Eucalyptus grandis* Hill ex Maiden tested by Hodgson (1976a). Temporally or spatially isolated individuals of species such as these could be expected to produce quantities of open-pollinated selfed seed. Bawa (1976) cites a number of cases of tropical tree species which are similarly, at least partially, self-fertile.

There have been relatively few studies of the breeding systems of *Acacia* species, though the Wattle Research Institute in South Africa has produced valuable information on *Acacia mearnsii* De Wild. and *Acacia decurrens* (Wendl.) Willd.. Since breeding commenced during the 1940s, many S₁ and S₂ progeny lines have been produced and evaluated (Moffett and Nixon, 1974), demonstrating that these species also fall into the "mixed but predominantly outcrossing" breeding system category. On the basis of relative fruit set following controlled self and outcross pollination, Kendrick (1986) concluded that five species from temperate Australia (*A. mearnsii*, *A. pycnantha* Benth., *A. retinodes* var. *uncifolia* J.M. Black, *A. terminalis* (Salisb.) McBride and *A. myrtifolia* (Sm.) Willd.) had effective self-incompatibility systems operating in at least some individuals, while sample trees of *A. paradoxa* D.C. and *A. ulicifolia* (Salisb.) Court were self-compatible.

All species discussed above exhibit co-sexuality in the sense that each individual can potentially act as both pollen and seed parent. Dioecious species, characterized by the presence of separate male and female plants, clearly cannot self but they are equally subject to the less severe forms of (neighbourhood) inbreeding which must occur widely in natural populations. In general, seed dispersal characteristics dictate that neighbouring trees are likely to be related, while pollen dispersal also favours mating between such spatially and genetically close individuals. Tropical forest species are particularly likely to have diverse population structures (Namkoong, 1985) since gene flow may be more restricted than in temperate zone forests (Bawa, 1976).

Results from recent studies on *Eucalyptus regnans* permit inference regarding the relative significance of neighbourhood inbreeding. A t value of 0.75 was estimated for a natural population of this species, but a seedling seed orchard in which individuals from 40 different open-pollinated families were planted in random mixture (i.e. there was no neighbourhood structure) gave a t estimate of 0.91 (Moran *et al.*, 1989). Although other differences related to tree size and spatial distribution could also have contributed to the result, it is considered that neighbourhood inbreeding was a major determinant of the lower t estimate in the natural stand.

EXPERIMENTAL DETERMINATION OF INBREEDING DEPRESSION

Inbreeding effects may be expressed as reduced yields of viable seed, poor germination, segregation of deformed or chlorophyll deficient seedlings and varying degrees of reduction in growth (Franklin, 1970). Other production traits may also be affected: for example, rubber yield (Simmonds, 1979) or tan bark yield in *Acacia mearnsii* and *A. decurrens* (Moffett and Nixon, 1974). Many strongly inbred plants will die or be culled out at the nursery stage, but this is less likely at lower levels of inbreeding, where effects are more quantitative but still highly undesirable in terms of plantation productivity. Mild inbreeding may thus represent the greater practical problem for seed production managers.

Most studies of inbreeding depression in tree species have been done on temperate conifers and have compared selfed progeny with outcrossed or open-pollinated controls (for example, see Franklin, 1970). Considering the range of species and test conditions, results are remarkably consistent in their demonstration of deleterious effects on survival and growth.

If the reduction of growth rate with increasing F is linear (an assumption validated for *Pinus radiata* D.Don. (Griffin *et al.*, 1986)), then across a range of species, decline in mean population growth rate per 0.1 increase in

F coefficient is in the order of 5%. That is, half-sib mating (F = 0.125), full-sib mating (F = 0.25) and selfing (F = 0.5) would reduce growth compared to an outcrossed control by about 6.2%, 12.5%, and 25% respectively. This is true even for species such as *Picea omorika* Purkyne which is highly self-fertile and considered to be an exception to the generalization that conifers are strongly outbreeding plants (Geburek, 1986).

To put these figures in perspective, first generation orchards of select, but untested, clones of *Pinus radiata* have yielded around 10% gain in growth rate (Pederick and Eldridge, 1983). All of this breeding effort would be more than negated in the event that seed from the select clones was produced by self-fertilization. Inclusion of sibling clones in the orchard (as could well happen following plus tree selection in unpedigreed plantations) would also result in a substantial reduction in realized gain.

There is limited evidence for inbreeding depression in hardwood species but the effects seem to be similar. The most detailed studies in the genus *Eucalyptus* have been carried out in *E. grandis* in South Africa and *E. regnans* in Australia. These species are from the major subgenera *Symphyomyrtus* and *Monocalyptus* respectively, and it is reasonable to assume that conclusions can be extended to other species of eucalypts which are grown as exotics in the tropics and subtropics. The only published record for the important commercial species *E. camaldulensis* Dehnh. (Mendonza, 1970) showed that seed from isolated trees had a lower germination percentage and more aberrant seedlings than controls – observations consistent with those from the more closely studied species.

A comprehensive series of experiments on *E. grandis* in South Africa has shown that inbreeding depression for height growth of S_1 progeny varied from 8 49% (Hodgson, 1976a). Van Wyk (1981) also found that progeny from full-sib matings of this species grew more slowly than outcrosses. Growth of seedling progeny of *E. regnans* produced by selfing, outcrossing and open pollination trees in the same natural population was studied in Australia (Griffin and Cotterill, 1988). Four years after planting the volume of outcrossed progeny was 37% greater than the selfs and 12% greater than the open-pollinated progeny. The open-pollinated seed was comparable to a normal commercial collection and it is important to emphasize that the 12% greater yield of the outcrosses was obtained simply through repackaging the sample of genes derived from this natural population into a less homozygous state, rather than through application of any selection pressure. Hodgson (1976a) reported a similar decline of 8–13% in yield of open-pollinated *E. grandis* progenies relative to controlled outcrosses from the same set of parents.

The progeny of each population type in the *E. regnans* experiment was located at random within blocks, thus simulating a stand derived from an unpedigreed mixture of inbred and outcrossed plants. The size class distribution of trees of each type, and hence their relative degrees of

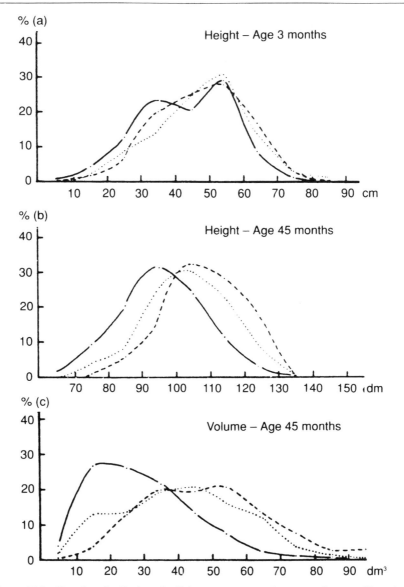

Figure 25.2 Size class distribution of selfed (— —), outcrossed (— — —) and open-pollinated (....) trees in a trial plantation of *Eucalyptus regnans*, at 3 months and 45 months after planting (after Griffin and Cotterill, 1988)

dominance within the stand, are shown in Figure 25.2. Three months after planting, there was little variation in mean height (Figure 25.2a). By 45 months, the selfs were falling behind for height growth (Figure 25.2b) and even more so on relative stem volume (Figure 25.2c), with few representatives in the dominant component of the stand. Forty-five per cent of the outcrosses were larger than the stand median volume tree, compared with 32% of the open-pollinated and only 9% of the selfs. Clearly, many selfs will not survive to produce usable trees. In an earlier experiment assessed to age 12.5 years, Eldridge and Griffin (1983) found that mean survival was 31% for four selfed families, 72% for their open-pollinated sibs, and 96% for the single outcrossed full-sib family evaluated. Representative plots from this experiment are shown in Figure 25.3.

Inbreeding depression for S_1 progeny of *Acacia mearnsii* and *A. decurrens* was very similar (Moffett and Nixon, 1974). By the third growing season, mean height of sets of S_1 progeny were 26% and 15% respectively less than open-pollinated controls. Literature for other hardwood species is sparse. Tan (1981) reported that *Hevea brasiliensis* Muell. (Euphorbiaceae) showed inbreeding depression for vigour following half-sib and particularly full-sib mating. He recommended avoiding such crosses in breeding programmes. Carpenter and Guard (1950) also showed that *Liriodendron tulipifera* L. (Magnoliaceae) conformed to the general pattern. After self-pollination, seed set was reduced relative to outcrossing, with inter-tree variation in apparent self-sterility. Controlled outcrossed progenies also grew faster than open-pollinated stock from the same trees.

EVIDENCE FOR INBREEDING DEPRESSION IN PLANTATIONS OF FOREST TREES

The above examples demonstrate that inbreeding depression is a potential source of reduced yield in plantations. What is the evidence that it *really* is a problem in practice? The general sequence of events involved in the initial domestication of tree species is shown in Figure 25.4. A number of features of this scenario lead to a high expectation of undesirable levels of inbreeding. As a first step, it has been usual to evaluate a range of species in small experimental test plots or even as arboretum entries, using very limited samples of seed. For those species which show promise, plans are made for introduction and testing of a wider range of germplasm. In the meantime, word of the promising new species spreads (well in advance of seed supply from the domestication programme) and the original introduction, which is now reproductively mature, is likely to be used as a seed source for widespread trial plantations and community woodlots.

Even if a genetic improvement programme is actively pursued, it is clear from Figure 25.4 that demand may outstrip improved seed supply for many years. There will thus be money to be made from sale of the seed readily available from the plantations. As international development assistance

Figure 25.3 Row plots containing selfed (right) and outcrossed (left) progeny from a tree of *Eucalyptus regnans*. Age 9 years (from Eldridge and Griffin, 1983)

networks become more active in disseminating information on species performance, the market for seed of tropical species will expand particularly rapidly. The net result of these activities is that control of the genetic quality

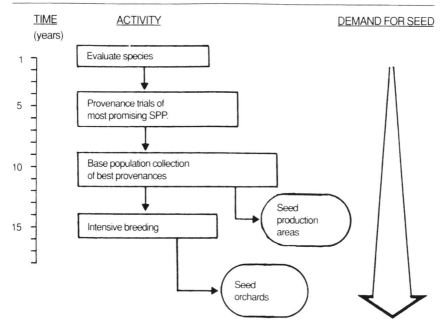

Figure 25.4 Generalized procedure for initial domestication of a fast growing tree species

of stock in general use passes (irretrievably?) from the hands of the scientists running the (more or less) well-planned, centralized, improvement and seed distribution programme.

I suggest that variations on the above theme are the rule rather than exceptions, and that world-wide, there are very significant losses in potential productivity because of the use of inferior seed. It is difficult to obtain strong documentary evidence since, inbreeding depression being a relative phenomenon, demonstration relies on designed experiments with material of genetically controlled origin. Without such experiments and adequate knowledge of pedigree, it is also difficult categorically to separate inbreeding effects from related genetic problems of sub-optimal choice of provenance, introgression with other species, or expression of maladaptation of the species to the test environment. The following examples are illustrative.

Development of land races of eucalypts have been discussed by Eldridge and Cromer (1987). These races have often developed from a very narrow genetic base. The origin of eucalypt seed used throughout the world, at least prior to the mid 1960s, is frequently not known; however, consideration of seed crops and commercial collecting practices lead to the view that many of the early introductions would have been made from a single tree (Larsen and

Cromer, 1970). For example, a single tree of *E. grandis* can yield about 3 million viable seeds, or sufficient to establish about 2500 ha of plantation. Since this species also seeds at an early age, further multiplication of the stock can be very rapid.

Inbreeding, as well as possible provenance variation, may provide an explanation for the fact that progeny from a seed orchard of select clones from South African plantations of *E. grandis* grew more slowly than samples from natural populations at Coffs Harbour, Australia (Ades and Burgess, 1982). Purnell (1986) has also reported that families from plus trees selected in South African stands of *E. nitens* Hook.f. were consistently poorer than unselected families from natural populations in Australia, and attributed this to inbreeding within the local plantations. Yet another example where inbreeding must be suspected is the Mysore gum (*E. tereticornis* Smith) which is grown widely in India and was derived from a 20-tree plantation of unknown Australian origin (Pryor, 1966). Mysore gum has performed consistently poorly relative to natural provenance samples in trials in Nepal (White, 1986). It has also been noted (Davidson, 1978) that demand for eucalypt seed in India is so great that it has often been necessary to collect all the seed that is available, irrespective of genetic quality. Finally, the impressive gains in productivity demonstrated in eucalypt breeding programmes in Brazil have been made relative to poor inbred and/or segregating hybrid stock, derived from early introductions (Brune and Zobel, 1981).

Unless the degree of inbreeding of control seedlots is known, it is also impossible to interpret reports of heterosis in interspecific eucalypt hybrids such as, for example, *E. tereticornis* x *camaldulensis* (Venkatesh and Sharma, 1977). A hybrid by definition is outcrossed, while a seedling which is true to type and produced by open-pollination might be inbred or outcrossed. The literature so far contains no reports of experiments with the appropriate intra-specific outcross controls necessary to resolve this point.

One example of genetic history that has been relatively well documented is that of the introduction of *Acacia mangium* Willd. to Sabah, Malaysia, from Australia (Sim, 1984). The first introduction was made in 1966, by way of open-pollinated seed from a single tree, with about 300 trees planted over two sites. There are now about 20 000 ha established with trees descended from this family, with a current annual planting rate of 4000 ha (Udarbe and Hepburn, 1987). The species is a precocious and prolific seed producer, and seedlings five generations removed from the original introduction are already being planted (Sim, 1984). Procedures used for selection of seed trees are not recorded but it is likely that, at least in the earlier generations, no great attention was paid to this practice. Indeed it was noted (FAO, 1982) that "atypical" segregants were favoured by seed collectors because they were small and tended to produce seed out of phase with the main crop.

In a nursery comparison of first, second, and third generation stock, mean height was 32.5, 20.7, and 18.1 cm respectively (Sim, 1984). The first generation was not significantly different from a mix of natural Australian provenance material. Seed from a seed production area developed by culling a second generation plantation showed some improvement at 35.2 cm. The extent to which higher quality seed is now being used in Sabah is unclear. However, because it is being marketed internationally at twice the price of routine plantation seed, it is highly likely that the latter is still an important source of planting stock.

Tree breeders in Sabah are well aware of the deficiencies of their material and have taken steps to broaden their genetic base (Sim, 1987), but it is difficult to visualize a short-term means of reducing the widespread use of the cheap inbred planting stock. The problem was created by genetic management decisions (or rather the lack of them) during the first generation of introduction, and we must look for solutions at this critical stage if we wish to prevent repetition of this scenario with other species or in other countries.

MEASURES TO MINIMIZE PROBLEMS WITH INBREEDING

In discussing tree domestication and improvement strategies it is helpful to consider four types of populations (Libby, 1973; Eldridge, 1984):

(1) The base population is a large, genetically unimproved, sample of the natural gene pool of the species;

(2) The breeding population is derived in the first instance by mating select base population individuals, and improved through successive generations of selection and testing;

(3) The propagation population is some subset of the current breeding population used for producing commercial quantities of seed. In the case of vegetatively propagated species it is the set of clones to be multiplied directly or mated to produce pedigreed families for mass propagation; and,

(4) The production population is the commercial plantation through which the grower realizes the benefits of the genetic improvement programme.

Not all improvement strategies require establishment of all four populations as physically separate entities. For example, the successful low-cost eucalypt improvement strategy developed in Florida (Franklin, 1986) uses the same plantation as breeding and propagation population, but the conceptual distinction is important.

Strategies for obtaining adequate base population samples and management of inbreeding in breeding populations have received considerable attention in the literature (e.g. Burdon and Shelbourne, 1971; Namkoong *et al.*, 1980), while inbreeding in production populations is of little consequence, since these should not be used as a source of seed.

The present discussion is primarily concerned with management actions designed to minimize, or at least control, inbreeding in the propagation populations of those species for which seed is the routine source of planting stock. Three major types of propagation population are available, or may be developed as a domestication programme proceeds: the natural forest, unpedigreed plantations, and seed stands or orchards.

The natural forest

In the early phases of domestication, the natural forest is likely to be the best available source of commercial seed. We have seen that such seed is likely to be inbred to some degree, and that this may be particularly so for the animal-pollinated species with mixed breeding systems which are common in tropical and subtropical forests. Continued reliance on this seed source is therefore undesirable. Even if there is no intention of embarking on an intensive breeding programme, some improvement in yield can be achieved by releasing the population from neighbourhood inbreeding effects. This can be done simply by collecting seed from the first generation plantations – provided of course that these are derived from a suitably broad genetic base. Additional benefits of this management decision are greater control over seed supply, together with the ability to improve the population by mass selection (choice of good phenotypes as seed parents) and to benefit from natural selection pressures operating in the new plantation environment.

It is clear from the continuing international trade in seed from natural populations, that this simple point is not appreciated in many quarters. For example it is estimated (J.C. Doran, pers. comm.) that commercial seed collectors in Australia still receive annual orders in excess of 10 tonnes for species such as *E. camaldulensis*, which have been under domestication for decades in many countries.

If the natural forest does represent the best available production population, then there are some simple measures which should minimize the extent to which seed is inbred, or is related and therefore going to inbreed further in the next generation: collect only from stands of trees after heavy general flowering; in particular, avoid isolated trees (which may carry heavy, easily collectable crops which are a great temptation to the commercial collector); also ensure that seed is collected from a recorded number of trees separated by a specific minimum distance (Willan, 1985).

Unpedigreed plantations

Plantings of trees of unknown genetic origin should be considered as a seed source of last resort – but in the short-term may be all that is available. Again, it is desirable to avoid collecting from isolated trees or small stands such as windbreaks, and to concentrate effort on heavy crops derived from the major annual flowering.

Maintenance of effective population size also requires increased attention. Consider, for example, hypothetical developments following the *Acacia mangium* introduction to Sabah described above. Figure 25.5 shows the rate of change in the coefficient of inbreeding for random mating populations of 20 and 1000 trees, with seed collected from either 1 or 20 trees. We know that a single female parent was involved in the original introduction. If we make the generous assumption that there were 20 unrelated fathers represented in the seed then, from Equation 2, $\Delta F = 0.13$ for that family relative to the (presumed) non-inbred parental population. If the first generation had been represented by 20 progeny, all of which interbred with each other to an equal extent, then seed collected from all of these ($N_f = N_m = 20$) would have become only slightly more inbred ($F = 0.14$). Repeating this calculation, $F = 0.16$ for the seed produced by fourth generation trees. If, on the other hand, a misguided attempt was made to improve the stand genetically by collecting seed only from the best one of the 20 phenotypes, then, with the same propagation and mating scenario, it would take only five generations to inbreed the population to a level equivalent to selfing trees in the original stand.

These calculations do not take selfing into account. If collection is made from a smaller number of trees, and these, by chance, happen to be highly self-fertile, then F may be substantially greater than predicted. It is clear (Figure 25.5) that the rate of build up in F is essentially the same if N_m is increased to 1000 – emphasizing the point that the critical factor is the number of individuals of the less frequent sex. The large population is, however, to be favoured on the grounds that one is able to select seed trees more intensively and thus obtain some improvement through mass selection.

Summary advice is therefore to collect seed from many trees in the largest available plantations, while actively planning a switch to better alternative seed sources at the earliest opportunity. Management action to reduce reliance on seed collected from unpedigreed production population trees may be particularly necessary in social forestry programmes, which are less likely to employ tree breeders. This improvement could be achieved by

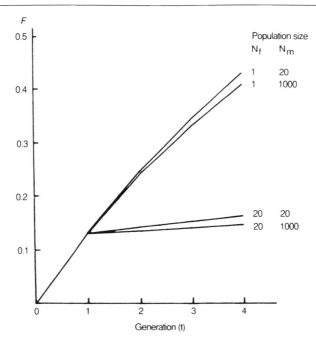

Figure 25.5 Increase in F over succeeding generations for populations of different sizes. Assume that all trees contribute pollen (N_m), and that seed is collected from some subset of the population (N_f)

ensuring an ongoing, cheap supply from a centralized seed production programme, or by including a seed production area as part of the management plan for each community (Eldridge, 1984).

Seed stands or orchards

These include seed production areas, seedling seed orchards, and clonal seed orchards which are specifically designed as propagation populations. Genetic history should be well defined, and assortative mating, rather than considerations of population size, is likely to be the major cause of any inbreeding problem. Natural outcrossing will be maximal when a large proportion of the orchard trees is in heavy flower at the same time. Useful management steps therefore include choice of planting sites conducive to flowering (and to high pollen vector activity for animal-pollinated tree species); use of planting designs which ensure that adjacent trees are of unrelated genotype; and matching of flowering times of the constituent genotypes.

It cannot be assumed that flowering characteristics will necessarily remain constant over a range of planting environments, and research on flowering biology is particularly important for species being domesticated as exotics. For example *Pinus caribaea* Morelet. flowers well in the lowland tropics, but not at high altitudes or latitudes (Slee, 1977). Chaperon (1978) and Geary (1984) commented on the variable and protracted nature of flowering of eucalypts in Congo-Brazzaville and Zambia respectively. In the Congo, a scarcity of pollen vectors was also noted as a hindrance to production of outcrossed seed.

Complete control over genetic quality of seed, and hence over inbreeding, can be achieved by investment in controlled-pollinated seed production programmes (Carson, 1986). This measure is most likely to be economic within the context of more elaborate breeding programmes, particularly where it is feasible to multiply each seedling through vegetative propagation.

More detailed discussion of the appropriate genetic constitution of a seed orchard is outside the scope of this contribution, but it is worth re-emphasizing that it is the degree to which the population of seed is inbred (rather than the parental trees from which seed is derived), which determines the extent of inbreeding depression in the production population. There may indeed be positive benefits in using unrelated S_1 clones in seed orchards (Lindgren, 1975).

IMPORTANCE OF RECORDING GENETIC HISTORY

Management of inbreeding, together with planning of the wider aspects of tree improvement, is heavily reliant on records of pedigree or genetic history of the various populations (for example, see Jones and Burley, 1973; Turnbull and Griffin, 1985; Willan, 1985). Provided the supplier is reliable, it is a relatively simple matter to record natural provenance and information on the number of trees sampled. This information must be kept associated with each of the plantations, derived from the seed, and added to with each successive generation of planting. This recording is a difficult task and will not occur unless managers are convinced of the economic benefits of such action. There is increasing international trade in seed of many tree species, and a corresponding need for agreement on conventions for recording genetic history. Existing schemes such as that of the OECD (1974) are not appropriate in all cases (Turnbull and Griffin, 1985).

CONCLUSIONS

Most discussion of alternative seed sources, including that presented here, emphasizes desirable seed stand characteristics and collection practices. I suggest that a very positive contribution to increased productivity can also be made by more clearly proscribing certain seed sources which may be expected to produce inbred seed.

To return to the example of *Acacia mangium* in Sabah, five generations of potential genetic improvement were lost by permitting plantation programmes to develop from an original narrowly based introduction. With hindsight, the forestry interests in Sabah would have been better served by felling the original plots at the first signs of flowering, and ordering a bulk seed collection from Australia for wider evaluation. In general, preliminary species trials should be viewed as invaluable sources of information, but quite the reverse with respect to seed production. Throughout the world, but particularly in the tropics, there are many species now at the stage of domestication of *Acacia mangium* 20 years ago. If we are prepared to learn from some of the examples outlined in this contribution and develop appropriate management procedures, it is possible to control and improve genetic quality of planting stock right from the outset of these programmes.

It will undoubtedly prove easier to prevent sub-optimal germplasm getting into general use than to replace it at some future date. It is therefore highly desirable to plan decisive action for obtaining adequate base population samples, co-incident with release of information on promising species.

REFERENCES

Ades, P.K. and Burgess, I.P. (1982). Improvement in early growth rate achieved by phenotypic selection of *Eucalyptus grandis*. *Australian Forest Research*, **12**, 169–73

Bawa, K.S. (1976). Breeding of tropical hardwoods: an evaluation of underlying bases, current status and future prospects. In Burley, J. and Styles, B.T. (eds), *Tropical Trees: Variation, Breeding and Conservation*, pp. 43–59. Linnean Society Symposium Series No. 2 (London: Academic Press)

Brune, A. and Zobel, B.J. (1981). Genetic base populations, gene pools and breeding populations for eucalyptus in Brazil. *Silvae Genetica*, **30**, 146–9

Burdon, R.D. and Shelbourne, C.J.A. (1971). Breeding populations for recurrent selection: conflicts and possible solutions. *New Zealand Journal of Forest Science*, **1**, 174–93

Burley, J. and Woods, P.J. (1976). *A Manual on Species and Provenance Research with Particular Reference to the Tropics*. Tropical Forestry Papers No. 10. (Oxford: Commonwealth Forestry Institute)

Carpenter, I.W. and Guard, A.T. (1950). Some effects of cross-pollination on seed production and hybrid vigor of Tulip tree. *Journal of Forestry*, **48**, 852–5

Carson, M.J. (1986). Controlled-pollinated seed orchards of best general combiners: a new strategy for radiata pine improvement. *New Zealand Agronomy Society Special Publication*, **5**, 144–9

Chaperon, H. (1978). Details of breeding improvement of Eucalyptus trees in the Congo-Brazzaville. In Nikles, D.G., Burley, J. and Barnes, R.D. (eds), *Progress and Problems of Genetic Improvement of Tropical Forest Trees*, Volume **2**, 1016–26 (Oxford: Department of Forestry)

Cunningham, T.M. (1960). *The Natural Regeneration of Eucalyptus regnans*. School of Forestry Bulletin No. 1. (Melbourne: University of Melbourne)

Davidson, J. (1978). Breeding tropical eucalypts. In Nikles, D.G., Burley, J. and Barnes, R.D. (eds). *Progress and Problems of Genetic Improvement of Tropical Forest Trees*, Volume **2**, pp. 941–56 (Oxford: Department of Forestry)

Eldridge, K.G. (1984). Breeding trees for fuelwood. In *Proceedings of the XV International Congress of Genetics*. Volume IV, pp. 339–50 (New Delhi: Oxford and IBH Publishing Co.)

Eldridge, K.G. and Griffin, A.R. (1983). Selfing effects in *Eucalyptus regnans*. *Silvae Genetica*, **32**, 216–21

Eldridge, K.G. and Cromer, R.N. (1987). Adaptation and physiology of eucalypts in relation to genetic improvement. In *Proceedings of Symposium on Silviculture and Genetic Improvement of Forest Species*, pp. 85–100 (Buenos Aires: CIEF)

Evans, J. (1982). *Plantation Forestry in the Tropics*. (Oxford: Clarendon Press)

Falconer, D.S. (1970). *Introduction to Quantitative Genetics*. (New York: Ronald)

FAO. (1982). *Variation in Acacia mangium Willd*. FAO/UNDP-MAL/78/009 Consultants Report No. 8. (Rome: FAO)

Franklin, E.C. (1970). *Survey of Mutant Forms and Inbreeding Depression in Species of the Family Pinaceae*. USDA Forest Service SE Forest Experimental Station Research Paper SE-61.

Geary, T.F. (1984). Flowering limitations on eucalyptus breeding strategies in Zambia. In Barnes, R.D. and Gibson, G.L. (eds), *Provenance and Genetic Improvement Strategies in Tropical Forest Trees*, pp. 516–17. (Oxford: Department of Forestry)

Geburek, T. (1986). Some results of inbreeding depression in Serbian Spruce (*Picea omorika* Panc. Purk.). *Silvae Genetica*, **35**, 169–72

Griffin, A.R., Moran, G.F. and Fripp, Y.J. (1987). Preferential outcrossing in *Eucalyptus regnans*. *Australian Journal of Botany*, **35**, 465–75

Griffin, A.R. and Cotterill, P.P. (1988). Genetic variation in growth of outcrossed, selfed and open-pollinated progenies of *Eucalyptus regnans* and some implications for breeding strategy. *Silvae Genetica*, **37**, 124–31

Griffin, A.R., Raymond, C.A. and Lindgren, D. (1986). Effects of inbreeding on seed yield and height growth of *Pinus radiata* D.Don. In *Proceedings of Joint Meeting of IUFRO Working Parties on Breeding Theory, Progeny Testing and Seed Orchards*, 603. IUFRO, Williamsburg, Virginia.

Hodgson, L.M. (1976a). Some aspects of flowering and reproductive biology in *Eucalyptus grandis* at J.D.M. Keet Forest Research Station. 2. The fruit, seed, seedlings, self-fertility, selfing and inbreeding effects. *South African Forestry Journal*, **98**, 32–43

Hodgson, L.M. (1976b). Some aspects of flowering and reproductive biology in *Eucalyptus grandis* (Hill) Maiden at J.D.M. Keet Forest Research Station. 3. Relative yield, breeding systems, barriers to selfing and general conclusions. *South African Forestry Journal*, **99**, 53–8

Jones, N. and Burley, J. (1973). Seed certification, provenance nomenclature and genetic history in forestry. *Silvae Genetica*, **22**, 53–8

Kendrick, J. (1986). A method for estimating self-incompatibility. In Williams, E.G., Knox, R.B. and Irvine, D. (eds) *Pollination '86*, pp. 116–20 (Melbourne: School of Botany, University of Melbourne)

Larsen, E. and Cromer, D.A.N. (1970). Exploration, evaluation, utilization and conservation of eucalypt gene resources. In Frankel, O.H. and Bennett, E. (eds), *Genetic Resources in Plants – Their Exploration and Conservation*. pp. 381–8. IBP Handbook No. 11. (Oxford: Blackwell Scientific Publications)

Libby, W.J. (1973). Domestication strategies for forest trees. *Canadian Journal of Forest Research*, **3**, 265–77

Lindgren, D. (1975). Use of selfed material in forest tree improvement. *Department of Genetics Research Notes*, **15**, 1–76 (Stockholm: Royal College of Forestry)

Mendonza, L.A. (1970). (The effects of self-pollination in *Eucalyptus camaldulensis* Dehn.). *IDIA Suplemento Forestal*, **6**, 41–5 (Es.)

Moffett, A.A. and Nixon, K.M. (1974). The effects of self-fertilisation on green wattle (*Acacia decurrens*) and black wattle (*Acacia mearnsii*). *Wattle Research Institute Report, 1973–74*, pp. 66–84

Moran, G.F. and Bell, C.J. (1983). Eucalyptus. In Tanksley, S.D. and Orton, T.J. (eds), *Isozymes in Plant Genetics and Breeding. Part B*, pp. 423–41 (Amsterdam: Elsevier)

Moran, G.F., Bell, C.J. and Griffin, A.R. (1989). The reduction in inbreeding in a seed orchard of *Eucalyptus regnans* compared to levels in natural populations. *Silvae Genetica*, **38**, 32–6

Namkoong, G. (1985). Genetic structure of forest tree populations. In *Proceedings of the XV International Congress of Genetics*. Vol IV, pp. 351–60 (New Delhi: Oxford and IBH Publishing Co.)

Namkoong, G., Barnes, R.D. and Burley, J. (1980). *A Philosophy of Breeding Strategy for Tropical Forest Trees*. Tropical Forestry Papers No. 16 (Oxford: Commonwealth Forestry Institute)

OECD. (1974). *OECD Scheme for the Control of Forest Reproductive Material Moving in International Trade*. Organization for Economic Cooperation and Development (Directorate for Agriculture and Food), Paris.

Pederick, L.A. and Eldridge, K.G. (1983). Characteristics of future radiata pine achievable by breeding. *Australian Forestry*, **46**, 287–93

Pryor, L.D. (1966). A report on past performance and some current aspects of the cultivation of quick-growing species (mainly *Eucalyptus*) in India. *Indian Forester*, **92**, 614–21

Purnell, R.C. (1986). Early results from provenance/progeny trials of *Eucalyptus nitens* in South Africa. In *Proceedings of Joint Meeting of IUFRO Working Parties on Breeding Theory, Progeny Testing and Seed Orchards*, pp. 500–13 (Williamsburg, Virginia: IUFRO)

Schemske, D.W. and Lande, R. (1985). The evolution of self-fertilisation and inbreeding depression in plants. II. Empirical observations. *Evolution*, **39**, 41–52

Sim, B.L. (1984). The genetic base of *Acacia mangium* Willd. in Sabah. In Barnes, R.D. and Gibson, G.L. (eds), *Provenance and Genetic Improvement Strategies in Tropical Forest Trees*, pp. 597–603 (Oxford: Department of Forestry)

Sim, B.L. (1987). Research on *Acacia mangium* in Sabah: a review. In Turnbull, J.W. (ed.) *Australian Acacias in Developing Countries*, pp. 164–6 (Canberra: ACIAR)

Simmonds, N.W. (1979). *Principles of Crop Improvement*. (London: Longman)

Slee, M.U. (1977). A model relating needleless shoots and dieback in *Pinus caribaea* to strobilus production and climatic conditions. *Silvae Genetica*, **26**, 135–41

Squillace, A.E. (1973). Comparison of some alternative second-generation breeding plans for Slash Pine. In *Proceedings of 12th Southern Forest Tree Improvement Conference*, pp. 2–13. (Louisiana: Baton Rouge)

Stern, K. and Roche, L. (1974). *Genetics of Forest Ecosystems*. (Berlin: Springer-Verlag)

Tan, H. (1981). Estimates of genetic parameters and their implications for *Hevea* breeding. In *Proceedings of 4th International SABRAO Congress*, pp. 439–46. Kuala Lumpur, Malaysia.

Turnbull, J.W. (1984). Tree seed supply – a critical factor for the success of agro-forestry projects. In Burley, J. and von Carlowitz, P. (eds), *Multipurpose Tree Germplasm*, pp. 279–97 (Nairobi: ICRAF)

Turnbull, J.W. and Griffin, A.R. (1985). The concept of provenance and its relationship to infraspecific classification in forest trees. In Styles, B. (ed.), *Infraspecific Classification of Wild Plants*, pp. 157–89 (Oxford: Oxford University Press)

Udarbe, M.P. and Hepburn, A.J. (1987). Development of *Acacia mangium* as a plantation species in Sabah. In Turnbull, J.W. (ed.), *Australian Acacias in Developing Countries*, pp. 157–9 (Canberra: ACIAR)

van Wyk, G. (1981). Inbreeding effects in *Eucalyptus grandis* in relation to degree of relatedness. *South African Forestry Journal*, **116**, 60–63

Venkatesh, C.S. and Sharma, V.K. (1977). Rapid growth and higher yield potential of heterotic *Eucalyptus* species hybrids FRI-4 and FRI-5. *Indian Forester*, **103**, 795–801

White, K.J. (1986). *Tree Farming Practices in the Bharbar Teri of Central Nepal*. Manual No. 2. Ministry of Forests, Kathmandu

Willan, R.L. (1985). *A Guide to Forest Seed Handling*. FAO Forestry Paper 20/2. (Rome: FAO)

CHAPTER 26

GENETIC STRUCTURE OF TROPICAL TREE SPECIES OF BRAZIL

Paulo Y. Kageyama

ABSTRACT

Knowledge on the genetic structure of tropical tree species is important for their rational utilization, both for the improvement of plantations and the management and conservation of natural communities. Focus is given to the association between the reproductive biology of tropical tree species and their genetic structure, as well as their form of occurrence in the forest, drawing upon the results of studies involving tropical tree species of the State of São Paulo highlands, Brazil. Silvicultural trials of indigenous species are also analysed, mainly in relation to the form of plantations of the different groups of species regarding their stage in secondary succession. The need to know the silviculture of the species is emphasized, not only in relation to their utilization, but also for the very installation of genetic trials. Results of progeny trials involving native species are analysed, and compared with the results of commercial species of Eucalyptus, *with the purpose of observing possible trends. The difficulty of standardization of genetic trials, the problems of errors in the sampling of the studied populations, and even the lack of knowledge about the population history regarding human intervention, are such that more questions than answers are presented. Electrophoresis studies hold promise in complementing the field trials to understanding the genetic structure of tropical tree species.*

INTRODUCTION

The genetic structure of a species is defined by the way in which genetic variation is distributed between and within populations. The genetic structure is the result of the joint action of mutation, migration, selection, and genetic drift, as pointed out by Loveless and Hamrick (1984). Furthermore, it depends on the genetic system of the species, which includes the breeding system, karyotypic structure, level of ploidy, and life cycle features (Brown, 1978). Frankel and Soulé (1981) also stress the importance of the history of the population, mainly in relation to frequency and severity of bottlenecks.

The understanding of the genetic structure of forest tree species is essential for sound management and conservation of natural populations, as well as for developing appropriate genetic improvement strategies for plantations. Several methods have been used to study genetic variation among individuals and populations: nucleic acid and protein sequencing, electrophoresis, immunology, cytology, morphometrics and breeding studies. No particular method is capable of unambiguously describing the total genetic diversity of a given species (Chambers and Bayless, 1983). Provenance and progeny trials are the usual methods that have been employed to measure genetic variation of forest populations, mainly for tree breeding purposes. More recently, electrophoretic techniques have gained more importance because of the rapidity with which they yield genetic data. This efficiency is particularly important for the long-lived tree species.

Here, I address several issues that provide the basis for an understanding of genetic structure of forest tree populations. I focus mainly on aspects of reproductive biology and experimental data of the Brazilian São Paulo State highland forests. The objective of our research in Brazil is to search for patterns of genetic variation that could group species, in order to organize and systematize the information about genetic structure in tropical forests. The questions and suggestions presented are preliminary, and are raised for the purpose of provoking discussions on this important and little known field. The experimental results presented cover only a few samples of the tropical forest and the incipient knowledge of tropical forest trees, leading to simplification and generalization of concepts and phenomena.

REPRODUCTIVE BIOLOGY AND GENETIC STRUCTURE OF TROPICAL TREE SPECIES

An intrinsic issue in the discussion of genetic structure of tropical tree species is reproductive biology, restricted here to breeding system and pollen and seed dispersal. Bawa and Krugman (1990) present a general discussion about the relation between specific reproductive characteristics of tropical forest trees and their effects on genetic variation within and between populations. Hamrick (1983) has shown that the mating system has a marked effect on the level and distribution of genetic variation of plant species. Autogamous species exhibit low variation within populations, but a high level of variation among populations; the reverse is true for the outcrossing wind-pollinated species. Outcrossing animal-pollinated species show levels of variation intermediate between the two groups. The seed dispersal mechanisms also influence genetic structure, mainly with respect to variation among populations. Species with winged or plumose seeds show a very distinct pattern in comparison to species with animal-attached seeds.

In relation to the breeding system, it appears that there is a strong tendency for predominance of cross-pollination in tropical tree species (Bawa, 1974; Bullock, 1985; Bawa *et al.*, 1985). The discussion about breeding system deserves closer attention because, being a genetically controlled characteristic, it can easily be modified (Allard, 1975). According to Roche (1978), such modification has been observed in *Theobroma cacao* and *Hevea brasiliensis*, with variation in the reproductive system from the centre of distribution towards the peripheral area of occurrence.

More important than the discussion of a possible dichotomy between self-pollination and cross-pollination is the discussion related to alternative reproductive mechanisms necessary for the maintenance of certain species, mainly those that are very rare in the forest. Bawa *et al.* (1985) raised the possibility that a higher frequency of self-compatibility could be prevalent in species that occur in low densities. Kaur *et al.* (1977) show that apomixis may be widespread in tropical forest trees. A certain rate of selfing, even in predominantly allogamous species, may be important for the reproduction of individuals in certain species.

Pollen and seed dispersal, by influencing gene flow, also affect genetic variation within and between populations. As an approximate rule, the differentiation between populations would be maximum in autogamous species and minimum in allogamous species (FAO, 1984). Animal-pollinated species should show greater differentiation than anemophilous tree species, but less than that in autogamous species (Hamrick, 1983). Particularly in tropical forest trees, there is great variation in flight distances of pollen and seed vectors. This variation could result in diverse patterns of genetic structure. For example, short distance pollen dispersal by flies in *Esenbeckia leiocarpa* (Crestana *et al.*, 1983) and *Theobroma cacao* (Roche, 1978) could lead to more genetic differentiation among populations than the long distance pollen dispersal by bats in *Hymenaea stilbocarpa* (Crestana *et al.*, 1984).

Pollinator foraging behaviour may be an evolutionary result of mutual adaptation between plants and pollinators which would be reflected in the tropical forest structure. Kageyama and Patiño-Valera (1985) have suggested that tree species pollinated by short-flying vectors may predominantly be clumped while the species pollinated by animals with long distance flight may be highly dispersed.

Considering the neighbourhood size and its effect on the genetic structure of populations, Hamrick and Loveless (1986*a*) state that the genetic structure within populations of species with short seed dispersal distances, gravity dispersal for example, should be pronounced. Such populations with reduced effective size of population should show greater family structure and inbreeding. In species with long distance seed dispersal, gene movement could prevent the divergence of populations, and intra-population variation should be high.

On the other hand, Levin (1979) states that the question of breeding structure of widely spaced plants is somewhat deceptive as, even though the plants are quite distant from each other, most pollination occurs between neighbouring plants, which are very likely to be related. This should not alter the breeding structure in comparison to the closely spaced species. Levin also states that the density-dependent pollen dispersal in animal-pollinated plants produces a neighbourhood size which is roughly constant, and a neighbourhood area which increases as plant density decreases.

In species with highly dispersed individuals, gene flow through pollen would be more effective if the larger neighbourhood area were coupled with a higher probability of crossing among non-related individuals. This would occur only if the long distance pollen movement was associated with the long distance seed movement. Namkoong and Gregorius (1985) discuss the effects of different combinations of migratory and sedentary behaviour of pollen and seed. Preliminary research with tropical tree species in the State of São Paulo, Brazil, appears to confirm the hypothesis that the dispersion pattern of individuals in a species is associated with the patterns of pollen and seed flight flow. Examples include:

(a) *Mimosa scabrella*: pollinated by small bees; seeds dispersed by gravity at small distances; pioneer species in a clumped form (Catharino *et al.*, 1982);

(b) *Esenbeckia leiocarpa*: pollinated by small flies; seeds dispersed by gravity at small distances; climax species in a clumped form (Crestana *et al.*, 1983);

(c) *Hymenaea stilbocarpa*: pollinated by bats; seeds dispersed by mammals at long distances; climax species in a very dispersed form (Crestana *et al.*, 1985); and,

(d) *Chorisia speciosa*: pollinated by birds; seeds dispersed by wind at long distances; late secondary species in a very dispersed form (Ramirez-Castilho, 1986).

If this hypothesis is correct, there could be a gradation in tree species, from those with typical clumped distribution and short pollen and seed dispersal distances to those with very dispersed distribution and wide ranging pollen and seed shadows. This gradation could allow the establishment of criteria for separation of species with regard to their probable patterns of genetic variability. However, long distance pollen dispersal can occur jointly with short distance seed dispersal, and vice-versa, as suggested by Namkoong and Gregorius (1985). This mechanism would increase the diversity of possible patterns of genetic variation. Hamrick and Loveless (1986a) have shown that there is association between both patterns of pollen and seed dispersal with genetic variation among populations in tropical tree species.

GENETIC STRUCTURE AND SILVICULTURE OF INDIGENOUS SPECIES

The tree species that have been planted in the tropics are mainly heliophile pioneer species of the early successional stages. The genetic trials that are being carried out in the tropics are related to this specific group of species with similar habitat preferences involving just a few genera, such as *Eucalyptus*, *Pinus*, *Tectona*, *Gmelina*, *Cedrela*, and *Leucaena*. The studies are mostly aimed at identification of the populations suitable for particular plantation sites. The characteristics normally evaluated in these trials are associated with wood production and wood quality rather than with adaptation. An important question is whether the genetic structures of the species in plantations are similar to those that occur in natural forests.

A common problem in native tree plantations in Brazil is that different light requirements of the species have not been considered in plantings. All species, in pure or mixed stands, have been planted in the same way as the commercial pioneer species. This procedure results in inadequate conditions for the development of the late secondary and climax species (defined according to Budowski, 1965), which require partial or even complete shading in their initial stages of development (Hubbell and Foster, 1986).

Kageyama *et al.* (1986), analysing spacing patterns, observed that the species reacted differently to the increase in available area per tree. Five types of response in tree growth to the increase in spacing were observed, making it possible to group the studied species in the following categories:

(1)   Positive effects for both height and diameter;

(2)   Positive effect for diameter and neutral for height;

(3)   Neutral effect for height and diameter;

(4)   Negative effect for height and positive for diameter; and,

(5)   Negative effects for both height and diameter.

These different responses to spacing can be interpreted as being, primarily, differential responses to light.

Genetic trials taking into account the differential light requirements of various species are being conducted in the Anhembi Experimental Station (ESALQ, University of São Paulo). Progenies of a pioneer species and of a climax species from a natural population are associated in a single experiment, with linear plots of the pioneer species crossing the also linear plots of the climax species. The individuals of the former group (planted first) shade the individuals of the latter (planted 1 year later). This trial illustrates the difficulty of treating different types similarly when one plants them in uniform environments, as in the case of genetic trials. The information about the distribution of genetic variability in the two groups of species could show how various species might respond to the same environment.

The association between successional stage and genetic structure put forth by Hamrick (1983) showed that genetic variation within and among populations of the early successional species was very distinct from that in the late successional species. This distinction is also likely to hold true for tropical forest tree species. The ephemeral pioneer populations in the form of islands would be very distinct from the more stable and continuous populations of climax species. The establishment of the former is dependent upon the existence of gaps, while the climax species, because of their tolerant habit, do not depend upon gaps. These two population types fit into the 'island' and 'isolation by distance' models of gene flow (according to Futuyma, 1986), possibly with a greater neighbourhood size for the climax species as compared with the pioneer species.

GENETIC STRUCTURE OF TROPICAL TREE SPECIES

The use of electrophoresis for genetic studies of indigenous species will certainly help to overcome the problems related to species diversity of the tropical forest, because it would be almost impossible to establish field trials even with only part of the total number of species. However, the use of isozymes for the study of genetic structure should involve a large number of enzymatic systems, such as reported by Hamrick and Loveless, 1986b). It would be interesting for isoenzyme studies to precede the genetic trials with quantitative data, which should then be used to check the laboratory results. Studies involving tropical tree species are few (Gan *et al.*, 1981; Hamrick and Loveless, 1986a, 1986b; Buckley *et al.*, 1986, as cited in Bawa and Krugman, 1990).

Variation between provenances of tropical pines and eucalypts

A general analysis of the provenance trials of pines and eucalypts, considering that the first species is pollinated by wind and the second one by animals, may help to clarify the importance of gene flow via pollen on the genetic structure of the species. Even though it is known that several other factors that influence genetic structure can be different in these two genera, the results of experiments with tropical species of these two genera in Brazil are discussed for the purpose of verifying the existence of certain trends.

The results of provenance trials of tropical pines (*P. caribaea* var. *hondurensis*, *P. caribaea* var. *caribaea*, *P. oocarpa* and *P. kesiya*), in different localities in Brazil (see Table 26.1), show that the between provenances coefficient of variation for wood volume varies from 3.3% to 12.3%, with an average of 7.7%. For the tropical eucalypts tested in Brazil (*E. grandis*, *E. urophylla*, *E. cloeziana*, *E. pilularis*, *E. citriodora*, *E. maculata*, *E. camaldulensis* and *E. tereticornis*), the average among provenances coefficient of variation is higher, about 26.6%, with a range from 12.4% to 72.3%. As the sampling used for the two genera was directed towards genetic improvement, it can be considered comparable. These results raise important questions such as: Are these differences due to the type of pollination vector? Can such differences be expected in other species pollinated by animals?

Genetic variation within populations of indigenous species of Brazil

The variation within and between progenies with respect to quantitative characters can be expressed by the coefficient of genetic variation among progenies (CV_g) and within progenies (CV_w). These genetic parameters for tropical eucalypt species, in comparison with the indigenous hardwood species of Brazil, can contribute to the present discussion on the efforts to establish trends in patterns of genetic variation in populations. Unfortunately, the characteristics often evaluated are growth, stem form, and wood density. Only growth will be discussed here, because it is the only character related to adaptation.

The results of progeny trials from non-improved populations involving the three main tropical eucalypt species planted in Brazil (*E. grandis*, *E. saligna*, *E. urophylla*) are shown in Table 26.2. The results of open pollinated progeny trials with indigenous species of Brazil are shown in Table 26.3. Although the data are not statistically comparable, some observations can be made in the analysis of the different species.

When one takes the average of the genetic parameters of native species, it can be observed that these data are compatible with those found for

Table 26.1 Results of among provenances coefficient of variation for wood volume, calculated from trials with tropical *Pinus* and *Eucalyptus* species in Brazil

Species	Age	Number of provenances	C.V.prov.* (%)	Author
P.kesiya	10	9	8.03	IPEF (1984)***
P.oocarpa	10	10	12.29	Garnica *et al.* (1983)
P.caribaea var. *hondurensis*	10	5	5.73	IPEF (1984)
P.caribaea var. *hondurensis*	8	5	9.46	Martins *et al.* (1983)
P.caribaea var. *caribaea*	10	5	4.95	IPEF (1984)
P.oocarpa **	6.5	7	17.12	Caser (1984)
P.caribaea var. *hondurensis*	6.5	5	10.14	Caser (1984)
P.caribaea var. *caribaea*	6.5	5	3.33	Caser (1984)
		Mean = 7.7		
E.urophylla	5.0	17	72.33	Pinto and Jacob (1979)
E.cloeziana	5.0	9	50.54	IPEF (1984)
E.grandis	7.0	12	14.05	Campinhos *et al.* (1983)
E.pilularis	5.0	9	12.43	Pasztor (1974)
E.citriodora	5.0	6	39.23	Assis *et al.* (1983)
E.maculata	5.0	13	25.82	Pasztor *et al.* (1983)
E.camaldulensis	3.0	21	18.59	Rodrigues *et al.* (1986)
E.tereticornis	3.0	8	25.70	Rodrigues *et al.* (1986)
		Mean = 26.62		

*   C.V.prov. = coefficient of variation among provenances
      calculated from the means of the trials
**   *P.oocarpa* including *P.tecunumanii* from Nicaragua and Belize
*** IPEF – Results of research trials – unpublished

Eucalyptus species, mainly for within progenies variation. This result could indicate a certain coherence for the referred genetic parameters.

There is a large coefficient of genetic variation in *Dipterix alata*, *Machaerium villosum* and *Araucaria angustifolia*. This result could reflect

Table 26.2   Genetic parameters in progenies of open pollination for plant height for *Eucalyptus* species

Species	Provenance	Age	Number of progenies	CV_g (%)*	CV_w (%)**
E.grandis	Coff's Harbour	2.0	81	4.7	19.2
E.saligna	Itatinga-SP	2.7	169	4.6	15.2
E.urophylla	Indonesia	3.0	42	6.9	20.8

* CV_g (%) = coefficient of variation among progenies

$$(= \frac{\sigma p}{X} \times 100)$$

** CV_w = coefficient of variation within progenies

$$(= \frac{\sigma d}{X} \times 100)$$

Sources: Kageyama (unpublished) for *E.grandis*, Patiño-Valera (1986) for *E.saligna*, Pinto (1984) for *E.urophylla*.

real genetic differences, possible sampling errors, or even effects of degraded populations. On the other hand, there is very little genetic variation, both between and within progenies, in such species as *Cariniana legalis* and *Gallesia gorarema*. This lack of variation indicates genetic uniformity within populations and could reflect inbreeding.

Cecropia cinerea and *Araucaria angustifolia* are both dioecious, the former being a hardwood species pollinated by animals, and the latter a conifer pollinated by wind. However, the genetic parameters of the two species are similar in terms of CV_g and CV_w.

The coefficient of variation within progenies, provided it is standardized, could be an indicator of allogamy, as suggested by Pires and Kageyama (1985). This suggestion is justified because, under random mating, the within progenies variation of allogamous species would contain three quarters of the additive variance, dominance variance, and almost all of the epistatic variance of the population, besides the within plot environmental variance. In autogamous species, the within progenies variance is represented mostly by the environmental variance.

Because we do not have precise information about the sampling procedure for the collection of seeds of the species used in these trials, and taking into account that it was not standardized, additional inferences about these results can be dangerous. However, it is evident that this type of genetic assay can be very useful for the quantification of the genetic variation of the populations, if sampling is standardized. Moreover, information about reproductive biology of the sampled species in the field could also provide important insights into the analysis of data.

Table 26.3   Genetic parameters in progenies of open pollination for plant height for some indigenous tree species of Brazil

Species	Provenance	Age (yrs)	CV_g (%)	CV_w (%)	Authors*
Mimosa					
scabrella	Several (4)	0.5	9.3	32.1	(1)
Cecropia					
cinerea	Piracicaba-SP	1.0	6.5	22.5	(2)
Araucaria	Ipuiúna-MG	2.0	5.8	32.4	(3)
angustifolia	Três Barras-SC	3.5	3.3	17.7	(4)
	Guarapuava-PR	3.5	7.4	18.4	(4)
	B.Jardim Serra-SC	3.5	0.0	19.3	(4)
Copaifera					
multijuga	Manaus-AM	4.5	6.2	11.7	(5)
Dipterix	Aquidauana-MS	5.0	4.6	18.0	(6)
alata	Campo Grande-MS	5.0	1.1	14.4	
	Iaciara-GO	5.0	0.0	14.8	
Machaerium	Campinas-SP	4.0	4.2	23.6	(6)
villosum	Porto Ferreira-SP	4.0	4.0	19.5	
	Camanducaia-MG	4.0	0.0	16.3	
Pterogyne	Baurú-SP	4.0	3.7	31.7	(6)
nitens	Ribeirão Preto-SP	4.0	2.1	34.5	
	Teodoro Sampaio-SP	4.0	1.1	29.2	
	Alvorada do Sul-PR	4.0	0.8	24.4	
Astronium	Novo Horizonte-SP	4.0	6.5	31.0	(6)
urundeuva	Rio Claro-SP	4.0	5.6	39.3	
	Penápolis-SP	4.0	5.0	32.8	
	Paulo de Faria-SP	4.0	2.7	41.5	
	Pederneiras-SP	4.0	2.3	52.1	
Cariniana	Campinas-SP	3.0	1.0	12.1	(6)
legalis	Piracicaba-SP	3.0	0.7	11.9	
	Vassununga-SP	3.0	0.5	11.4	
Gallesia	Ribeirão Preto-SP	3.0	1.7	11.1	(6)
gorarema	Baurú-SP	3.0	0.4	8.8	
	Campinas-SP	3.0	0.2	11.7	
Peltophorum	Alvorada do Sul-PR	3.0	7.2	14.7	(6)
dubium	Baurú-SP	3.0	3.7	9.9	
Mean			3.25	22.29	

* (1) Fonseca (1982); (2) Kageyama (unpublished); (3) Gianotti *et al.* (1982); (4) Kageyama and Jacob (1980); (5) Vastano (1984); (6) Siqueira *et al.* (1985)

CONCLUSIONS

Genetic structure of tropical tree populations is largely unknown. In plants, the mating systems and gene flow greatly influence patterns of genetic variation within and between populations. Because tropical forest trees show a great diversity of breeding systems, pollination and seed dispersal mechanisms, we might expect tropical tree populations to exhibit diversified population structures. Genetic variation in trees has been analysed by detecting polymorphism in quantitative traits and for loci coding for isozymes. There is a need to link isozyme studies with those concerned with morphometric traits. Furthermore, it is important to examine the patterns of genetic variation in the context of reproductive biology, successional status and geographical distribution – the parameters which greatly affect population genetic structure.

ACKNOWLEDGEMENTS

Thanks are due to several colleagues who contributed to the elaboration of this work, especially: to Prof. Carlos F.A. Castro, Prof. Keith Brown, and Prof. Virgilio M. Viana, for critical review of the manuscript; to Prof. K. Bawa for his advice in finalizing the paper; to Prof. Walter P. Lima for help in the translation to English. This work was partially funded by the FAO/Brazil Research Project on *In Situ* Genetic Conservation.

REFERENCES

Allard, R.W. (1975). The mating systems and microevolution. *Genetics*, **79**, 115–26

Assis, T.F., Brune, A. and Euclides, R.F. (1983). Ensaio de procedências de *Eucalyptus citriodora*. *Silvicultura em São Paulo* (São Paulo), **28**, 162–4

Bawa, K.S. (1974). Breeding systems of tree species of a lowland tropical community. *Evolution*, **28**, 85–92

Bawa, K.S. and Krugman, S.L. (1990). Reproductive biology and genetics of tropical trees in relation to conservation and management. In Gómez-Pompa, A., Whitmore, T.C. and Hadley, M. (eds.) *Rain Forest Regeneration and Management*, Man and the Biosphere Series vol.6, pp. 119–136. (Paris: Unesco and Carnforth: Parthenon Publishing)

Bawa, K.S., Perry, D.R. and Beach, J.H. (1985). Reproductive biology of tropical lowland rain forest trees. I. Sexual systems and incompatibility mechanisms. *American Journal of Botany*, **72**, 331–45

Brown, A.H.D. (1978). Isozymes, plant population genetic structure and genetic conservation. *Theoretical and Applied Genetics*, **52**, 145–57

Budowski, G. (1965). Distribution of tropical American rain forest species in the light of successional process. *Turrialba*, **15**, 40–2

Bullock, S.H. (1985). Breeding systems in the flora of a tropical deciduous forest in Mexico. *Biotropica*, **17**, 287–301

Campinhos, E., Jr., Martins, F.C.G. and Ikermori, Y.K. (1983). Teste de procedência de *Eucalyptus grandis* em Aracruz (es). *Silviculturea*, (São Paulo), **28**, 221–5

Caser, R.L. (1984). Variações Genéticas e Interações com Locais em *Pinus* Tropicais e suas Associações com Parâmetros Climáticos. Thesis of Magister Scientias. ESALQ/USP, Piracicaba.

Catharino, E.L.M., Crestana, C.S.M. and Kageyama, P.Y. (1982). Biologia floral da Bracatinga *Mimosa scabrella* Benth. *Silvicultura em São Paulo*, (São Paulo), **16A**, 525–31

Chambers, S.M. and Bayless, J.W. (1983). Systematics, conservation, and the measurement of genetic diversity. In Schonewald-Cox, C.M., Chambers, S.M., MacBryde, B. and Thomas, W.L. (eds), *Genetics and Conservation. A Reference for Managing Wild Animal and Plant Populations*, pp. 349–63, (Menlo Park, California: Benjamin Cummings Publishing Company)

Crestana, C.S.M., Kageyama, P.Y. and Souza Dias, I. (1983). Biologia floral do Guarantã. (*Esenbeckia leiocarpa* Engl.). *Silvicultura*, (São Paulo), **28**, 35–8

Crestana, C.S.M., Souza Dias, C.S.M. and Mariano, G. (1985). Ecologia de polinização de *Hymenaea stilbocarpa* Hayne, o jatobá. *Silvicultura em São Paulo*, (São Paulo), **19**, 31–7

FAO. (1984). *Conservación in situ de los Recursos Fitogeneticos. Bases Cientificas y Tecnicas*. (Rome: FAO)

Fonseca, S.M. (1982). Variações Fenotipicas e Genéticas em Bracatinga – *Mimosa scabrella* Benth. Thesis of Magister Scientias. ESALQ/USP, Piracicaba.

Frankel, O.H. and Soulé, M.E. (1981). *Conservation and Evolution*. (Cambridge: Cambridge University Press)

Futuyma, D.J. (1986). *Evolutionary Biology*. Second Edition. (Sunderland, Massachusetts: Sinauer Associates)

Gan, Y., Robertson, F.W. and Soepadmo, E. (1981). Isozyme variation in some rain forest trees. *Biotropica*, **13**, 20–8

Garnica, J.B., Nicolielo, N. and Bertolani, F. (1983). Teste de procedência de *Pinus oocarpa* na Região de Agudos-São Paulo. *Silvicultura*, (São Paulo), **28**, 296–7

Gianotti, E., Timoni, J.L., Mariano, G., Coelho, L.C.C., Fontes, M.A. and Kageyama, P.Y. (1982). Variaçao genética entre procedências e progênies de *Araucaria angustifolia*. In Anais do Congresso Nacional sobre Essências Nativas. *Silvicultura em São Paulo*, (São Paulo), **16 (A)**, 970–5

Hamrick, J.L. (1983). The distribution of genetic variation within and among natural forest populations. In Schonewald-Cox, C.M., Chambers, S.M., MacBryde, B. and Thomas, W.L. (eds), *Genetics and Conservation. A Reference for Managing Wild Animal and Plant Populations*. pp. 335–48 (Menlo Park California: Benjamin Cummings Publishing Company)

Hamrick, J.L. and Loveless, M.D. (1986*a*). The influence of seed dispersal mechanisms on the genetic structure of plant populations. In Estrada, A. and Fleming, T.H. (eds), *Frugivores and Seed Dispersal*, pp. 211–23 (The Hague: Dr. W. Junk Publishers)

Hamrick, J.L. and Loveless, M.D. (1986*b*) Isozyme variation in tropical trees: procedures and preliminary results. *Biotropica*, **18**, 201–7

Hubbell, S.P. and Foster, R.B. (1986). Commonness and rarity in a neotropical forest: implications for tropical tree conservation. In Soulé, M.E. (ed.) *Conservation Biology: The Science of Scarcity and Diversity*, pp. 295–31 (Sunderland, Massachusetts: Sinauer Associates)

Kageyama, P.Y. and Jacob, W.S. (1980). Variaçaõ genética entre e dentro de populações de *Araucaria angustifolia* Bert O. Ktze. In *Forestry Problems of the Genus* Araucaria, pp. 83–6 IUFRO Meeting Proceedings. FUPEF, Curitibia.

Kageyama, P.Y. and Patiño-Valera, F. (1985). Conservación y Manejo de Recursos Geneticos Forestales: Factores que Influyen En La Estructura y Diversidade de Los Ecosistemas Forestales. Invited paper presented to IX World Forest Congress. México, July 1985.

Kageyama, P.Y., Brito, M.A. and Baptiston, I.C. (1986). Estudo do mecanismo de reproduçaõ das espécies da mata natural. In *Estudos para Implantação de Matas Ciliared de Proeção na Bacia Hidrográfica do Passa Cinco*. pp. 102–228. DAEE/USP/FEALQ, São Paulo.

Kaur, A., Ha, C.O., Jong, K., Sands, V.E., Chan, H.T., Soepadmo, E. and Ashton, P.S. (1977). Apoximis may be widespread among trees of the climax rain forest. *Nature*, **271**, 440–2

Levin, D.A. (1979). Pollinator foraging behavior: genetic implications for plants. In Solbrig, O.T., Jain, S., Johnson, G.B. and Raven, P.H. (eds), *Topics in Plant Population Biology*, pp. 131–53 (New York: Columbia University Press)

Loveless, M.D. and Hamrick, J.L. (1984). Ecological determinants of genetic structure in plant populations. *Annual Review of Ecology and Systematics*, **15**, 65–95

Martins, F.C.G., Ikemori, Y.K., Campinhos, E. Jr. and Maciel (1983). Teste de procedência de

Pinus caribaeae em Aracruz (ES). Resultados preliminares. *Silvicultura*, (São Paulo), **28**, 336–9

Namkoong, G. and Gregorius, H.R. (1985). Conditions for protected polymorphisms in subdivided plant populations. 2. Seed versus pollen migration. *American Naturalist*, **125**, 521–34

Pasztor, Y.P.C. (1974). Teste de procedência de *Eucalyptus pilularis* SM. na região Mogi Guaçú. *IPEF* (Piracicaba), **8**, 69–93

Patiño-Valera, F.F. (1986). Variação Genética em Progênies de *Eucalyptus saligna* e sua Interaçoã com o Espaçamento. Thesis of Magister Scientias. ESALQ/USP, Piracicaba.

Pinto, J.E., Jr. (1984). Variabilidade Genética em Progénies de uma Populaçao de *Eucalyptus urophylla* S.T. Blake da Ilha Flores-Indonésia. Thesis of Magister Scientias. ESALQ/USP, Piracicaba.

Pinto, J.E., Jr. and Jacob, W.S. (1979). Comportamento de Procedências de *Eucalyptus urophylla* na Região de Aracruz (ES). Boletim Informativo. IPEF, Piracicaba.

Ramirez-Castilho, C.A. (1986). Dispersão Anemocórica de Sementes de Painerira (*Chorisia speciosa*), na Região de Baurú-SP. Thesis of Magister Scientias. ESALQ/USP, São Paulo.

Roche, L.R. (1978). Antecedentes biológicos. In *Metodologia de la Conservación de los Recursos Genéticos Forestales*, pp. 5–18 (Rome: FAO)

Rodrigues, L.C., Vastano Jr., B. and Silva, A.P. (1986). Manejo e melhoramento de florestas de *Eucalyptus* em areias quartzosas na região noredeste do Estado de São Paulo. *Silvicultura*, (São Paulo), **41**, 104–10

Siqueria, A.C.M.F., Noguira, J.C.B., Kageyama, P.Y., Zanatto, A.C.S., Morais, E., Mariano, G. and Salles, L.M. (1985). Conservacion de Recursos Genéticos de Algunas Espécies Nativas de Brazil. Paper presented to the IX World Forestry Congress. Mexico, July 1985.

Vastano, B., Jr. (1984). Estudos de Aspectos da Estrutura Genética de uma População Natural de *Copaifera multijuga* Hayne Leguminosae — Caesalpinoideae na Região de Manaus. Thesis of Magister Scientias. INPA, Manaus.

CHAPTER 27

USE OF SMALL, CLONAL PLANTS FOR EXPERIMENTS ON FACTORS AFFECTING FLOWERING IN TROPICAL TREES

K.A. Longman, R.M. Manurung and R.R.B. Leakey

ABSTRACT

The factors which control flowering in tropical forest trees have seldom been studied experimentally because of the overwhelming problems of large size, long life-cycle and great variability. Methods to overcome these restrictions are described, including the use of model species that start flowering early in life, selection of precocious and heavily flowering genotypes for clonal propagation, growth under controlled environments, and injection with plant growth regulators. By combining these techniques in different ways, a logical progression of inductive treatments can be applied to develop an understanding of the physiological factors involved. Experiments with clonal, potted plants in heated glasshouses and growth cabinets in Scotland showed that flowering in Tabebuia pallida *(which can occur in successive stem modules) can be delayed or accelerated by modifying the cycles of shoot growth and branching. Treatments which altered these patterns included partial defoliation, day length, gibberellin, 'Ancymidol' and 'Alar'. In other experiments,* Triplochiton scleroxylon *(which would normally only flower as a large tree) on several occasions produced reproductive spurs in some ramets of certain clones, especially when plants had been grown in various root environments. Carbohydrate and nitrogen analysis showed possible relationships with flowering, amounts changing when root systems were transferred to lower temperatures. Stimulation of cloning has also been achieved in* Callitris *spp. by injection with GA_3, a reliable technique in the Cupressaceae. Other potential species for reproductive studies are also discussed.*

INTRODUCTION

Studying reproduction in forest trees is amongst the most difficult of botanical investigations, and this is especially true with the scale of genetic, spatial and temporal variation found in tropical biomes. The problems are, of course, further magnified because many tropical trees only commence flowering when they are tall enough to reach the canopy. Thus, it may seem premature to discuss experimental approaches to the subject, but it is in fact possible to commence such investigations, as it is possible to find tree species in which flowering occurs, or can be induced, while the plant is still small enough to be grown as a potted plant in controlled environments. Consequently, progress can be made, using clonal material, adequate replication, and known conditions to allow the powerful techniques of experimental biology to be employed. The physiological knowledge gained provides a sound basis for ecological interpretation of the factors operating within the tropical forest, as well as being of practical importance for seed production and for genetic improvement programmes in forestry.

FLOWERING EXPERIMENTS

Tabebuia pallida

Studies with the small Central American tree species *Tabebuia pallida* Lindl. were carried out in tropical glasshouses at the Institute of Terrestrial Ecology at Penicuik near Edinburgh, using four clones obtained from young plants of West Indian origin (Manurung, 1982). These clones were multiplied vegetatively by stem cuttings, using the techniques described by Leakey *et al.* (1982), with an average of about 75% rooting in 6 weeks. Plants were grown at $28°C \pm 2°C$ in a potting compost (7:3:1 peat/sand/loam) with added slow release fertilizers (4.2 g kg^{-1} 'Enmag', 2.6 g kg^{-1} John Innes base, and 0.3 g kg^{-1} trace elements). Pots were watered daily, with weekly application of soluble fertilizers (23% N; 19.5% P; 16% K). Where growth regulators were applied, they were injected into the xylem by the technique of Dick and Longman (1985), using a 'Precision Petite' miniature drill and an 'Agla' micrometer syringe.

Unless otherwise stated, day length was maintained constant at 19.5 h by supplementing natural daylight with mercury vapour lighting, giving irradiances in excess of 150 µmol m^{-2} s^{-1}. In some experiments, plants were grown in modified Fison's 140 G2 growth cabinets, under 8 or 16 h days. Irradiance (PAR) during the core 8 h period was about 480 µmol m^{-2} s^{-1} (83% warm white fluorescent and 17% tungsten). Photoperiod extension for the 16 h day was provided at less than 10 µmol m^{-2} s^{-1} for 4 h before and after the core

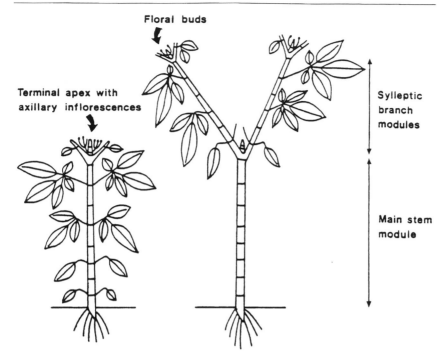

Figure 27.1 Diagrammatic representation of the growth habit of *Tabebuia pallida* (some leaves omitted for clarity)

period. Temperatures were maintained at 26°C day/21°C night ± 1°C. The tips of plants were kept about 5 cm below the top of each cabinet by progressive lowering of the shelf. When plants became too big for the cabinets (7–9 weeks), they were transferred to a glasshouse under 19.5 h day length.

Rooted cuttings typically produced a 7–8 node unbranched shoot, with opposite and decussate leaves, which changed in shape and size from simple and small to 5-lobed and large (Figure 27.1). At about the eighth node, sylleptic branching occurred, producing a forked plant that conformed to Leeuwenberg's architectural model (Hallé *et al.*, 1978). Growth of the two branches was similar, and the symmetrical modular habit was soon repeated, this time after 5–6 nodes. Following each branching event, the terminal bud of the main axis developed as a short shoot, and either:

(1) remained inactive;

(2) became reproductive, or

(3) was abscised;

depending on various physiological and environmental factors.

When reproductive, the terminal shoot typically formed large inflorescences with flower buds borne on short axillary shoots.

A series of experiments was carried out (see Manurung, 1982, for details) in which the vegetative growth habit of this species was manipulated, either by environmental or growth regulator treatments. Many of these treatments markedly affected the flowering process, and their effects are summarized here. In general, those treatments which delayed branching, by increasing the number of nodes on the main axis, also delayed flowering. Conversely, those which induced earlier branching, on shoots with fewer nodes, enhanced flowering.

Treatments that reduced the incidence of flowering included:

(1) Partial defoliation (successive removal of lower foliage, to retain only four leaf pairs per plant). Although this treatment did not affect the number of nodes produced on the main stem, plant height was reduced and flowering delayed. Seventy-five per cent of intact controls had flowered by week 7, whereas only 42% of the partially defoliated plants had done so in the same time;

(2) Application of gibberellins (between 20 and 200 µg per plant of GA_3, or between 100 and 500 µg per plant of $GA_{4/7}$, dissolved in a small volume of methanol) generally enhanced vegetative growth, especially stem elongation, in this set of experiments (Table 27.1). There was also a tendency towards an increased number of nodes on the main stem module, and consequently some delay to branching and flowering, although the latter effects did not always reach statistically significant values.

Treatments which increased the incidence of flowering included:

(1) Application of growth retardants ('Ancymidol' and 'Alar', dissolved in dimethyl sulphoxide) significantly reduced the height of the main stem prior to branching. 'Ancymidol' (100 µg per plant) reduced the number of nodes occurring before branching, rather than shortening the internode length (Table 27.2). Flowering started earlier and continued longer, presumably by enhanced retention of flowers, since their number was not significantly altered. In contrast, 'Alar' (1000 µg per plant) shortened internode length without markedly affecting the number of nodes per module. Here there was a considerable increase in the number of flower buds formed (Table 27.2), although the timing of flowering was not altered, in comparison with the controls;

Table 27.1   The effects of GA₁ and GA₄₋ on the vegetative growth and flowering of *Tabebuia pallida* (* = significantly different from control)

	Mean height of mainstem module (mm)	Mean stem extension rate (mm day⁻¹)	Mean number of nodes	Mean internode length (mm)	% plants branching	% plants flowering
Control	166	3.2	8.0	41.8	—	12
GA₃ 200 μg	306*	6.0	8.4	51.1	—	0*
Control	122	3.0	7.3	39.5	75	25
GA₄₇ 100 μg	158*	3.8	7.5	43.6*	63	25
GA₄₇ 500 μg	274*	6.7	8.6*	51.7*	50	13*
Control	97	2.2	8.6	35.5	83	68
GA₃ 100 μg	230	5.6	10.4*	43.9*	55*	48*

Table 27.2   The effects of 'Alar' and 'Ancymidol' on the vegetative growth and flowering of *Tabebuia pallida* (* = significantly different from control)

Treatments		Mean height of mainstem module (mm)	Mean number of nodes	Mean internode length (mm)	Number of flower buds per flowering plant	Mean number of days to first branching	Mean number of days to first flowering
Control		175	6.4	27.7	6.4	44	47
'Alar'	200 μg	156	6.1	25.6	8.0	42	42
	1000 μg	136*	5.9	23.1	10.0*	42	40
'Ancymidol'	20 μg	180	6.1	29.5	5.0	40	39
	100 μg	148*	5.6*	26.4	8.0	39	36*

(2)   Short days (8–10 h, compared with 16–19.5 h) seldom affected the rate of main stem extension significantly (Table 27.3), but induced earlier branching (Figure 27.2). The number of main stem nodes was usually considerably reduced by short days compared with long days (Table 27.3). Flowering also occurred sooner in plants that had been given short day treatment (Figures 27.3 and 27.4), and more than twice as many flowers were produced (6.4 ± 0.6 compared with 2.8 ± 0.3). It therefore appears that *T. pallida* is a facultative short-day plant, although further testing is needed to determine whether it is sensitive to the relatively small changes in natural photoperiod of the tropics (Longman and Jeník, 1987).

These experiments confirm that flowering in *T. pallida* is closely related to the factors controlling branching, which clearly include environmental and hormonal regulation. Other aspects of the physiological status of the plant may

Table 27.3 The effects of daylength on the vegetative growth of *Tabebuia pallida* (* = significant effect of daylength)

Daylength		Mean height of mainstem module (mm)	Main stem extension rate (mm day⁻¹)	Mean number of nodes	Mean internode length (mm)
(a) Growth cabinets/glasshouse					
8h	19.5h	230	4.5	7.7	47.3
16h	19.5h	253*	5.0	9.3*	45.3
8h	19.5h	191	3.8	7.3	41.1
16h	19.5h	213	4.3	7.8	43.0
(b) Glasshouse only					
10h		165	4.0	7.3	45.9
19.5h		204	5.0	8.3*	44.1
10h		97	2.4	8.4	32.4
19.5h		118	2.9	9.0*	33.0

well play a role (Reich and Borchert, 1982), but there is clearly a strong 'endogenous' component determining the strictly modular habit of development. Assuming that there is usually a predetermined number of nodes on a module, earlier branching would result in a greater number of flowering nodes situated in the terminal 'short shoot'. Such buds at lower nodes may therefore have a greater propensity to flower, provided they have the 'opportunity-to-flower' as a result of the induction of early branching. In this connection, it is interesting to note that a feature of older, profusely flowering trees of *Tabebuia*, and many other species, is to have a short growth flush.

Triplochiton scleroxylon

Previously reported observations on the precocious flowering of small, clonal plants of *Triplochiton scleroxylon* K. Schum. in tropical glasshouses at the Institute of Terrestrial Ecology at Penicuik (Leakey *et al.*, 1981) showed that flowers, fruits and viable seeds could be formed on rooted cuttings that were only 18 months old. In comparison with the usual onset of flowering in the field at 15–30 years, this period represents a considerable reduction in the duration of the reproductive cycle. Three generations have in fact been produced in 7 years by controlled pollination, using deep-frozen pollen. The smallness of the potted plants in this glasshouse study made it simple to observe the sequence of events leading to the formation of short reproductive spurs from inactive axillary buds. Flower buds opened successively over a period of weeks, and their subsequent development

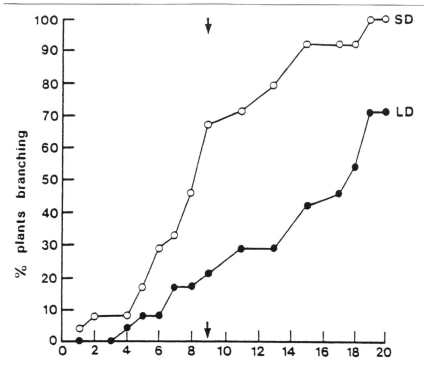

Figure 27.2 The effects of day length on the occurrence of branching by the main stem module of *Tabebuia pallida*. (Open symbols - 8 h; closed symbols - 16 h; the arrows indicate the date of transfer from controlled environment cabinets to a 19.5 h regime in a glasshouse)

depended on successful cross-pollination and on their position within the spur and within the plant. The first formed fruits appeared to be highly competitive, because many later flower buds and smaller fruits were aborted.

This research also suggested that early floral induction might have been linked with watering with cold water (down to 2°C in winter) of plants kept at an air temperature of 28°C. The subsequent installation of water pre-heaters in the glasshouses appeared to eliminate flowering. To investigate this possible connection further, deep-freeze cabinets were modified to operate at a range of temperatures between 5°C and 25°C. The lids were replaced with a board containing 14 holes to accommodate 125 mm diameter plant pots, allowing soil and air temperatures to be controlled independently.

Fourteen plants of clones 8057 and 8106 were grown in these root environment boxes for 5 or 10 weeks at soil temperatures of 8°, 12°, 16°, 20° and 28°C (control), all with an air temperature of 28°C and all watered with cold water. One or two plants flowered in each treatment, except that at 8°C, in which the plants eventually died. Numbers of flowers ranged from 10 to 100, with most at 16°C, but the differences were not significant. Although

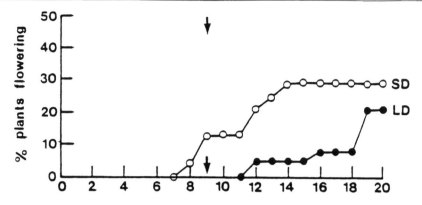

Figure 27.3 The effects of day length on the occurrence of flowering on the main stem module of *Tabebuia pallida.* (Open symbols - 8 h; closed symbols - 16 h; the arrows indicate the date of transfer from controlled environment cabinets to a 19.5 h regime in a glasshouse)

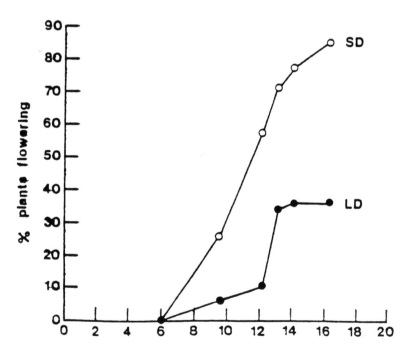

Figure 27.4 The effects of day length on the occurrence of flowering in glasshouse grown *Tabebuia pallida.* (Open symbols - 10 h; closed symbols - 19.5 h)

Table 27.4 The effects of chilling the root system for 10 weeks on the soluble carbohydrate and total nitrogen contents (as % of dry matter) of leaves, stems and roots of *Triplochiton scleroxylon*. Air temperature was 28°C

	\% soluble carbohydrates					\% total nitrogen				
	8°	12°	16°	20°	28°C Control	8°	12°	16°	20°	28°C Control
Leaves	10.0	10.0	9.4	8.0	6.5	2.0	2.4	2.4	2.7	2.8
Stems	12.0	14.0	17.0	14.0	8.9	1.3	0.9	1.5	1.2	1.2
Roots	3.0	4.2	7.7	6.7	5.4	2.0	2.2	2.3	2.2	2.2

the results of this experiment were inconclusive, it is possible that other combinations of temperature and treatment duration might prove stimulatory, or perhaps the plants need to be in a more appropriate physiological condition. Nevertheless, It is interesting to note that, in the present experiment, the soluble carbohydrate and total nitrogen contents of the stems, leaves, and roots were markedly altered by the root temperature treatments (Table 27.4). Carbohydrate accumulated in the leaves to a greater extent as the temperature decreased, whereas its accumulation in both stems and roots was greatest at 16°C, decreasing progressively when root temperatures were either lower or higher than this. Patterns of total nitrogen content were fairly similar to those for carbohydrates in the stem and root, but, in leaves, the levels fell as temperatures were decreased.

In cocoa (*Theobroma cacao* L., another member of the Sterculiaceae), the cauliflorous flowering has been associated with high stem carbohydrate content. It is possible that further work with *Triplochiton scleroxylon* may indicate that checks to root activity, such as water stress or low temperature, could promote floral initiation through changes in carbohydrate nutrition. In this connection, it may be noted that a correlation has been found in Nigeria between the incidence of mast years and the occurrence beforehand of a more pronounced short dry season than normal.

Other species

Stimulation of reproduction is readily achieved in the coniferous family Cupressaceae by application of gibberellic acid (GA_3) (Longman, 1984, 1989; Longman *et al.*, 1982; Manurung, 1982). Most species that have been tested produce large numbers of male and female cones, including the subtropical *Callitris glauca* and *C. intratropica*, as well as the important plantation tree *Cupressus lusitanica*. Even small cypress plants respond, including those of seedling origin, as well as rooted cuttings derived from older trees. This response provides unique opportunities for miniaturized flowering studies. However, the potential for detailed investigation of the effects of environmental factors on flowering has so far only been utilized in the temperate genera *Thuja* (Longman, 1989) and *Tsuga* (Pollard and Portlock, 1984). In the latter case, the gibberellin mixture $GA_{4/7}$ was used as it is more effective in the Pinaceae.

An alternative method of obtaining plants small enough to study floral initiation easily is the selection of unusually precocious clones of tree species in which reproduction commences early in life (Longman, 1984, 1985; Longman and Jeník, 1987). For example, using controlled environment cabinets, the temperate zone conifer *Pinus contorta* has been shown to be a facultative short-day plant for female cone formation (Longman, 1982). A promising broad-leaved tree for potted plant studies of flowering appears to be teak (*Tectona grandis* L.f.), which can produce, in some seed-lots, its first terminal inflorescence within a few years, or even months, of sowing. Other possible species include the South-east Asian *Dipterocarpus oblongifolius*, the West African *Hildgardia barteri* and the New World *Pinus caribaea* var. *hondurensis*. Several common colonizers and mangroves could be added to the list of potential tree species for flowering studies in controlled environments.

DISCUSSION

This contribution has summarized some of the research on flowering of tropical tree species at the Institute of Terrestrial Ecology near Edinburgh. The key features enabling such experimental studies to be carried out are:

(1) The vegetative multiplication of clonal material to produce many similar replicate potted plants of manageable size;

(2) The use of controlled environments, such as glasshouses and growth cabinets, to maintain more uniform conditions and allow testing of the effects and interactions of individual factors;

(3) The choice of appropriate species, provenances and cultivars which show a particular propensity to flower; and,

(4) Where available, the use of reliable flower induction treatments (such as plant growth regulators) to predispose the experimental plants towards reproductive activity.

The results indicate that successful study and manipulation of flowering is possible in tropical forest trees, despite their heterogeneity and longevity.

It is clear that species vary greatly, not only in flowering morphology and ecology, but also in their physiological mechanisms and responses to changes in environment. In *Tabebuia* and *Triplochiton*, factors which altered the relative sink activity of different organs may perhaps have been particularly important in controlling the balance between vegetative and reproductive activity. Fuller elucidation of such questions will follow as research becomes better co-ordinated, for instance by developing 'model' cultivars and standard techniques that can be adopted internationally, as has been done with various herbaceous species.

REFERENCES

Dick, J.McP. and Longman, K.A. (1985). Techniques for injecting chemicals into trees. *Arboricultural Journal*, **9**, 211–14

Hallé, F., Oldeman, R.A.A. and Tomlinson, P.B. (1978). *Tropical Trees and Forests: an Architectural Analysis*. (Heidelberg: Springer-Verlag)

Leakey, R.R.B., Chapman, V.R. and Longman, K.A. (1982). Physiological studies for tropical tree improvement and conservation. Factors affecting root initiation in cuttings of *Triplochiton scleroxylon* K. Schum. *Forest Ecology and Management*, **4**, 53–66

Leakey, R.R.B., Ferguson, N.R. and Longman, K.A. (1981). Precocious flowering and reproductive biology of *Triplochiton scleroxylon* K. Schum. *Commonwealth Forestry Review*, **60**, 117–26

Longman, K.A. (1982). Effects of gibberellin, clone and environment on cone initiation, shoot growth and branching in *Pinus contorta*. *Annals of Botany*, **50**, 247–57

Longman, K.A. (1984). Tropical forest trees. In Halevy, A.H. (ed.), *CRC Handbook of Flowering*. Volume I. pp. 23–39 (Boca Raton, Florida: CRC Press)

Longman, K.A. (1985). Variability in flower initiation in forest trees. In Cannell, M.G.R. and Jackson, J.E. (eds) *Attributes of Trees as Crop Plants*. pp. 398–408. Institute of Terrestrial Ecology, Monks Wood, Huntingdon.

Longman, K.A. (1989). *Thuja*. In Halevy, A.H. (ed.) *CRC Handbook of Flowering*, Volume VI, pp. 610–24. (Boca Raton, Florida: CRC Press)

Longman, K.A. and Jenik, J. (1987). *Tropical Forest and its Environment*. Second edition. (Harlow: Longman Scientific and Technical)

Longman, K.A., Dick, J.McP. and Page, C.N. (1982). Cone induction as an aid to conifer taxonomy. *Biologia Plantarum (Praha)*, **24**, 195–201

Manurung, R.M. (1982). Environment and growth substances affecting gibberellic acid-induced coning of *Thuja plicata* D. Don. and flowering in *Tabebuia pallida* Lindl. Ph.D. Thesis. University of Aberdeen, Aberdeen

Pollard, D.F.W. and Portlock, F.T. (1984). The effects of photoperiod and temperature in gibberellin $A_{4\,7}$ induced strobilus production of Western hemlock. *Canadian Journal of Forest Research*, **14**, 291–4

Reich, P.B. and Borchert, R. (1982). Phenology and ecophysiology of the tropical tree, *Tabebuia neochrysantha* (Bignoniaceae). *Ecology*, **63**, 294–9

Section 8

Conclusions

CHAPTER 28

REPRODUCTIVE ECOLOGY OF TROPICAL FOREST PLANTS: CONCLUDING REMARKS

T.N. Khoshoo

INTRODUCTION

This book, and the workshop on which it was based, focus on a particular aspect of the functioning of tropical forest ecosystems, that is the reproductive ecology of plants. The tropics themselves have been the cradle of origin and evolution of biota, including humankind itself. One wonders how long it would have taken for Charles Darwin to finalize his theory of organic evolution if he had not visited the tropics. Even his less known competitor, Alfred Wallace, who came to similar conclusions about organic evolution, also worked in the tropics.

Naturalists have continued to be fascinated by the diversity and richness of tropical ecosystems, and by the complex webs of interactions and interrelationships that are characteristic of these systems. Nowhere is this better reflected than in the field of the reproduction of tropical forest plants, addressed in this book. The topic is of considerable interest and importance, from policy, management and scientific standpoints. This interest and importance were well reflected in the three trend-setting addresses which opened the workshop by YB Datuk Amar Stephan K.T. Yong, the Minister of Science, Technology and Environment of Malaysia, by Dr. Salleh Mohd Nor, Director of the Forest Research Institute, Malaysia, and by Dr. K.S. Bawa, Professor at the University of Massachusetts at Boston, and head of the scientific programme committee for the workshop. The addresses conveyed, respectively, the perceptions of a political leader and a policy-maker; of an implementer responsible for seeking bridges in translating policy into action; and of a scientist who does things for the sake of knowledge.

These three strands provided a backdrop to the Bangi workshop and the various contributions which make up the present volume. These contributions range from those dealing with basic questions on reproductive rhythms to those on applied tree breeding, and many other topics in between. As an ensemble, they shed light on the widening frontiers of reproductive ecology, from flushing, flowering, and fruiting to resource and ecosystem management. This work involves specialists from not only botany, physiology and genetics, but also from pathology, entomology, biochemistry, and other disciplines.

Many of the contributions are based on the use of simple techniques and tools that are within the reach of anyone, but require a keen mind and an observant eye. Several of the syntheses point out the need for a commonality of approach in data collection, so that results are comparable. In the first instance, such collaboration might be attempted on the topic of phenology. Today, we find a number of more or less parallel lines which do not seem to converge. While not stifling innovation, convergence is as important as divergence. We can achieve commonality in approach in several ways, for example by organizing regional training courses, and by evolving standardized protocols, using unified approaches adopted by several contributors to this volume.

About 150 persons attended the Bangi workshop, many of them fully familiar with the tools and techniques of study. If the message is to be spread and the gigantic tasks that lie ahead are to be accomplished, we need an army of trained people to take up this work at the earliest possible time. Therefore, protocols and training are needed for many other persons who were not able to attend the workshop. We have to utilize simple tools and techniques which can be used by students in high schools, colleges, and universities. The late Professor J.B.S. Haldane was asked, in India during the 1950s, if some relevant and excellent work could be done without any equipment. His choice was reproductive biology. Many in India, including me, started such work. I ended up finding versatile reproduction (sexual, semi-sexual and apogamous) in a weed, *Lantana camara*.

By the turn of the century, there will be a 90% increase in the population in the tropics, and escalating demands and expectations for food, shelter, and clothing and other essentials for a better standard of living. Thus, the economic planner and the ecologist face the dilemma of reconciliation of short-term benefits and long-term ecological costs, because there is an inverse correlation between projects that are economically sound in the short-term, and those that are sustainable in the long-term. So far, everyone has followed the path of least resistance, and we have reaped the short-term economic benefits and passed long-term ecological costs to succeeding generations, creating a serious ecological deficit or environmental drag.

Sustainable use of forests requires that biologists be aware of the problems faced by forest managers. It also requires that economists be familiar with ecological constraints on the exploitation of biological resources. The following commentary highlights some of the major problems in tropical forestry and the ways that foresters, biologists, and economists may interact to resolve problems in the sustained use of tropical forest resources. Examples from India are frequently drawn upon to emphasize particular points, because of my familiarity with that country.

INTERFACE WITH FORESTERS

For effective collaboration with foresters, we need critically to examine the data and extract specific items of relevance to forestry. For this, we need to understand the mandate of foresters, which is to produce practical results on the ground. Therefore, no off-the-cuff or *ad hoc* suggestions will help. Only cases where adaptability and production trials have been made should be taken up. Failing this test, the whole exercise may be counterproductive. In fact, what is needed is a periodic dialogue with the foresters to identify areas amenable to collaborative work with research workers. Having done this, a conference/workshop could be specially designed for the purpose of rendering help in the sustainable management of wood and non-wood resources in forests. This conference would encompass the areas of forestry (wood-based), biodiversity, and minor forest products referred to by Dr Salleh in his opening remarks to the Bangi workshop as VIPs (Very Important Products) or VISPs (Very Important and Special Products). The management of all these resources has to be for the benefit and the welfare of the people, thus necessarily involving socio-economic and environmental goals.

Forest cover and forestry are critical to the well-being of the people of a country, particularly those whose society and economy is biomass-based. In the Indian context (or, for that matter, for most less-industrially developed countries), there are four principal goals as far as forestry is concerned:

(1)   Affording long-range ecological security for the conservation of climate, water, soil and biodiversity;

(2)   Meeting the needs of goods and services, including firewood, charcoal and fodder, for rural/tribal communities and urban poor;

(3)   Meeting the wood needs of the people and industry of timber, pulp, fibre and silvi-chemicals; and,

(4)   Amelioration of soils of the degraded areas and wastelands to enhance the productive capacity of such derelict lands as well as improve general aesthetics.

Emanating from these principal goals are four mutually supportive types of forestry: conservation forestry, agroforestry, industrial forestry, and environmental/revegetation forestry.

Conservation forestry

This is most relevant to all water regimes/watersheds/catchment areas; representative ecosystems and biosphere reserves (located in different biomes); centres of diversity; national parks and sanctuaries; and fragile ecosystems. In these regions, there should be no exploitation of wood and non-wood resources unless warranted on scientific and technical grounds for maintaining the well-being of the concerned forest, but not exceeding the mean annual increment.

Restoration and repair of such areas has to be done with local and indigenous species, and, on no account, must exotic species be introduced into conservation areas. Conservation forestry benefits all people, and linked with it is the stabilization of microclimate, conservation of soil, water and biodiversity, sources of non-wood products, and other amenities.

Agroforestry

Here the objective is the integration of agriculture, forestry and animal husbandry to meet food, fuel, and fodder needs through a well worked-out Agri-Silvi-Pastoral or Agri-Silvicultural or Silvi-Pastoral model of development. It is essentially a polycultural system with low inputs but reasonably high outputs, and, if established on proper lines, can be sustainable. The basic idea is to aim at the intensification and diversification of biomass production in rural areas.

There should be no objection to the use of exotic trees in agroforestry, if their use is warranted on the grounds of land-use and end-use, is in line with the location-specific edaphic conditions, and meets the demand of local people. The beneficiaries of agroforestry in India are the vast number of the rural and urban poor in its 576 000 villages, who meet their needs of fuelwood and small timber and also earn some money. Agroforestry, practised sustainably, would ultimately relieve pressure on natural forests and thereby help in forest conservation.

Industrial forestry

Industrial forestry is an economic and commercial venture based on wood quality as well as on the economic input–output considerations. The objective here is to meet the needs of timber, pulpwood and fibre industries. Industrial forestry has to be related to land-use and end-use considerations. Since these are commercial ventures, the chief consideration is that of production/productivity, and, if warranted on other grounds, fast growing exotic species are most welcome.

A country like India can be oblivious to the needs of industrial/ commercial forestry only at the cost of ecological security. Industrial forestry should not be ignored on account of emotional considerations. Realism demands that there be a crash programme on industrial forestry in order to save our forest wealth.

Suggestions such as those favouring the import of timber, firewood, and pulpwood, etc., can only help to avert the most critical and immediate situation. They do not offer a permanent solution, for more reasons than one. First, they only serve to shift forest degradation to other countries (most probably developing ones), which is not ethical. Second, the kind of money required for imports of wood products may not always be available. Third, wood can become a "political weapon" like food and oil, and thus its price could well keep on soaring. The best strategy is to give very high priority to industrial forestry and take steps to extend all help by suitable modification of land laws, etc.

Environmental/revegetation forestry

The objective here is to ameliorate and finally restore waste and derelict lands. The process can be started by creating natural wilderness areas and by using the principles of plant colonization based on genetic considerations. As a result of litter fall, a decomposer chain will start, and will be followed by soil conservation and water retention. These first steps would go a long way in improving the quality of these lands. Starting with plantations of tolerant species, leading to some improvement of soil, there would be a distinct possibility of growing less tolerant species, and of further improvement in soil characteristics. In the succeeding cycles, it would be possible to grow increasingly less and less tolerant species. A stage may come when community respiration is less than community production.

There is an ever-widening gap between production and demand. In India, there is a gap of over 200 million m^3 of fuelwood, of over 34 million m^3 of timber, and of about 5 million m^3 of wood for pulp and paper. If India is to be self-sufficient in wood requirements, around 658 million ha of land is needed, with a harvesting rotation of 10 years for fuelwood, 50 years for timber, and 8 years for pulpwood, and a production of 7.5 and 10 million m^3 per hectare per year, respectively, from the so-called wastelands, which are the only lands available for the purposes of forestry. The area demand (658 million ha) is about twice the area of India (329 million ha). Clearly, the only way to augment production and productivity is to make recourse to forestry based on science and technology.

So far, forestry in India has revolved around teak, sal, pines, and deodar. However, for agroforestry, a lot of new species have become relevant, for which we have no information as far as reproductive ecology is concerned. Furthermore, most of the forestry has been based on the species of wild origin with a high degree of genetic variability following the principles of organic evolution in forest stands which approach the situation found in nature. However, to enhance production/productivity, the use of advanced breeding technology, based on detailed reproductive ecological studies is needed. In addition, there is a need to optimize the silvicultural and nutritional requirements of the use of fertilizers, irrigation, and bacterial and mycorrhizal cultures. There is also need of improved understanding in disease and pest management.

The four major uses of forestry outlined above, are not mutually exclusive, nor is one at the expense of the other. They are mutually supportive. Furthermore, it may be pointed out that the slogan of wasteland development, though laudable, cannot be expected to be the panacea for all our food, fodder, timber and fuelwood needs. Wastelands are essentially derelict lands, and, for several years to come, these will have low productivity. While forestry is expected to perform miracles on wastelands, the prime agricultural land continues to be used for non-food and non-forestry uses, like human settlements, industries, road and rail systems, airports and similar ventures. The land laws are either too weak or non-existent to permit such ventures on prime agricultural land.

Traditional forestry and agroforestry differ in many ways. In the former, the escalating population of people and livestock creates pressure on forests because the people have no direct stake. There is illegal grazing, cutting and clearing, and a continuous need for surveillance for protection. However, people do have a stake in agroforestry, which responds favourably to people's pressure. It produces small timber, crops, fuel and fodder and involves no unfamiliar technology. Besides, no surveillance is needed. People themselves act as a social fence.

BIODIVERSITY

The Indian region has been the source of over 160 economic plants, whose centres of diversity fall in this region. Moreover the region is a secondary centre for a number of important, plants. Notable among the plants belonging to this region are several woody species, like bamboos, conifers, teak, sal, dipterocarps, etc.; rice, sugarcane, small millets, brassicas, bananas, citrus, mango, jute, pepper, eggplant, jasmine, musk melon, cucumber, and several minor plants. These can be conserved at minimal cost *in situ* in respective ecosystems, provided they are conserved in relation to a minimum area demand based on the principles of reproductive ecology. The conservation of such ecosystems can be sustainable, because, if the extraction of species is based on considerations of their reproductive ecology, these ecosystems can be exploited in perpetuity.

Ex situ conservation can be undertaken in botanic gardens, arboreta, provenance plots, seed/tissue/pollen banks, etc. Again, considerations of reproductive ecology can be brought to bear in determining the sizes of population, seed sample, etc. If such principles are not taken into consideration in both *in situ* and *ex situ* conservation, serious genetic drift could result, and the whole exercise could become counterproductive.

MINOR FOREST PRODUCTS

Minor forest products constitute important non-wood elements of vegetation in tropical forests which are a source of non-edible oils, gums, resins, dyes, tannins, lac, tussar, fibres, canes, aromatics and medicines, and herbal drugs, etc. Many of these products are well-established in national and even international trade. With the tremendous resurgence of interest in anything that is natural, there has recently been considerable increase in both domestic and international trade.

Since 1975, WHO has provided support for the traditional system of medicines which are a part of all the old and well-established civilizations. There are many single drugs and compound formulations in trade used as herbal medicine, and even as toothpastes, shampoos, bath oils, soaps, sherbets, etc. During 1984–85, the estimated production of crude drugs and active principles in India alone has been of the order of US $178 million. The export of medicinal plants was about US $45 million. This sum is exclusive of the internal market. Obviously, the somewhat inappropriately named "minor forest products" constitute an important element in national and international trade in India. This importance holds true for most tropical countries as well.

While demand is considerable, and only some well-established medicinal and aromatic plants (mentha, aromatic grasses, atropa, digitalis, psyllium pyrethrum, bulgarian rose, etc.) are cultivated, the bulk of the remaining drug plants are collected from nature with no consideration of their reproductive ecology. This exploitation has resulted in a situation where many such species (like podophyllum, coptis, valerian, violas, dioscorea, aconites) are today endangered by over-collection.

The collection of drug plants (including gums, herbal drugs, and even seeds of trees like teak, etc.) has been a family vocation for several village/tribal communities. Prohibiting the collection of many of these useful but endangered species has not served the desired end. There is, therefore, a need to cultivate these plants, rather than to continue the present hunting-gathering practice. Results of a project aimed at utilizing seed and tissue culture approaches have been remarkable. There is not only a possibility of cultivating some of the species, but also of restocking the endangered populations in nature. Telescoping the domestication process, as well as restocking in nature, can be sustainable if based on reproductive ecological data. This possibility also applies to several Malaysian species like rambutans, garcinia, durio, etc. Most of the minor forest product species, as well as the underutilized local fruits, have not undergone any intensive process of domestication based on recombination and selection. The response of such hitherto unselected species to such treatments may well be dramatic. An important example is the actinidia which, from simple techniques of local cultivation evolved in China, has now become a world class fruit through the use of technology evolved in New Zealand. Even its name has changed from Chinese Gooseberry to Kiwi Fruit. In all such cases, sustained production is the best form of conservation.

ECOLOGY-ECONOMICS INTERFACE

Work on reproductive ecology is very significant in establishing an interface between ecology and economics. Recently, I used the studies of Bawa and his group in my address to a select group of economists and environmentalists. I talked of breeding biology as a model to optimize the fitness of resource allocation, investment costs and benefits, and population recruitment. I was able to convince the group that it was very necessary for the biologists to examine the economic principles that underlay biological phenomena such as reproduction. Conversely, it was equally important for economists not to run away from the biological principles that underlay the well-being of all living creations. I concluded that there is an imperative need to look into the economic principles underlying biological systems, and the biological principles underlying economic systems. The commonality of such principles should actually govern our judgements about bio-resource management. After all, economics (financial housekeeping) and ecology (environmental housekeeping) have the same root, (*oikos*, meaning a house). Finally, I pointed out to the meeting that, for proper evaluation of such indices as the annual rate of economic growth and gross national product, such data should be accompanied by data on the annual rate of ecological degradation of environmental assets. Thus, annual estimates of the Gross National Product and Gross Ecological Product would give a true picture of a country's economic situation. Such a notion underlines the critical importance of biomass for the survival the human race, and, for that matter, the whole biosphere.

CONCLUSIONS

There are, ultimately, two basic questions which need attention. They are:

(1)   To what extent will tropical forests meet the escalating demands of the increasing population without ecological degradation? and,

(2)   How can the present and future knowledge be marshalled for the purposes of sustainable use of these areas for the five 'f's (food, fodder, fuel, fertilizer and fibre) and the medical and spiritual welfare of the people in tropics, and yet not degrade the environment?

I am sure that intimate knowledge on the reproductive ecology of tropical forest plants will play a major role in addressing these two challenges.

INDEX

413